Beyond the Battlefield

This volume draws together an international team of scholars to explore the experience and significance of early modern European continental warfare from an interdisciplinary perspective.

Individual essays add to the lively fields of War and Society and the New Military History by combining the history of war with political and diplomatic history, the history of religion, social history, economic history, the history of ideas, the history of emotions, environmental history, art history, musicology, and the history of science and medicine. The contributors address how warfare was entwined with European learning, culture, and the arts, but also examine the ties between warfare and ideas or ideologies, and offer new ways of thinking about the costs and consequences of war. In addition to its interdisciplinarity, the volume is distinctive in including chapters focused not only on Western and Central Europe but also the often-ignored European peripheries, such as the Baltics and the Russian frontier, Scandinavia, and the Habsburg-Ottoman borderlands of Southeastern Europe. As a whole, the volume offers readers interesting alternatives and threads for reconsidering the place and meaning of warfare within the larger history of early modern continental Europe.

This book will be valuable for general readers, undergraduate and graduate students, and scholars interested in military, early modern, and European history.

Tryntje Helfferich is Associate Professor of History at The Ohio State University, Lima. She studies the history of war, religion, and politics in early modern Europe. Her publications include *The Essential Thirty Years War* (2015) and *The Iron Princess* (2013).

Howard Louthan is Professor of History and Director of the Center for Austrian Studies at the University of Minnesota. His scholarship focuses on the intellectual and cultural history of early modern Europe. His publications include *Theuerdank: The Illustrated Epic of a Renaissance Knight* (2022) and *Converting Bohemia* (2009).

Beyond the Battlefield

Reconsidering Warfare in Early Modern Europe

Edited by Tryntje Helfferich and Howard Louthan

LONDON AND NEW YORK

Designed cover image: Minerva protects Pax from Mars (Peace and War) - Peter Paul Rubens, circa 1630. Niday Picture Library/Alamy Stock Photo

First published 2024
by Routledge
4 Park Square, Milton Park, Abingdon, Oxon OX14 4RN

and by Routledge
605 Third Avenue, New York, NY 10158

Routledge is an imprint of the Taylor & Francis Group, an informa business

British Library Cataloguing-in-Publication Data
A catalogue record for this book is available from the British Library

ISBN: 978-0-367-74419-9 (hbk)
ISBN: 978-0-367-74417-5 (pbk)
ISBN: 978-1-003-15770-0 (ebk)

DOI: 10.4324/9781003157700

Typeset in Times New Roman
by KnowledgeWorks Global Ltd.

Contents

Tables

Figures

Maps

Contributors

Mary Elizabeth Ailes is Professor of Early Modern European History at the University of Nebraska at Kearney.

Heidi Hausse is Assistant Professor of History at Auburn University.

Tryntje Helfferich is Associate Professor of History at The Ohio State University at Lima.

Sabine Jesner is a Research Fellow at the section of Southeast European History and Anthropology at the University of Graz, Austria.

Dariusz Kołodziejczyk is Professor of Early Modern History at the University of Warsaw, Poland, and member of the Polish Academy of Sciences.

Mary Lindemann is Professor Emerita in the Department of History at the University of Miami.

Howard Louthan is Director of the Center for Austrian Studies and Professor of History at the University of Minnesota.

Thomas Marks is Choral Director at the River School in Weston, Massachusetts, and former Postdoctoral Fellow at Yale University.

Kristoffer Neville is Professor and Chair of the Department of the History of Art at the University of California, Riverside.

Daniel Riches is Associate Professor of History and Director of Graduate Studies at the University of Alabama.

Brian Sandberg is Professor of History at Northern Illinois University.

Mindaugas Šapoka is Researcher at the Lithuanian Institute of History, Lithuania, and a former fellow of Yale University's MacMillan Center.

Jeffrey Chipps Smith is Professor of Art History and Kay Fortson Chair in European Art at the University of Texas at Austin.

Acknowledgments

This project began in November 2018 with a conference sponsored by the Center for Austrian Studies and the Center for German and European Studies at the University of Minnesota, followed by further presentations and discussion at the 2019 Sixteenth Century Studies Conference in St. Louis, Missouri. We are extremely grateful for the ongoing support offered by both of these Centers, with special thanks for the organizational work provided by Jennifer Hammer and James Parente. Our thanks to our contributors as well, who, despite the difficulties that arose due to the emergence of a global pandemic, steadfastly pressed forward as the volume began to take shape. We had a marvelous cartographer, Johnathan Hardy, who patiently worked through a variety of drafts with us. As we moved into the editing process, a gifted team of students at the University of Minnesota helped in various phases of the project. Here, we are especially thankful for the assistance of James Gresock, Emmey Harris, Briana Heidkamp-Vu, Jake Henke, Claire Mavis, and Tim McDonald. Thanks also to Elizabeth Dillenburg, who played an important role throughout the entire process and the critical transition from initial conference to finished book.

Map 0.1 Western Europe (sixteenth to eighteenth centuries)

Map 0.2 Central and Eastern Europe (sixteenth to eighteenth centuries)

Going Beyond the Battlefield

Tryntje Helfferich

Introduction

> As the Larke began to peep, the seventh of September 1631, having stood all night in battalion a mile from Tilly's Army, in the morning, the Trumpets sound to horse, the Drums calling to March, being at our Arms, and in readiness, having before meditated in the night, and resolved with our Consciences; we began the morning with offering our souls and bodies, as living Sacrifices unto God, with Confession of our sins, lifting up our hearts and hands to Heaven, we begged for reconciliation in Christ, by our public prayers, and secret sighs, and groans; recommending ourselves, the success, and event of the day unto God, our Father in Christ; which done by us all, we marched forwards in God's name....[1]

Thus did Sir Robert Monro, an officer in a Scottish regiment serving King Gustavus Adolphus of Sweden, describe how he and the rest of the combined Swedish-Saxon army began their advance to the sound of drums and trumpets. And it was on this day at Breitenfeld, near the city of Leipzig, that these forces met, and smashed, the Catholic League army under Johann Tserclaes, Count of Tilly. The Battle of Breitenfeld resulted in as many as 12,000 dead and thousands more captured or injured, but it also cemented the Swedish king's reputation as a tactical virtuoso. So too did it encourage other Protestant princes, among them the duke of Brandenburg and the landgrave of Hesse-Cassel, to declare themselves for the Swedish-led alliance, a move that ensured that the war would continue for many years to come.

Given the significance of this battle for the progress of the war, as well as its extreme bloodiness, it is not surprising that it captivated contemporary news accounts and artistic imaginations, and that many first-person accounts of the battle subsequently circulated, including that by Monro, which appeared within his much longer account of his long years fighting on the continent. The Battle of Breitenfeld has also fascinated historians, who have generally focused on the political implications of the Swedish victory,

DOI: 10.4324/9781003157700-1

but also on its military significance, especially its importance in demonstrating the era's tactical and organizational innovations and technological developments.

Yet the Battle of Breitenfeld is only one of many other early modern European battles that have received significant scholarly attention and interest over the centuries. Mohács (1526), for example, has been fully dissected and analyzed, as have Dreux (1562), Lepanto (1571), Lützen (1632), Vienna (1683), Poltava (1709), Austerlitz (1805), and so on. So many famous battles occurred, of course, because this era also produced so very many wars. The Schmalkaldic War, the French Wars of Religion, the Great Northern Wars, the Habsburg-Ottoman Wars, the wars of Louis XIV, and the French Revolutionary Wars, as just a few notable examples, have all received in-depth studies, with scholars having fleshed out and mapped the details of their major campaigns, explored the conduct of military operations, investigated the various armies' organizational patterns, institutions, financing, and command structures, traced advances in military technology, argued the skill or incompetence of the generals and analyzed their victorious or losing tactics and strategies, laid out adherence to or deviations from contemporary military theories and principles, and followed the back and forth of war-time diplomatic efforts to gain advantage or make peace. In the process, military historians have created, enriched, and reinforced the grand narratives and standard analytical frameworks that have traditionally guided the study of war in history.

New Military Histories

Beginning in the late twentieth century, however, scholars also began looking at these wars and conflicts in new ways, using approaches termed either the "new military history" or "war and society." Although there are differences of opinions on how one defines these terms, and they are sometimes conflated, the former, new military history, is usually described as asking many of the same questions posed by the "old" military history, such as how armies functioned, campaigns proceeded, and wars were won, but it does so from the perspective not of generals or rulers, but of ordinary soldiers, civilians, or other non-elite participants.[2] One of the pioneering works in this field is John Keegan's *The Face of Battle* (1976), which, strongly influenced by the techniques of social history, explored the brutal reality of the experience of war for those who fought. A passage from his analysis and description of the 1415 Battle of Agincourt is illustrative:

> The distance between horses and archers narrows. The archers, who have delivered three or four volleys at the bowed heads and shoulders of their attackers, get off one more flight. More horses—some have already gone down or broken back with screams of pain—stumble and fall, tripping their neighbours, but the mass drive on....[3]

While the new military history thus often attempts to make the business and practice of warfare more personal, real, and visceral, the approach known as war and society is often understood as being more innovative in its methodology, as it sets aside what have been the fundamental questions of the actions of soldiers, the nature of tactics and strategies, or the causes of victory and defeat. Instead, it mines the vast quantities of records housed in archives to draw out more fully the interrelationship between warfare, society, and culture, sometimes to such an extent that battlefield combat and violence become an afterthought. The scholar André Corvisier's comparative 1976 study, *Armées et sociétés en Europe de 1494 à 1789*, set the tone for this approach to military history, as he highlighted a growing divergence between military and civilian life, and also argued that warfare strongly influenced the nature of early modern state politics, economics, society, and ways of thinking.[4]

Of course, one does not have to look very far to see that many of the analytical categories claimed by the new military history and by war and society have also appeared in various guises throughout the history of writing about war, and so are less new than they are happily rediscovered. The fourth-century Roman scholar Ammianus Marcellinus's description of the Battle of Adrianople (378) between the Goths and the Byzantines, for example, shares with Keegan the gripping narrative point of view of the common fighting men on the battlefield:

Then you might see the barbarian towering in his fierceness, hissing or shouting, fall with his legs pierced through, or his right hand cut off, sword and all, or his side transfixed, and still, in the last gasp of life, casting round him defiant glances. The plain was covered with carcasses, strewing the mutual ruin of the combatants; while the groans of the dying, or of men fearfully wounded, were intense, and caused great dismay all around.[5]

Similarly, most are familiar with the famous passage in Thucydides's *History of the Peloponnesian War* (fifth century BCE) in which he critiqued the evil effects of warfare on the values, culture, and even language of Greece:

In peace and prosperity states and individuals have better sentiments, because they do not find themselves suddenly confronted with imperious necessities; but war takes away the easy supply of daily wants, and so proves a rough master, that brings most men's characters to a level with their fortunes..... Words had to change their ordinary meaning and to take that which was now given them. Reckless audacity came to be considered the courage of a loyal ally; prudent hesitation, specious cowardice; moderation was held to be a cloak for unmanliness; ability to see all sides of a question, inaptness to act on any. Frantic violence became the attribute of manliness; cautious plotting, a justifiable means of self-defense.

The advocate of extreme measures was always trustworthy; his opponent a man to be suspected.... Thus every form of iniquity took root in the Hellenic countries by reason of the troubles. The ancient simplicity into which honour so largely entered was laughed down and disappeared; and society became divided into camps in which no man trusted his fellow....[6]

In the modern era, and despite some initial resistance, both the new military history and war and society approaches have been embraced by many within the academy, and it is not at all controversial for current scholars to assert that war must and should be considered as a complex human enterprise inextricably intertwined with the societies that wage it. The significance of this shift is reflected in its appearance even in general surveys of military history. In his *What is Military History*, for example, Steven Morillo advocates for what he terms "a broad definition of military history" as something that:

encompasses not just the history of war and wars, but that includes any historical study in which military personnel of all sorts, warfare (the way in which conflicts are actually fought on land, at sea, and in the air), military institutions, and their various intersections with politics, economics, society, nature, and culture form the focus or topic of the work.... Indeed the best history, military and otherwise, necessarily crosses many of these abstract boundaries in order to present as rich and rounded a view of the past as possible.[7]

Even when modern histories follow a more traditional narrative path, or maintain a focus on wartime operations or military activities writ large, a more theoretical war and society framework may also be included. While the chapters within the recent popular military history textbook *The Cambridge History of Warfare,* for example, provide students with helpful though generally standard chronological overviews, they also work together to advance a larger argument that—as explained in the volume introduction—emphasizes the cultural and social foundations of war-making. "Every culture develops its own way of war," the volume editor Geoffrey Parker notes, and there has been since antiquity a specifically Western way of war that "rests upon five principal foundations: technology, discipline, an aggressive military tradition, eclecticism, and finance."[8]

While this way of thinking about the study of warfare has thus spread throughout the larger field of military history—and yielded not just numerous monographs, journals, and book series, but also undergraduate courses, scholarly centers, and entire academic programs—the trend has been less linear within the narrower field of pre-Napoleonic early modern European military history. After a flurry of scholarship spurred by the innovative arguments about ties between military developments and the rise of the centralized state that emerged in relation to the so-called Military Revolution debate (sparked in the 1950s by Michael Roberts), many early modern military historians then moved away from big theoretical models and new approaches. Some embraced micro-historical studies of particular wars, campaigns, battles, or generals, while others retreated to the comfort of the "old" military history, often termed (even by its fondest supporters) "drums and trumpets" history.[9]

By the 1980s and 1990s, however, the tide again began to shift, with increasing scholarly interest in topics that explored the intersection of early modern European war and society. This trickle, led in particular by British historians, began to resemble more of a flood by the 2000s, and we are now seeing a surge of interest in these newer approaches throughout the field of early modern European history. These works have not only drawn or expanded on earlier scholarship on the Military Revolution, they have explored the political, diplomatic, social, and economic context of early modern warfare, they have investigated the experiences and values of soldiers and their interactions with civilians, and, in sum, they have expanded our knowledge of the ways in which warfare serves as a social and cultural phenomenon deeply embedded into all levels of human existence.[10]

So too is there now an ever-growing body of cross-disciplinary works by scholars interested in applying to early modern military history the perspectives offered by different methodologies and thematic topics. Recent examples of this latter expansion in military history include everything from examinations of warfare imagery or military architecture and urbanism;[11] to gender studies that explore women's roles in early modern warfare or literary representations of masculinity and martial identities and culture;[12] to environmental historians looking at how weather and agricultural cycles placed limits on military operations or how warfare was influenced by the climactic trends of the Little Ice Age.[13] Literary scholars too have advanced the field of early modern military history by exploring the use and significance of warfare in plays and other works of literature, but also by publishing and analyzing eye-witness accounts of war by soldiers, officers, clergy, and ordinary civilians. In addition, we now have a large historiography of works that bring early modern warfare into a global and more comparative perspective.[14]

In this volume, we aim to advance these trends and further demonstrate some of the exciting possibilities offered to early modern studies by both the new military history and war and society approaches. This book project grew out of a research initiative sponsored by the Center for Austrian Studies and the Center for German and European Studies at the University of Minnesota. In Minneapolis in November 2018, we assembled an interdisciplinary team of leading academics and rising scholars from North America and Europe interested in the many conflicts of the long seventeenth century. We then met again in St. Louis at the Sixteenth Century Society & Conference in October 2019, where we offered a series of engaging panels that further pushed scholars to break out of the traditional disciplinary or theoretical categories that often channel our studies of early modern warfare. We asked participants to think imaginatively, consider alternate or interdisciplinary viewpoints, or reimagine what war and conflict may have meant or how it was experienced. We also asked our contributors to consider how war functioned or was experienced across the traditional geographic, national, or conflict-specific boundaries that have almost always defined military history.

The resulting essays, which we have collected for this volume, explore the social and cultural consequences of early modern warfare, trace some of the ways it functioned or was justified, and illuminate how it was intertwined with culture, society, and daily life. The goal, as mentioned above, is to attack these issues in innovative or unusual ways, and thereby suggest to the reader some interesting alternatives and threads for reconsidering the place and meaning of conflict and warfare within the larger history of early modern continental Europe. Indeed, one of the most important features of this book is its interdisciplinary nature, as we bring together military history with political and diplomatic history, the history of religion, social history, economic history, the history of ideas, the history of emotions, environmental history, art history, musicology, and the history of science and medicine. Most of our contributors, in fact, are scholars who do not consider themselves military historians but instead seek to understand the ways in which conflict and violence resonate more broadly across continental European culture and society. The volume is also distinctive in emphasizing and highlighting multiple geographic areas within the continent. Essays thus focus not only on Western and Central Europe, but also the often-ignored European peripheries, such as the Baltics, Scandinavia, and the Habsburg-Ottoman borderlands of Southeastern Europe. Many of the chapters, moreover, like early modern armies, do not limit themselves to a single geographical area, but instead eagerly and profitably range across Europe's political boundaries.

All essays engage with the work's central theme of reconsidering or thinking in different ways about the experience, significance, or resonance of early modern European continental warfare, even while their topics reflect the contributors' own unique interests and expertise. To highlight interesting connections and provide a logical structure, we have divided the essays into three major sections representing broad thematic categories, although the ideas and arguments raised in each section also resonate with those in others. In Part I, essays examine how warfare and conflict influenced or were entwined with European learning, culture, and the arts; in Part II, essays address the interrelationship and ties between warfare and ideas or ideologies; and in Part III, essays offer new ways of thinking about the costs and consequences of war.

Part I: Learning, Culture, and the Arts

Part I of this volume, on learning, culture, and the arts, opens with Tryntje Helfferich's chapter on learning the art of war in early modern Europe. Drawing on contemporary military manuals and treatises, Helfferich describes how the era's significant changes in military practice, strategy, and technology required a new type of learning and expertise among the noble officer class, one that simultaneously embraced the newest developments of the gunpowder era and the classical methods and values of Roman antiquity. This shift in the requirements for successful military leadership, she argues, opened up new paths for merit-based advancement, one eagerly seized by ambitious young noblemen interested in the possible social and financial advancement offered by a successful military career. Perhaps less interested than their ancestors in defending a particular sovereign (or sometimes even religion), these men traveled to study or practice the art of war under the most renowned European captains, and also perfected their skills and knowledge through the flood of military prescriptive literature or treatises emerging from Europe's printing presses. Finally, Helfferich argues that as they pursued their individual ambitions, the era's noble military men also laid the basis for a new pan-European officer class and helped develop a shared international military culture and identity. Through such connections and exchanges, enterprising noblemen across Europe were not only trained in the practicalities of contemporary military action and leadership, but also absorbed a shared understanding of elite military and masculine virtues that combined traditional noble values with those advanced by ancient Roman military authors.

In Chapter 2 Heidi Hausse continues to examine the issue of learning and literature, but shifts our focus from the art of war to the art of military medicine. Utilizing case histories recorded by the early modern medical practitioner and author Johannes Scultetus (1595–1645), as well as archival records and chronicles from the German city of Ulm during the last part of the Thirty Years War (1618–48), Hausse argues that while historians have often treated learned medicine, surgery, and military surgery as separate categories of study, all three uniquely overlapped in Scultetus's career as both an academically trained doctor of anatomy and surgery and, from 1625 to 1645, the town physician of Ulm. She notes that the injuries and wounds he described in his case studies were partly a result of marauding soldiers, sieges, and open battles, but also included bodily damages from duels and everyday accidents. Scultetus brought these cases together in his widely circulated treatise, *Armamentarium Chirurgicum* (1655), a work, Hausse argues, that showed his careful thinking about not only how to do surgery, but also what objects are needed for surgery, and how to communicate such medical knowledge to others. It thus provided physicians, instructors, and eager general readers across Europe with clear descriptions and illustrations of his specialized surgical instruments, and cross-referenced these with information on the techniques and procedures he used to treat the people of Ulm. In this chapter, Hausse not only exposes the many-layered ways of performing surgery in a time of war, she also offers a new way of understanding the stormy history of Ulm during the mid-seventeenth century. Furthermore, she suggests the multifaceted relationships among bodies of surgical knowledge learned in university lecture halls, in Latin and vernacular treatises, and in hands-on work with patients—often patients injured through violence or conflict.

From the physical suffering of patients in Ulm during the Thirty Years War, in Chapter 3 Thomas Marks then moves us to consider the emotional suffering of German Lutherans during the same era. In this essay, Marks draws from approaches in the history of emotions as he first illustrates how the sigh in Lutheran theology was often considered a form of prayer that emerged especially during difficult times of suffering, as may also be seen in Robert Monro's description, provided above, of how soldiers "begged for reconciliation in Christ, by our public prayers, and secret sighs, and groans."[15] Marks then highlights the relationships between this affective gesture and a growing corpus of sigh-prayer devotional literature published throughout the first part of the seventeenth century. Faced with wartime dangers and fears, German Lutheran composers within the Empire also began to compose with increasing frequency a type of sacred vocal compositions they called *Seufftzer* or *suspiria* (sighs). While these pieces were quite diverse, Marks argues that they conformed stylistically and in their composition to the larger genre of Lutheran sigh-prayers and typically addressed the harsh realities of the war and the necessity for peace. In this essay Marks also traces the various strategies composers employed when setting to music these wartime *Seufftzer,* such as those found in the highly popular collection of wartime prayers by Josua Stegmann (1588–1632), and suggests that their performance not only taught Lutherans emphatic ways to pray, but also offered an emotional vocabulary by which they might put words—and sounds—to their otherwise indescribable experiences of war.

From music, we then turn to art and architecture, and to two chapters that demonstrate very different approaches to the intersection of the history of art and the history of warfare. First, in Chapter 4, Jeffrey Chipps Smith addresses the impact of the Thirty Years War on the Society of Jesus, which, he argues, was both a prime target for Protestant forces and a source of curiosity. Drawing on diaries, letters, and other sources, including visual records, Chipps Smith focuses particularly on the Jesuits' artistic heritage in southern German towns such as Bamberg, Eichstätt, Landshut, Munich, and Neuburg, and traces how members of the Society of Jesus, their students, and their patrons simply tried to survive the horrors of war as their colleges and churches were occupied or destroyed, and their artistic treasures plundered or damaged. The terror and uncertainty of war, he writes, halted virtually all of the Society's building campaigns and major artistic projects. The ultimate restoration of Jesuit churches and the renewal of their artistic decorations only came decades after the war ended with the Treaty of Westphalia in 1648. Even then, Chipps Smith argues, the scars of their wartime experiences during the 1630s and 1640s lingered. This essay thus emphasizes in vivid form the destructive power of warfare, and the long-term cultural, artistic, and even institutional losses it often generates.

As an indication of the complexity of warfare as a social phenomenon, we then move on in Chapter 5 to Krisoffer Neville's essay, in which he argues that within the destruction of the Thirty Years War and subsequent conflicts, there was also a form of cultural florescence often overlooked by scholars. While not denying the war's losses and destructive power, Neville argues that the enormous wealth and stature acquired by certain ambitious seventeenth-century military commanders, such as Albrecht von Wallenstein, Carl Gustaf Wrangel, and Prince Eugene of Savoy, provided them both the financial means and the social impetus to become important patrons and commission significant works of art. These three men, driven by the desire for titles and wealth, were able to rise socially and financially through their military skill and then attempted to demonstrate and reinforce their new-found social influence through cultural extravagance and patronage. Neville notes that Wallenstein and Prince Eugene each competed with their employers—the imperial Habsburg family—in the scale and quality of their palaces and other artistic commissions. Wrangel, who worked for the Swedish crown, was less grandiose in his resources and ambitions, but in addition to serving as a major patron, he also became a cultural arbiter who often determined the development of cultural life at court. To some extent, Neville argues, these generals' cultural efforts and spending on art and architecture offset the parallel decline among some older noble families, whose own fortunes and status as art patrons were ruined by war and other calamities over the course of the seventeenth and early eighteenth centuries. This chapter also reminds the reader of the prevalence of wartime looting, a phenomenon that facilitated a significant movement of artistic works and books across the continent.

Part II: Ideas and Ideologies of War

Part II begins with Daniel Riches's essay, which focuses more closely on the ideas and ideologies wrapped up with, explaining, or justifying early modern warfare. This chapter also shifts our focus from Central Europe to France, and to the era of the wars of Louis XIV (1638–1715). Riches notes that European opposition to the Sun King brought together groups with fundamentally different interests, confessional alignments, and governmental forms. Riches's central argument is that creating and preserving such alliances required not only the mobilization of enormous political resources, but also the development of a sufficiently ecumenical language of common cause. In the process, Europeans unintentionally articulated a set of concepts that contributed in important ways to emergent understandings of what Europe itself was and how its various parts should relate to one another. Drawing especially on the extensive pamphlet literature that emerged in response to Louis's wars, Riches focuses on a series of themes that contributed to what he terms a new "language of Europe." This included appeals to the "Liberties of Europe," intense discussions of a European balance of power, transformations and redeployments of the concept of Christendom as a fundamentally cultural entity anchored in cross-confessional values and standards of conduct, as well as debates over the plight of refugees and the collective responsibility to assist them. Taken collectively, Riches argues, the wartime discussion of these issues helped produce changed understandings of Europe that persisted long after Louis's reign itself was finally over.

Continuing with the theme of ideology and adding to a robust body of scholarship on the significance of religion in the wars of the early modern era, in Chapter 7 Howard Louthan addresses the role of Reformed Protestant militantism in contributing to the outbreak of the Thirty Years War. He argues that although many historians have portrayed the rash acceptance of the Bohemian crown by the young and inexperienced Elector Frederick of the Palatinate as a major trigger for the conflict, Frederick's court preacher and advisor, Abraham Scultetus, bears much of the blame as the *éminence grise* of the Palatine affair. Although Scultetus is now ignored or at least underappreciated in most modern scholarship, he was, as Louthan explains, a significant theologian, historian, and preacher who was active in many of the era's most important religious events, and who also stood at the center of an extensive Calvinist network that reached from England to Poland. Although rejecting the idea that Calvinism in this era was inherently more militant than other confessions, since many of the convictions that could lead to violence were equally shared among Catholics, Lutherans, and Calvinists, Louthan nevertheless argues that it was in large part Scultetus's vision of a unified Protestant front against the threat of resurgent Catholicism, as well as his absolute confidence that divine providence favored the Reformed, that helped push Frederick to make a decision that upset the fragile confessional balance of Central Europe and brought about the devastating cataclysm of the Thirty Years War. Louthan's work thus not only provides a more nuanced view of the religious currents swirling around Europe during this pivotal period, it also offers an interesting intersection between the new military history and the new diplomatic history, a field that attempts to highlight the importance of individual advisors, courtiers, and diplomats in the process of war- and peacemaking, and thereby complicates traditional historical narratives that privilege the role of rulers.

The focus on the importance of religious ideology and shared confessional identities in shaping or sparking early modern conflict is further developed and supported by Brian Sandberg in Chapter 8, which explores the French religious wars of the late sixteenth and early seventeenth century. Drawing on an extraordinarily wide array of manuscripts and printed sources, Sandburg closely analyzes the intersections between the practice of religious warfare and the ways in which this conflict was described and represented by contemporaries. Like Riches, therefore, Sandberg asks us to look more closely at the importance and power of language, and how it helps us better understand the meaning and import of early modern war. In France, he argues, written descriptions of war, and especially the language of "troubles," show how religious divisions and confessional animosities not only supplied the root causes of the Wars of Religion, they also motivated many nobles and commoners to join military units, shaped the war aims and strategies of the belligerents, and guided peace negotiations. Religious identities and confessional politics, he argues, thus became embedded in the everyday practices of waging war and making peace, such that not only did contemporaries see the mere existence of religious difference as sparking violence, many also specifically blamed the opposing confession for causing the troubles. In sum, Sandberg demonstrates that the French Wars of Religion are appropriately named, since the language used by early modern French combatants and civilians indicates that while political, economic, social, and cultural factors certainly figured in the conflicts, they were almost always mediated through a religious lens.

Like both Louthan and Sandberg, in Chapter 9 Dariusz Kołodziejczyk also addresses the role of religious ideology in warfare. His aim, however, is to examine to what extent religious, political, and dynastic tensions, which are typically seen as significant factors driving early modern wars, also influenced individual decisions of young noblemen to enter military service or join a particular campaign. To address this question, which once again demonstrates the important insights the new military history can provide about the ideas and experiences of fighting men, Kołodziejczyk examines a selection of memoirs and diaries left by noble career soldiers who served in the Polish-Lithuanian army, both foreigners and natives. Unlike Louthan and Sandberg, however, he finds that to these men, the hope for material enrichment and social promotion was perhaps more important than religious motives or the proclaimed duty to defend one's fatherland and monarch. In particular, he argues, the popularity among the noble soldiery of the various wars against the Ottomans and Crimean Tatars owed less to the soldiers' religious fervor or anti-Muslim sentiment than it did to the widespread expectation of rich spoils, valuable captives, and prestige. Such motivations, he writes, help explain why wars against Muscovites or Swedes were far less popular. These conflicts offered fewer spoils and a greater risk of permanent captivity or death in a pitched battle, since the Russian and Swedish armies were largely recruited among peasant infantrymen who did not share either the elite warrior honor code of the Polish-Lithuanian nobility or their tradition of ransoming. The least popular wars, Kołodziejczyk notes, were those waged against Ukrainian Cossacks, who were viewed by most noble officers as nothing but rebellious serfs. Such disdain for these enemies, he argues, may partly account for the prevalence in these wars of acts of savage cruelty and massacres by the Polish soldiery.

The violence and brutality demonstrated in the wars against the Cossacks is a reminder, if one were needed, of the sheer destructiveness of war. Indeed, such a topic is hard to avoid in any scholarly consideration of the early modern era, as almost all estimates of the numbers of dead, wounded, and displaced in the era's wars—as well as the disease, famine, and economic crises that usually accompanied them—stretch into the tens of millions, not to mention the trauma and suffering survivors experienced from marauding troops, the constant insecurity, the loss of friends and family, and widespread damage to or destruction of homes, barns, crops, orchards, mills, bridges, towns, and even entire cities. Given the extent of such losses, it is not surprising that descriptions of wartime destruction and violence appear frequently throughout contemporary chronicles and other sources. One contemporaneous account of the 1631 sack and destruction of the city of Magdeburg, for example, reports that:

> the soldiers began to beat, frighten, and threaten to shoot, skewer, hang, etc., the people, so that even if something had been buried under the earth or locked away behind a thousand locks, the citizens would still have been forced to seek it out and hand it over. Through such enduring fury—which laid this great, magnificent city, which had been like a princess in the entire land, into complete burning embers and put it into such enormous misery and unspeakable need and heartache—many thousands of innocent men, women, and children were, with horrid, fearful screams of pain and alarm, miserably murdered and wretchedly executed in manifold ways, so that no words can sufficiently describe it, nor tears bemoan it.[16]

Part III: The Costs of War

In the final section of this volume, Part III, our contributors address this issue of wartime damages or costs, and consider some of the ways individuals and institutions attempted to cope with them. Perhaps even more than in the previous two sections, the essays here show a striking variety of approaches, a fact that helpfully reinforces this volume's goal of demonstrating the many ways in which the history of warfare intersects with other forms of historical inquiry or offers fruitful avenues for cross-disciplinary studies.

First, in Chapter 10, Mary Elizabeth Ailes uses the tools of social and gender history to investigate early modern women's experiences of warfare, with a particular focus on how warfare affected the family lives of officers' wives in seventeenth-century Sweden. Unlike in modern militaries, most early modern armies were accompanied by large numbers of camp followers, including women sutlers, servants, companions, and wives. These women sometimes participated in battles or sieges, but also faced such daunting conditions as frequently moving from place to place, being fired upon or attacked by enemy forces, tending to children while on campaign, and giving birth in military camps or in the open on the march. Because of the constant warfare that the Swedish kingdom experienced during this era, its officers spent long periods abroad, and thus, as Ailes notes, their wives did too. While sources that focus specifically on the lives of these women are hard to find, Ailes carefully extracts details from official records and personal correspondence in order to explore such issues as how the women decided to stay home or go abroad on campaign with their husbands, and what strategies they and their husbands used to create stable homes for their families in the midst of conflict. She notes as well the difficulties faced by husbands and wives in shielding their children from danger, a concern that sometimes meant sending them away to be raised by relatives. In addition, Ailes points out the significance of social networks for these women and their husbands, and argues that in times of trouble, especially when facing the death of a spouse, both military wives and officers developed and called upon extensive and long-lasting social ties that included not only other military colleagues but also extended family and friends.

Where Ailes explores the social strategies used by Swedish officers' wives to cope with harms to family life, Sabine Jesner, in Chapter 11, considers the institutional strategies used by Habsburg governing authorities to address conflict-driven pestilence and disease. Based on extensive archival research and focusing particularly on a bubonic plague outbreak that spread from the Ottoman Empire into the Habsburg principality of Transylvania during the Habsburg-Ottoman War (1737–39), Jesner analyzes how the Habsburg civil and military apparatuses responded to the problem of infectious disease—something that threatened not only the army's military effectiveness, but also the wellbeing of the civilian population. Rather than only one response, she explains, these institutions tried multiple overlapping strategies, including an intensification of central control over the movements of people and goods; a repurposing of old bureaucratic structures and creation of new institutions, such as provincial Sanitary Commissions and quarantine stations; and the introduction of new military procedures and sanitary regulations for medical staff. In the long run, she argues, and despite some lack of enforcement by officials and resistance from local residents, these measures were reasonably effective, helping end the outbreak by 1740.

The essays of Ailes and Jesner thus both demonstrate some of the ways that early modern people successfully managed or mitigated wartime damages. The volume's final two essays, however, explore some of the ways in which the costs of war were less easy to avoid or had far longer-term impacts or affects. First, in Chapter 12, Mindaugas Šapoka offers an important examination of the lasting influence of the events surrounding the Great Northern War (1700–21) on geopolitical relations in Eastern Europe. He notes that while Russia only established its protectorate over the Republic of Poland-Lithuania in the late 1720s, it was in the previous two decades that the shift in power relations between Poland-Lithuania and Russia began. Drawing on correspondence and government records, Šapoka demonstrates that during this war, the vast majority of the bitterly divided Polish-Lithuanian ruling elite and nobility accepted the Russians as their allies. Concerned by the power of the Republic's ruler, Poland's Saxon king Augustus II, whom many in the nobility feared was attempting to strengthen his royal powers at the expense of the Republic's constitution and traditional political liberties, the elites acquiesced to Tsar Peter the Great's growing interference in the internal affairs of the Republic. In other words, fearing one autocrat, Šapoka argues, the local elites unwisely placed their (partial) trust in another. The result was the growth of Russia in the early eighteenth century as a major European military power and, at the same time, the decline of the Republic of Poland-Lithuania and its increasing slide under Russian influence and dominance. This shift, moreover, was not simply a short-term phenomenon, but an event of epoch-making significance in Eastern European history, and one that provides important background to current geo-political events, including Vladimir Putin's 2022 invasion of Ukraine and ongoing attempts to extend Russia's power into Eastern Europe.

In Chapter 13, the volume's final essay, Mary Lindemann focuses on the Electorate of Brandenburg, an area whose rulers, like those of Russia, made major military and diplomatic moves to enlarge their territory and influence during the late seventeenth and early eighteenth centuries. Brandenburg is also notable, however, for the extent to which it, more than many others in Central Europe, fell prey to the vast destructive fury of the Thirty Years War, suffering a massive population decline of fifty to ninety percent due to combat, disease, and migration. Lindemann notes that while these human losses have been well studied, far less attention has been paid to the costs of war on the landscape, both built and natural, and to the longer-term disruptions (the aftermath of war) that continued for decades after the Peace of Westphalia. Furthermore, she argues that such hidden costs—including the damage to the land that fighting and troop movements caused—shaped a recovery that profoundly altered social, economic, communal, and even political life, while problems associated with the decline of populations—including not only the need to recast social, communal, economic, and especially agricultural relationships but also the complications of war-conditioned neglect—proved every bit as decisive and wrenching as immediate damage. Based on her study of archival and other sources, she also demonstrates how such neglect persisted for decades and affected the landscape, altered the vegetational profile, skewed the animal world, and retarded the reconstruction of structures that served as markers and anchors of social life in countryside and towns alike. Rebuilding was thus an extended process, not an event, and required a range of people—nobles, villagers, local administrators, merchants, and tradespeople—often ignored in the history of the region. So, while much of the story of these years has been cast in the larger narrative of the rise of Prussia, she argues, just as critical were local efforts and ones by no means directed by the Hohenzollern electors and their ministers. Even in matters that lay particularly close to the heart of the Great Elector, such as the *Peuplierungspolitik* (repopulation policy), and especially in the first years after 1648, the efforts from below proved more consequential.

Lindemann's focus on the long-term physical wounds or impacts of war is an important one, since most traditional military histories lose interest once the battles are done, the soldiers have all returned home, and the weapons are once again hung on the wall. But after wars are over, one must also count the dead and take stock of the many things that have been lost, or are missing or harmed: lives, homes, and farms ruined; literature, art, architecture smashed; entire communities gone. People rally to cope with such crises, as we see for example in the essays by Ailes and Jesner, but the losses add up nevertheless. Sometimes these long-term costs are more subtle, but for that no less damaging. Children not born, crops not planted, opportunities missed. Social disruption and trauma are perhaps just as hard to measure, but can also be long lasting, as we see not only in Lindemann's essay but also in Sandberg's, since the bitterness and division he described as emerging from religious conflicts would resonate within French history for many years to come. Cultural losses create another painful gap, as Chipps Smith's description of the damage to the Jesuits' artistic heritage demonstrates, and political damage too can have a long tail, such as the decline of the Palatine Electorate after the Thirty Years War as suggested by Louthan, or that of Poland-Lithuania after the Great Northern War as described by Šapoka.

Yet, interestingly, most of the essays in this volume focus not on the horrors and destruction of war, but on the ways in which it is transformative, productive, or even creative. We see the development of new military professionalism and forms of training in Helfferich's essay, new surgical tools and methods in Hausse's, and new musical forms in Marks's. Kołodziejczyk's chapter, along with those of Helfferich and Neville, also shows how war could serve as a tool to advance or benefit both individuals and families. More concrete, physical echoes of the productive nature of early modern warfare might also stretch not just for decades, but even into the modern day. We see this, for example, in Neville's chapter, since modern tourists who today visit the Swedish castle of Skokloster are walking down halls built with money that Carl Gustaf Wrangel mostly acquired through wartime extractions, and are admiring paintings and textiles his troops may have looted. Similarly, Jesner's essay not only describes the creation of new institutions, but also points to the Holy Trinity Column on the *Piața Unirii* in Timişoara, built as an enduring physical testament to the lethal epidemic that struck Habsburg lands during the Habsburg-Ottoman War. Other long-lasting mementos of early modern wars appear more intangibly in such things as a new conception of a united Europe, a topic explored in Riches's essay. One might also note that Šapoka's essay, in describing the decline of Poland-Lithuania, is simultaneously describing the rise of Russia, a phenomenon of lasting import in the history of Europe.

These are only a few of the ways in which the essays in this volume trace not only the nature or practice of early modern European warfare, but also how its influence has stretched well beyond the battlefield and has lodged itself inextricably into the weft and warp of European society and culture both then and now. Contributors have brought in a wide range of different perspectives and disciplinary approaches—including political and diplomatic history, religious history, social history, economic history, the history of ideas and of emotions, environmental history, art history, musicology, and the history of science and medicine—and have thereby put forward a more nuanced and complex picture of early modern European warfare that enriches our understanding and expands our fields of view. Taken together, the essays in this volume make no pretense of being a systematic or comprehensive presentation of the new military history and war and society approaches. However, by demonstrating some of the fascinating possibilities that arise when one is willing to explore early modern European warfare through different disciplines, methodologies, and frames of inquiry, they do help to reinforce the idea that the current field of study is not only thriving, but also wide open to innovation and exciting interdisciplinary analysis.

Notes

1 Robert Monro, *Monro his Expedition with the Worthy Scots Regiment Called Mac-Keyes* (London: William Jones, 1637), II, 63.
2 See Robert M. Citino, "Review Essay: Military Histories Old and New: A Reintroduction," *American Historical Review*, 112, no. 4 (2007): 1070–90.
3 John Keegan, *The Face of Battle: A Study of Agincourt, Waterloo, and the Somme* (London: Penguin, 1983), 95.
4 André Corvisier, *Armées et sociétés en Europe de 1494 à 1789* (Paris: Presses Universitaires de France, 1976), published in English as *Armies and Societies in Europe 1494–1789*, trans. A.T. Siddal (1979).
5 Ammianus Marcellinus, *The Roman History of Ammianus Marcellinus During the Reigns of The Emperors Constantius, Julian, Jovianus, Valentinian, and Valens*, trans. C. D. Yonge (London: G. Bell & Sons, 1911), 609–18.
6 Thucydides, *History of the Peloponnesian War* 3: 82–83, in Thucydides, *The Complete Writings*, trans. Robert Crawley (New York: Modern Library, 1951).
7 Stephen Morillo and Michael F. Pavkovic, *What Is Military History?*, 2nd ed. (Cambridge, UK: Polity, 2012), 4.
8 Geoffrey Parker, ed. *The Cambridge History of Warfare* (New York: Cambridge University Press, 2005), 1. Note that Victor Davis Hanson, who also contributed to Parker's volume, has advanced a very different definition of the Western way than Parker advanced argued here. Victor Davis Hanson, *The Western Way of War: Infantry Battle in Classical Greece* (Berkeley: University of California Press, 2009).
9 More colorfully, in a review article, Colin Jones described this traditional mode of writing military history as something "which, in the least competent hands, all too swiftly descended to the level of a chronicle of one damn battle after another, with interludes for descriptions of uniforms." Colin Jones, "New Military History for Old? War and Society in Early Modern Europe," *European Studies Review* 12 (1982): 97–108 (97). Nevertheless, few could argue against the usefulness of having a clear, chronological narrative history of early modern European wars, detailed studies of their military leaders, or explanations of how warfare as an art and science developed and changed over that time. See here Dennis E. Showalter, "A Modest Plea for Drums and Trumpets," *Military Affairs* 39, no. 2 (1975): 71–74. For the Military Revolution, see Geoffrey Parker, ed. *The Military Revolution: Military Innovation and the Rise of the West, 1500–1800*, 2nd ed. (Cambridge, UK: Cambridge University Press, 1996); and Clifford J Rogers, ed., *The Military Revolution*

Debate: Readings on the Military Transformation of Early Modern Europe (Boulder, CO: Westview Press. 1995).

10 Examples of the new military history include Charles Carlton, *Going to the Wars: The Experience of the British Civil Wars, 1638–1651* (New York: Routledge, 1992); and Christopher Duffy, *The Military Experience in the Age of Reason* (London: Routledge and Kegan Paul, 1987). For examples of war and society approaches, see John R. Hale, *War and Society in Renaissance Europe, 1450–1620* (Baltimore: Johns Hopkins University Press, 1985); M. S. Anderson, *War and Society in Europe of the Old Regime, 1618–1789* (New York: St. Martin's Press, 1988); Frank Tallett, *War and Society in Early Modern Europe, 1495–1715* (London: Routledge, 1992); J.B. Wood, *The King's Army: Warfare, Soldiers and Society During the Wars of Religion in France, 1562–1576* (Cambridge, UK: Cambridge University Press, 1996); M. P. Gutmann, *War and Rural Life in the Early Modern Low Countries* (Princeton: Princeton University Press, 1980); Jan Glete, *War and the State in Early Modern Europe: Spain, the Dutch Republic and Sweden as Fiscal-Military States, 1500–1660* (New York: Routledge, 2007); and Christine Shaw and Michael Mallett, eds., *The Italian Wars, 1494–1559: War, State and Society in Early Modern Europe* (London: Pearson, 2012).

11 Pia F. Cuneo, ed., *Artful Armies, Beautiful Battles: Art and Warfare in Early Modern Europe* (Leiden: Brill, 2002); Martha Pollack, *Cities at War in Early Modern Europe* (Cambridge, UK: Cambridge University Press, 2010).

12 See especially John A. Lynn, *Women, Armies, and Warfare in Early Modern Europe* (Cambridge, UK: Cambridge University Press, 2008). See also Ulinka Rublack, "Wench and Maiden: Women, War and the Pictorial Function of the Feminine in German Cities in the Early Modern Period," *History Workshop Journal*, no. 44 (1997): 1–21; Stefan Dudink and Karen Hagemann, "War and Gender: From the Thirty Years' War and Colonial Conquest to the Wars of Revolution and Independence—an Overview," in *The Oxford Handbook of Gender, War, and the Western World since 1600*, eds. Karen Hagemann, Stefan Dudink, and Sonya O. Rose (Oxford: Oxford University Press, 2020), 36–73.

13 For example, see Dagomar Degroot, "'Never Such Weather Known in These Seas': Climatic Fluctuations and the Anglo-Dutch Wars of the Seventeenth Century, 1652–1674," *Environment and History* 20, no. 2 (2014): 239–73; Janken Myrdal, "Food, War, and Crisis: The Seventeenth Century Swedish Empire," in *Rethinking Environmental History*, eds. A. Hornborg, J. R. McNeill, and J. Martinez-Alier (Lanham, MD: AltaMira Press, 2007), 79–98.

14 For an example of the former, see Matthew Woodcock and Cian O'Mahony, eds., *Early Modern Military Identities: Reality and Representation* (Cambridge, UK: D.S. Brewer, 2019). For the latter, see Jeremy Black, *European Warfare in a Global Context 1660–1815* (London: Routledge, 2006).

15 Monro, *His Expedition*, 63.

16 Tryntje Helfferich, *The Thirty Years War: A Documentary History* (Indianapolis: Hackett Publishing, 2009), 109.

Part I

Learning, Culture, and the Arts

1 "Skilfull Captaines in Warlike Affaires"

Learning the Art of War in Early Modern Europe

Tryntje Helfferich

The Knowledge of Warfare is thrown away on a General who dares not make use of what he knows. I commend it only in a Man of Courage and of Resolution; in him it will direct his Martial Spirit; and teach him the way to the best Victories, which are those that are least bloody, and which tho' atchiev'd by the Hand, are manag'd by the Head. Science distinguishes a Man of Honour from one of those Athletick Brutes whom undeservedly we call Heroes. Curs'd be the Poet, who first honour'd with that Name a meer Ajax, a Man-killing Ideot (John Dryden, 1700).[1]

The early modern era saw the European art of war transformed, with military leaders and theorists reviving certain ancient Greco-Roman military practices, especially the use of massed infantry formations, while simultaneously modernizing to meet the challenges brought by the new gunpowder weapons. These extraordinary changes not only influenced all aspects of war, they also affected the demands placed upon the noble officer class. In addition to their traditional requirement of personal skill at arms, commanders now also needed to learn how to read maps, craft written orders and reports, oversee new methods of military training and drill, implement complex tactical maneuvers, and so on. Precision, self-discipline, technical proficiency, plus expertise in the use of artillery, a knowledge of siegecraft, and a keen eye for logistics were now the hallmarks of the successful officer. Thus, as J.R. Hale has noted, in this era "it was no longer enough simply to be brave and a gentleman, to know how to ride and to use a lance and a sword" since "the potential officer needed to know more about a more complex and a more disciplined craft of war than had his late medieval predecessor."[2] The historian Keith Thomas, looking at early modern English military elites, puts a greater emphasis on the continued significance of masculine virtue and bravery as the gentleman soldier's "most essential qualification," but has also argued that both longer-term changes in the nature of the English aristocracy and the new technical demands of warfare meant that by the end of the sixteenth century, "military success depended on the collective actions of well-trained armies, not on dazzling feats of individual

DOI: 10.4324/9781003157700-3

heroism. Disciplined obedience was what mattered most; sheer animal vigour became less important."[3]

As such scholars have shown, early modern contemporaries were themselves keenly aware of the need for an officer class that was both supremely disciplined and well-trained in the new art of war. John Dryden's scorn, for example, of the unlearned officer as nothing but an "Athletick Brute" and "Man-killing Ideot," is echoed in the anonymous *The A, B, C, of Armes* (1616), which stated that if a nobleman "should at any time, either for their own or the countries good, attempt to undertake command, or expect preferment by the wars," they should not then "disdain to be taught, or shame to seem to learne."[4] Likewise, the seventeenth-century English author Robert Ward argued that every man who planned to "beare the honourable name of a Souldier, well beseeming a generous person of that Noble qualitie," "must and ought to be diligent, and carefull to learne the art of Warre."[5]

This emphasis on a modern, educated noble officer class, and on military leaders who could offer more than mere "animal vigour" and an aristocratic lineage, made learning the art of war a more complicated process than in prior centuries, but it also offered ambitious young men new paths for merit-based advancement. Eager to make a career and to win titles and fortunes—and not always concerned with defending a particular sovereign or religion—the era's noblemen traveled to learn the newest techniques and tactics under the most renowned European captains, perfected their skills and knowledge through the flood of military prescriptive literature or treatises emerging from Europe's printing presses, or even attended one of the new knightly academies. As they fed their individual ambitions, moreover, these men also laid the basis for a new pan-European officer class and helped develop a shared international military culture. Italian specialists in military engineering, for example, had avid readers in France, while French models of knightly academies were imitated in Germany, German works on gunnery were eagerly read in England and Holland, and Dutch methods of teaching the elements of military drill were imitated everywhere. Through such connections and exchanges, enterprising noblemen across Europe were not only trained in the practicalities of contemporary military action and leadership, but also absorbed and furthered a common understanding of elite military and masculine virtues that combined traditional chivalric noble values with those advanced by ancient Roman military authors.

Field Experience

Since the early medieval era, elite fighting men had begun their military training at home or fostering at another noble court, where in addition to the exercises common to their class, such as riding and hunting, young men might sign on to serve as pages or squires to older knights. In this position, they would learn to ready a horse for battle, handle a sword and lance, and maintain armor and other military gear. They could also accompany their lords into battle, where they might perform as an aide-de-camp, groom, or flag bearer, and might even fight. In this way, young men often gained some experience of war even before they themselves became full knights. This traditional method of learning by apprenticeship continued in one form or another through the early modern era, even as the heavy mounted knight himself was becoming increasingly sidelined by massed, close-order infantry formations as the primary means for the successful prosecution of battles.[6] Many young noblemen who wished to become warriors thus still learned the basics of combat at home or at another court,[7] and then served under an experienced officer, usually a relative or patron. After some years, with luck, talent, money, and/or the right connections, a nobleman might then be promoted to a position of command, such as captain of a company of infantry or cavalry. All the while his education would continue in the field, especially if he were serving under an able general who could help him develop and hone his martial skills.

Such field experience offered ambitious young men some of the greatest opportunities for professional advancement, since not only could they thereby gain familiarity with actual battle conditions and tactics, they could also make important contacts, cultivate patrons, and build their reputation among both their peers and any potential future employers. Frequently, moreover, such employers were no longer exclusively a man's own immediate lord or king, as had often been the case in past centuries. Instead, it was increasingly common in the early modern era for soldiers to fight and gain expertise in the military arts abroad. And while earlier eras had offered soldiers the educational opportunities of service in the crusades of the Levant, the Baltic, or Iberia, in the late fifteenth and early sixteenth centuries it was the Habsburg-Valois wars that raged throughout Italy that beckoned adventurers. These Italian Wars also provided the most modern developments in gunpowder weaponry, siegecraft, and military tactics that were then being developed and implemented, and so European noblemen of the time eagerly traveled to Italy to participate in, and learn from, its many conflicts. There was no better or more useful act of self-improvement, the scholar Justus Lipsius noted, than for a gentleman to "further, and increase his learning by peregrination," and "Italy, that queen of countries," he wrote, was an excellent place to visit, for the Italians were "renowned in all histories, both old and new, for their mighty wars, waged with the whole world, for their martial discipline in war, and politic government in peace."[8]

Once the Italian wars had cooled in the mid-sixteenth century, however, the wars in the Netherlands became Europe's new draw for noblemen seeking military experience. In particular, Protestants from Germany, France, and the British Isles now flocked to the Low Countries in large numbers to serve under such generals as Prince Maurice of Orange-Nassau and his cousins, Willem Lodewijk of Nassau-Dillenburg and Johann VII of Nassau-Siegen. There they learned how to implement Dutch tactical innovations, studied their renowned siegecraft and logistical organization, and followed their stress on the crucial importance of drill for pike and musket armies, a practice that, like many other aspects of the Dutch military reforms, was drawn almost directly from the works of classical Roman authors such as Aelian. Many successful young officers then used their experience in the Dutch armies as a springboard for further military service and advancement, perfecting their education in the armies of England, Italy, France, Sweden, Spain, and the Holy Roman Empire, and thereby further spreading the Dutch method of war across Europe. Thus the Dutch theorist Hugo Grotius noted in his history of the Low Countries that "hither have been brought all the Politike Inventions, both of Pristine and Modern Warfare, the long continuance of the War having drawn from all Parts Foreign Spectators, as to a publike School of War."[9] Similarly, the English-language preface to a 1616 edition of Aelian's *Tactics* noted that the Netherlands "at this day are the Schoole of war, whither the most Maritiall spirits of Europe resort to lay downe the Apprenticeship of their service in Armes."[10]

The training that recruits gained at this Dutch "Schoole of war" might be innovative, but it could also become tiresome. While serving in the Netherlands in 1631 as a young officer, for example, Henri de la Tour d'Auvergne, vicomte de Turenne, complained that rather than seeing action, "we are still at the same place and do nothing. All we do is drill the majority of the troops each morning."[11] Some military units offered a more engaging educational experience, however, and seemed to have been more narrowly targeted toward officer education. The English captain Horatio Vere, Lord Tilbury, for example, commanded a company in the Netherlands in the 1620s that, according to the modern scholar Roger Manning, "functioned much like a military academy; besides military exercises, the young cadets also received instruction in mathematics, fencing, and dancing."[12] Certain French regiments also became known as models and training grounds for young noblemen. *Mestre de camp* Pierre Arnauld d'Andilly, for example, who had himself fought in Holland, wrote that his regiment drew "young gentlemen," "from all points of the compass to Fort-Louis [where Arnauld was governor] to learn their profession there." While at Fort-Louis, young men both studied and practiced Arnauld's methods of martial discipline and drill, performed, as in the Netherlands, in the ancient Roman style.[13]

To improve the ability of the French king to call up men ready for battle, the Huguenot general François de la Noue also recommended that gentry and noblemen maintain and improve their martial skills by meeting once a year for military practices lasting "eight or ten daies," during which the commanders would instruct the men "by the discourses & writings of skilfull Captaines in warlike affaires." Such regular military training, he argued, would "prepare them for time to come: exercise them, and by lively exhortations print in their harts the goodly portraiture of honour, to the ende afterward they might doe things worthie their fame." In the process, Noue suggested, "as well the commaunder as the commanded should be the better learned therein. For everie man coming to this martiall schoole would bring in the best that hee had collected out of the deeds of our auncestours, which by continual conference, adioined to practise, wold be both seene & grauen in memory."[14]

While such practice was useful, contemporary authors often described actual battle experience as vital in preparing a man to become a successful soldier or officer, advice that provided a further spur for ambitious men to travel to wherever war might be found. In his *Briefe Discourse of Warre*, for example, Roger Williams argued that "it is an errour to think that experimented [experienced] Souldiers are sodeinlie made like glasses, in blowing them with a puffe out of an yron instrument. There can be no Leaders of good conduct, unles they have beene in foughten Battailes, asseiged and defended Townes of warre."[15] Williams also noted that the rapidly changing nature of modern war made on-the-ground experiences especially valuable for advancing a soldier's military education. "A campe continually maintained in action is like an University continually in exercises," he noted, and "when famous Schollers [scholars] die, as good or better step in their places: Especially in armies, where there be every day new inventions, stratagems of wars, change of weapons, munition, & all sort of engins newly invented, & corrected dayly."[16]

The author Robert Barret also emphasized the importance of field experience, due both to the rapid changes of contemporary war and to what he saw as contemporary "imperfections of manie training captaines." "Time altereth the order of warre, with new inventions daily," he noted, "then was then, and now is now; the wars are much altered since the fierie weapons first came up: the Cannon, the Musket, the Caliver and Pistoll." This desperately called out for well-prepared, experienced officers, but here some significant tension between old and new forms of training are evident. "There be many points in a souldier, and more in a Captaine," Barret wrote, "which cannot be attayned by reading, but by practise and experience." Those who only know their profession through books, he wrote, are like a captain who tries to drill his troops in forming infantry squares, but "is puzzled," and must call out to the men: "*stand still untill I have looked in my Booke!*"[17] This advocacy for field experience and seeming disdain for theoretical or book-gained knowledge was even more succinctly phrased by the sixteenth-century military architect Giovanni Battista Belluzzi, who stated simply that "measurements and books don't fight."[18]

Just as one can see tension concerning the value of traditional field experience versus the newer theoretical knowledge, one can also observe how the era's religious conflicts complicated the efforts of young noblemen to use foreign military service as their training grounds in the modern art of war. The count of Souvigny, for example, described in his memoirs how he had recalled two of his brothers from their military service in Holland after Cardinal de la Rochefoucauld advised him that "they could not gain salvation as long as they served the Dutch heretics against king of Catholic Spain." Souvigny noted, however, that when they returned, he "found that they had greatly profited on their journey."[19]

The cardinal's scruples were not unusual, as an emphasis on the strict religious obligations of military men was also commonplace within contemporary military advice books. The author Robert Ward, for example, urged all who claimed "that Noble qualitie" of soldier to observe the "defence of true Religion," "to the last breath and drop of bloud."[20] So was it then ever appropriate and morally correct for a Protestant to study in a Catholic military academy, train and fight under a Catholic general, or serve in the army of a Catholic ruler; or for a Catholic to study from, fight for, or serve under a Protestant? The Venetian historian Paolo Sarpi addressed this very issue in one of his works, which appeared posthumously in English translation as *The Free School of War* (1625). Here he recounted a story of some Italian gentlemen who had been denied absolution by their Catholic confessor at The Hague because they had, "for the better attainment of militarie skill, served in the army of the States in the Low Countries."[21] Sarpi argued that the confessor's refusal was inappropriate since the men had joined the army of the Prince of Orange merely to take advantage of "the learnedst Schoole for that kind of discipline, that at this time is in all Europe, yea in the whole world."[22] Because the men's motive had not been "Heresie or false Doctrine," but only military experience and education, they had committed no sin or crime. "The onely and true cause why these gentlemen served under the States of Holland," Sarpi continued,

> was a meere intent to learne the art of Warre under that Captaine unto whom without controversie all the Commanders of our time must give place, without having relation or regard unto the Doctrine and Religion professed in those Countries; Nay, it appearing in fact, in that... they persevered in the Catholicke Romane Religion; neither doe they favor or fight out of respect of the Religion that disagreeth from the Romane faith.[23]

While it was not a universal opinion, early modern European nobility also generally tended, like Sarpi, to favor pragmatism over religious precision when it came to training, career-building, and the pressing demands of waging war. And even as confessional conflict raged throughout the sixteenth and seventeenth centuries, cross-confessional military alliances and coordinated military campaigns were also common—Catholic France, Lutheran Sweden, and Reformed Hesse-Cassel, for example, frequently fielded joint, coalition armies throughout the 1630s and 40s. Even within individual armies, moreover, religious uniformity was not the rule, since while most troop units (company, regiment, or army) were dominated by one or another religious persuasion, few were monolithic. Troops were recruited or impressed from areas of a different religion or from defeated enemy armies; rulers employed generals of a different faith simply because they seemed the best men for the job; and young noblemen seeking to further their military educations continued to sign on with the best captains, wherever they might be found.

Military Literature

In addition to the all-important field experience, the early modern era also provided noblemen another extremely important means of knowledge acquisition—books.[24] The historian David Lawrence has noted that "contemporaries argued... that experience alone was not enough to make a complete soldier and those bound for the wars should also take time to study and read about the *arte militarie*, gleaning what they could from the wisdom and experience of others."[25] This sentiment can indeed be seen in numerous works of the time, such as one by Clement Edmonds, who argued that "Reading and Discourse, are requisite to make a souldier perfect in the Arte militarie, how great soever his knowledge may be, which long experience and much practice of Armes hath gayned."[26] Similarly, in the diary of his military service on the continent, the Scottish Colonel Robert Monro exhorted his reader "to beleeve what profit the diligent and serious Souldier doth reape by reading, and what advantage he gaineth above him, who thinketh to become a perfect Souldier by a few yeares practise, without reading... for I dare be bold to affirme, that reading and discourse doth as much or rather more, to the furtherance of a perfect Souldier, than a few years practise without reading."[27]

Manuals on the art of warfare had of course circulated for centuries in Europe. The knightly reference work *L'Arbre des batailles* [The Tree of Battles] (c. 1382–87), for example, by the Provençal monk Honorat Bovet, would remain popular among European nobility into the early modern era. It would also serve as a major source for another popular work, *Livre de faits d'armes et de chevallerie* [Book of Feats of Arms and Chivalry], Christine de Pisan's famous 1410 treatise for knights, which also served to compile more ancient military authorities. It was in the late fifteenth and sixteenth centuries, however, with the rise of a print culture and increase in literacy rates, that books began to become standard tools for the commanding officer and a regular part of the recommendations by theorists for building a proper military education. In addition to medieval texts, printing houses then also began to issue multiple new editions of works by ancient and Byzantine-era authors in modern Latin and in vernacular translation. Among the most published of these revived classical works were Sextus Julius Frontinus's *Stratagems,* Modestus's *Libellus de Vocabulis rei Militaris* [Book on Military Vocabularies], Maurice's *Strategikon* [Handbook of Strategy], Polybius's *History*, and Aelian's *Tactics,* a comprehensive manual of Hellenistic warfare that, as mentioned above, strongly influenced Dutch military practices, but was also used to train English infantrymen during the Civil War.[28] *De Re Militari* [Concerning Military Matters] by Flavius Vegetius was another text that was well-known in Europe even in Carolingian times, and not only appeared in innumerable editions and multiple languages, but was also liberally cribbed by many early modern military theorists.[29]

Perhaps even more popular than any of these, however, was Julius Caesar's *Commentaries on the Gallic War.* This work received numerous printings in all major European languages and also appeared in popular critical editions, such as Henri de Rohan's *Le parfait capitaine: autrement l'abrégé des guerres des commentaires de César* [The Perfect Captain, or an Abridgement of Caesar's Commentaries] (1642), and Sir Clement Edmonds's *Observations, upon the first books of Caesars Commentaries* (1600). Latin versions also abounded, mostly because the work's clear structure and simple grammar made it a popular text for teaching the language in schools, a role it plays even today. Unlike Vegetius and most other classical texts in wide circulation during the early modern era, Caesar's work was not a theoretical treatise on the art of war, but a military memoir in which he described (in the third person) his own military exploits in Gaul in the sixth century B.C.E. and then in a subsequent civil war against Pompey. As the modern scholar Matthew Woodcock has noted, readers thereby learned through example, absorbing lessons about Caesar's military practices and tactics, but also seeing him as a man who married military genius and bravery with both self-control and literary skill—a combination of characteristics that were then frequently held up as an example for early modern officers to emulate. Captain John Bingham's introduction to Aelian's *Tactics*, for example, took the opportunity to compare his contemporary Prince Maurice of Orange to ancient military leaders, lauding him as "a Prince borne and bred up in Armes and… for skill, experience, judgement, and military literature comparable to the greatest Generals, that ever were."[30]

Knowledge of the military wisdom of historical Greek and Roman armies along with assertions of the superiority of both the ancient way of war and the character of its warrior class were also spread throughout Europe by synthetic or integrative texts such as the humanist Petrus Ramus's Latin work *De Militia Caesaris* [On Caesar's Army], which appeared in P. Poisson's French translation of 1583 as *Traité de l'art militaire ou usance de guerre de Jules César* [Treatise on the Military Art or Julius Caesar's Use of War], and in a similarly titled German edition in 1614. Another analytical volume was Justus Lipsius's *De militia Romana* [On the Roman Army] (1595). In this work, which was written in the form of a dialogue, Lipsius argued that a study of the ancient Roman army would not only benefit scholars of history but also provide modern commanders with important and useful suggestions on how to improve their recruiting, battle array and organization, armament, and discipline.[31]

Similarly, Machiavelli, most famous today for his *Prince*, thought his most important work was instead his *Art of War* (1521), a popular treatise which, in the form of a humanistic discourse and drawing heavily on classical texts (including entire uncredited sections from Vegetius), recommended a return to Roman military practices in such areas as recruitment, logistics, encampments, siege warfare, and training. In describing an imaginary army created on the Roman model, for example, he asked his reader to "see with what *virtù* our men charge. The expertness they have acquired by long drilling and discipline inspires them with confidence." In his preface, Machiavelli noted that his work was designed "for the improvement of others desiring to imitate the ancients in warlike exploits." Although martial virtue was rare in the modern era, he argued, through studying the ancient art of war "it would not be impossible to revive the discipline of our ancestors and, in some measure, to retrieve our lost *virtù*."[32]

The era's military books thus frequently matched practical subjects with encouragement for the modern military officer to moderate his behavior and to consider honor, loyalty, honesty, courage, Roman-style stoicism, and disciplined self-control as complementary and intertwined. Thus the French general François de la Noue's discourses on war, according to its English editor, "doth at large discourse of martiall discipline, and withal teacheth how to use and well employ our weapons," but also "exhorteth every one in his vocation to embrace pietie & to honor Justice: It teacheth Princes, Lordes, and generally all gentrie the true path & high way to climbe to vertue, and to recover the auncient honor of France, as also how to eschue the danger of shame and miserie."[33] Along with the practical training men received in the field, such works were thus useful not merely for teaching young men skill at arms, how to manage a battle or campaign, or how to command men, but also served a prescriptive and moralizing function that aimed to inculcate the era's standards of noble, masculine warrior behavior.[34] It is hard to know, of course, to what extent students and other readers absorbed such lessons, but the scholar David J.B. Trim has argued that early modern "men's actions were shaped by what was valorised in the texts," which not only "prescribe an ideal course of action and ideal virtues, but also describe how the ideals were practiced and implemented."[35]

Many of the works, moreover, like Machiavelli's, held up ancient elites as far more virtuous, noble, and masculine than those of the modern era due not just to their strength, courage, and discipline, but also to their greater familiarity with war. Robert Barret, for example, despaired that "our long continued peace… wherein we have not knowne what the name of warre hath meant" had not only led to the "neglect of Martiall discipline" in England, but also "hath metamorphosed manly minds,"[36] since "the profession of armes" he wrote "is the verie source, mother, and foundation of Nobilitie."[37] Barret was not unusual in this view, for as Keith Thomas has noted, English authors repeatedly argued that both individual and national virtue could be demonstrated through successful military action, and "that war was a necessary healing and purging ritual for an 'effeminate' state, given over to softness and luxury. Warfare was invigorating, peace debilitating, and the fall of the Roman Empire a warning of what would happen if martial discipline and 'manly exercises' were neglected."[38]

In addition to classical texts and works that summarized or recommended ancient practices, another important form of military literature comprised books of military advice written by contemporary authors based on their own personal experiences in battle or as commanding soldiers in the field. Such works by more modern "skilfull Captaines in warlike affaires," as François de la Noue termed them, could be useful to complement the education young noblemen gained elsewhere. The modern scholar F. González de Leon has argued that Spanish military officers, in particular, were prolific authors of such military treatises intended for their fellow noblemen, but works of military science and advice were published by the hundreds in every language in the early modern era.[39] Many were also then translated from their original vernaculars and republished elsewhere, such that foreign texts in translation were widely disseminated from and throughout the entire continent and the British Isles, creating an international body of shared, hard-won practical knowledge that soldiers and prospective soldiers across Europe could draw upon.[40]

Béraut Stuart, seigneur d'Aubigny, in his *Traité sur l'art de la guerre* [Treatise on the Art of War] (c. 1508), for example, drew heavily on Greco-Roman models, especially as seen in Vegetius, but then modified or interpreted these lessons and recommendations by matching them to his own military experiences in the Italian wars during the 1490s. This was thus not just a work of theory, but a practical handbook, which emphasized, as David Potter has noted, "the professional approach to the command and administration of armies: decisions for war, the role of experienced commanders, the disposition of reserves, reconnoitering and mapmaking, spies."[41] Another best-seller was *Das Kriegsbuch* [The Book of War] (1573) by the Bavarian soldier, armorer, and author Leonhardt Fronsperger. Based on his own expertise gained in service to three successive Holy Roman Emperors and to the free imperial city of Ulm, he provided a comprehensive survey of up-to-date information on military theory and law, organization and structures, and fortification and military defense.[42] Duke Philippe Eberhard of Cleves's *Instruction de toutes manières de guerroyer tant par terre que par mer* [Instruction on all Manner of Warfare Both by Land and by Sea] (1558), as another example, used his own extensive experiences of war in Flanders and the Mediterranean in order to provide information and advice on the real-world functioning of an army and navy.[43]

Perhaps the most influential and widely studied late seventeenth-century author on military affairs was the brilliant Italian general Raimondo Montecuccoli, who wrote numerous military works that drew on his own extensive experience fighting for the Austrian Habsburgs during the Thirty Years War and thereafter. His works, such as *Trattato della Guerra* [Treatise on War] (written c. 1641) and *Dell' Arte Militare* [On the Military Art] (written c. 1653), circulated in manuscript form during his lifetime, and were then published posthumously in French, German, and Latin editions that cemented his lasting influence as a military theorist. In his books, Montecuccoli offered deep analyses of fortifications, siegecraft, and tactical and strategic considerations generals should make in war, but also explored the psychology of combat and sought to teach his readers attitudes and behaviors to improve their chance of military success. For example, in an early unpublished work, he advised that in victory, "modest behavior, imputation of success to the grace of God, and tempering the effects of victory are necessary because even great captains become overbearing when their affairs have gone well and prove incapable of digesting their triumph. They ought to remember that *fortuna* is like glass, that while it may glitter, it is also fragile.... The victor must ask, 'What is there to fear?' The answer is that which he does not fear."[44]

Often, the authors of these works not only used their personal military adventures as a selling point for their books, but also suggested it could provide a corrective for readers' own lack of battlefield experience. The retired soldier Robert Barret, for example, noted in the introduction to his book that he had "spent the most part of my time in the profession of Armes, and that among forraine nations, as the French, the Dutch, the Italian, and Spaniard," and suggested that his vast field experience and "such Martial points, as I have noted, gathered, or learned from them" could then benefit "the younger and unexperienced sort" and particularly "such young Gentlemen, and others, my willing countrie men, as have not, as yet, entered within the bounds of Mars his bloudy field."[45] Later in the work, which took the form of a dialogue, Barret addressed the apparent conflict with his prior critique of officers who depended too much on books. While he may have seemed to have disdained the educated "booke Captaine," he wrote, one should not "mistake me" when it came to the importance of texts, since "I ever allow and honour the learned souldier: for what famous Commaunders have there yet bene, unlearned and without letters?" "Thus may you see," he argued, "how many good partes are requisite to a perfect souldier; not learned by hearesay, nor gayned with ease: but with care, diligence, industrie, valour, practice, and continuance; and most of all perfected with learning, annexed with long exercise and use."[46]

Still, even while such authors often touted their personal knowledge or experience, many of these books (as with earlier volumes, such as Christine de Pisan's) were nevertheless strongly derivative of classical works on warfare by Greco-Roman authors, drew on classical examples, or blatantly copied large sections from Vegetius, Aelian, or other ancient writers. Montecuccoli, for example, interspersed his advice with extensive citations of military practices both from his own day and from the ancient world. Barret combined information based on his own exploits in the armies of his day with quotes from Vegetius, discussions of Roman practices, as well as additional historical examples from both the ancient world (such as Greece and Persia) and earlier European wars. Such histories, he argued, were beneficial for modern students, since they taught not only "the courses and changes of times, and ages," but also "the conductions, and strategemes of battels wonne and lost; the carriage of brave men, and baseness of bad persons; the virtue and fame of the valiant, the shame and infamie of the vile: the use of auncient discipline, and manner of our Moderne warres."[47] Other popular works modeled the literary style or pattern of Caesar's *Commentaries* by wrapping their advice and suggestions within personal accounts of battles, campaigns, or entire military careers. The sixteenth-century

French nobleman Blaise de Monluc, for example, explicitly credited this model for his own narrative of his war experiences as Marshal of France. He noted that "the greatest man who ever lived in the world, Caesar, showed us the way, having himself written his Commentaries, writing by night what he executed by day." Monluc's own personal narrative, he argued, was similarly "not a book for intellectuals, they have enough historians, but for a soldier, a captain, and perhaps a lieutenant of the king, who might find something to learn from it."[48]

In addition to military memoirs and works offering broad discussions of military theory, practice, and recommended behavior, other works focused on providing instruction and advice on narrower, more targeted or technical topics, such as mathematical aiming guidelines and formulae for gunpowder artillery, or practical instruction on the tactical application and everyday use of other military equipment and devices of war. One especially influential text in this genre was the nicely illustrated treatise by the German artillery master Franz Helm, titled *Buch von den probierten Künsten* [Book of the Practical Arts] (c. 1530), and also printed in Frankfurt am Main in 1625 as *Armamentarium principale oder Kriegsmunition und Artillerie-Buch* [Main Armory or Munitions of War and Artillery Book]. This text followed the contents of an earlier anonymous gunner's treatise by offering readers expert advice (with illustrations) on the use of cannons, explanations of how to use and manufacture gunpowder, up-to-date information on modern guns and advanced ammunition, as well as the use of mines, grenades, and other various explosive devices, fire arrows and smokescreens, catapults and bomb-throwing machines, and battering rams and additional useful siege warfare tools.[49] Jacob de Gheyn's successful *Wapenhandlinghe van Roers, Musquetten ende Spiesson* [The Exercise of Arms with Arquebus, Musket, and Pike] (1607), was another notable technical treatise that explained the sequence of steps for loading, handling, and firing the arquebus and musket and how troops should be trained in these weapons. The work had a Dutch explanatory foreword, but otherwise consisted almost entirely of sketches demonstrating the ancient techniques of military drill as reconstituted for the modern gunpowder era by Maurice of Nassau and his cousins. As such, it was helpful for officers training the usually illiterate common infantryman, but was also easily translated into other languages, such as the 1608 French edition published in Amsterdam as *De l'utilisation du mousquet* [On the Use of the Musket].[50]

Experts in the science of military architecture, whose popular works circulated throughout Europe, were particularly aware of how the art of war had been revolutionized by gunpowder artillery, a fact that many such authors then argued limited the usefulness of the wisdom and teachings of their forebearers.[51] In his late fifteenth-century treatise *Trattato di architettura civile e militare* [Treatise of Civil and Military Architecture], for example, Francesco di Giorgio Martini pointed out that such ancient authors as Vetruvio were far less relevant for modern military practitioners than were contemporary engineering treatises (such as his own), since "the ancients did not know our artillery."[52] This critique of the ancients could also be found in other military volumes of the day, as authors attempted to balance their admiration of classical military forms with a recognition that modern weapons had now caused irrevocable changes to the art of war. Although Roger Williams praised the Roman-style training, drill, and tactics then popular in contemporary armies, for example, he also argued that a slavish imitation of ancient and medieval military practices was ill-advised. "Let us not erre in our ancient customes," he wrote, for "although the ground of ancient Discipline is the most worthiest and the most famous; notwithstanding, by reason of fortifications, stratagems, ingins, arming, with Munition, the discipline is greatly altered; the which we must follow and be directed as it is now: otherwise we shall repent it too late."[53]

Despite some older academic skepticism, recent scholarship suggests that some, and perhaps many military elites and even common soldiers did indeed use military advice books, analytical treatises, and manuals not just to learn their craft, but also subsequently, as ongoing guides or tools of the trade. Drill books, mathematical tables, and lists of square roots, in particular, were frequently carried on campaign or in field camps by officers, but so were books on tactics, gunnery, logistics, and siege warfare, especially those graced with easy-to-understand diagrams. Presses across Europe certainly found a ready market for all forms of military treatises and general works on the modern art of war, which also suggests their continued popularity throughout the sixteenth and seventeenth centuries. As a result, as David Lawrence has noted, "the proliferation of military books across Europe acted as one of the best means of spreading innovation and codifying practice."[54] Of course, such books also opened up increased opportunities for military advancement and social mobility among literate non-noble soldiers who would not have access to the noble patronage networks required for plum military apprenticeships.

These military treatises and the values they advocated could be read by interested individuals hoping thereby to improve their skills and knowledge, but they might also appear as part of a more formal educational program—whether merely as basic Latin primers, as mentioned above in the case of Caesar's *Gallic War*, or as texts for use in university courses on such topics as literature, history, mathematics, or engineering. In addition, military texts and prescriptive literature also found a place in the curricula of the new knightly academies that began to appear across Europe in small numbers in the late sixteenth century. Such schools were especially popular in France, where institutions designed to teach or train young noblemen in the general arts of war were founded on both the model of the popular Italian riding schools that had begun to emerge earlier in the century, as well as the educational theories of François de la Noue, which were published in 1587 as *Discours Politiques et Militaires* [Political and Military Discourses], a work that was quickly translated into both English and German.[55]

While these new knightly academies were usually intended to train the local nobility or upper bourgeoisie for service at princely or royal courts, many of them also drew in and enrolled foreign students from across the continent and the British Isles.[56] As an example of the pan-European appeal of such military schools, in 1658 the young duke of Brunswick-Lüneburg-Bevern, as part of his Grand Tour, visited the academy at Lyon, remaining there for five months and training in "princely-knightly practices, such as riding, fighting, jumping, dancing, learning languages, most often fortifying himself with the art of song and other useful arts, as also the art of arithmetic."[57] The duke's account of his education at Lyon is also illustrative of the fact that while describing themselves as "knightly academies" or even "schools of war," most of these institutions' roots were firmly grounded in the era's far larger humanist educational movement, which was itself an offshoot of a general Europe-wide nostalgia for the Greco-Roman past. Thus while these early military schools did indeed often train students in the practical arts of weapons, fencing, military mathematics (or fortifications), and map-reading, they also looked to the educational model of the ancient world found in the writings of men such as Plato, whose own school, the *Akademia*, provided the usual name of these new institutions.[58] This meant that in many respects these academies served more as noble finishing schools, teaching such topics as foreign languages, religion, history, and politics, as well as the aristocratic physical graces and courtly skills of horsemanship, dance, and music.[59] This broad program was advised by many educational reformers of the day, including Noue, as important for inculcating contemporary noble youths with both military skills and idealized knightly cultural values. Noue thus recommended that young noblemen interested in a military career should, as the English translation stated, "be taught many kinds of exercises as wel for the bodie as the mind." This included not just weapons training and all the things that "make a man strong & nimble," but also such topics as mathematics, geography, languages, and "lectures out of the auncient writers ye intreat of moral vertues, policy, & war."[60]

While France had some of the most well-known academies, Germany and Scandinavia also saw their fair share (also frequently modeled after the French style and teachings of Noue), such as the Collegium Illustre of Tübingen, founded in 1594; the Danish Kongelige Adelige Akademi of Sorö, established in 1623; and the influential Wolfenbüttel Ritter-Akademie, founded in 1688, a prestige project of Dukes Anton Ulrich and Rudolf August of Brunswick-Lüneburg that drew students from across the Empire, as well as from Sweden, Denmark, the British Isles, and Russia.[61] One of the most famous, if short-lived, academies in Germany, however, was the Kriegs- und Ritterschule [School of War and Knights], established in Siegen in 1617. Like many other knightly academies on the continent, Siegen's also offered a broad humanist education, but in keeping with the major army reform efforts spearheaded by its founder and patron, Johann VII of Nassau-Siegen, the school placed a larger emphasis than many such academies on actual military training. The statement appointing the first director of the school, Johann Jacobi von Wallhausen, who was himself the author of numerous military treatises (many used as academy textbooks), also indicated that the school was intended to serve "persons belonging to princely, comital, and noble estates of the common fatherland and German nation who intend to take part, today or tomorrow, in military campaigns or in the defense of the fatherland."[62] Even before his appointment, Wallhausen himself praised the school's planned emphasis on teaching students a full range of useful and necessary military skills following the example of both the ancient Roman art of war and more modern military theory and practice.[63]

Even in the second half of the seventeenth century, when more military academies were appearing across Europe, focused training in topics essential for modern military command was still usually supplemented by education in the arts and sciences, noble values, and courtly skills that would help noblemen prepare for a successful modern career that bridged the dual worlds of battlefield and court. The nine specialized training companies for cadets established by the famous French military reformer the marquis de Louvois in 1682, for example, were designed to "draw young nobles into military service….[and] give the nobility the means and the training to fulfil the duties associated with their social Order."[64] As such, they not only taught topics of practical use for a military career—such as shooting, riding, conducting drills and other military exercises, map-reading, siegecraft and theories of fortifications, geography, mathematics, and writing—but also offered instruction in music, art, and literature, as well as such courtly skills such as dancing and fencing, and training to improve the cadets' manners, behavior, and morals.

Conclusion

In the early modern era, a young nobleman desirous of becoming a military commander faced new challenges not seen by his medieval predecessors. Military theorists, driven by the same fascination for classical culture and institutions shown by many of their contemporaries in other fields, were then eagerly embracing and widely implementing Greco-Roman military practices, organization, and training methods, as well as recommending classical models of ideal officership and noble masculine warrior values. At the same time as this turn toward antiquity, however, early modern armies were also integrating ever more sophisticated gunpowder weapons, thereby utterly transforming the early modern art of war and generating entirely new tactical, strategic, and logistical demands and challenges. These two developments, moreover, the ancient and the modern, did not sprout from an empty field, but joined a deeply rooted traditional system of medieval military practices, chivalric values, and patterns of military education, many of which persisted in some form throughout the early modern era.

In the face of this complexity, a young nobleman of the sixteenth or seventeenth century who hoped to follow Robert Barret's advice "diligently to learne the Art he professeth, which is warre,"[65] would still usually follow traditional forms of learning at home and at court, and then supplement his knowledge of the practicalities of contemporary warfare through apprenticeship and actual military service. This field experience, however, was now far less tethered to the traditional bonds of lordship, and instead offered greater opportunities for soldiers to find honorable engagement in foreign armies or under foreign captains. This meant that men who were open to such geographic flexibility could learn the most up-to-date methods abroad, where they especially gravitated toward generals and armies that most successfully integrated the modern and classical arts of war. Such contemporary military practices, along with the courtly arts, might also be learned in one of the era's handful of new knightly or military academies, or from the hundreds of volumes of military literature that flowed in every language from the era's printing presses. Indeed, written by both ancient and modern authors, these popular handbooks and treatises offered readers such a treasure-trove of information that it sparked an argument among contemporaries over whether book-learning could complement or perhaps even replace the knowledge gained through actual field experience. In addition, the international nature of the era's military literature, along with the free flow of Europe's nobility within and between armies, and the multinational student enrollment in universities and military academies, all helped to reinforce a broadly shared European understanding of the ideal masculine noble warrior virtues that reflected these major shifts in early modern military practices. Though the heroic model remained potent throughout the era, therefore, the perfect military leader was now clearly less a passionate, impulsive Roland gloriously engaging the enemy from horseback, and more a tactically and strategically sophisticated Julius Caesar—still brave, strong, and daring, but now also able to oversee every aspect of his troops' discipline and training, arrange his army's supplies and transport, and then carefully and efficiently organize and array his forces and artillery.

Yet whatever military education at home, at a foreign court, or in an academy that a nobleman received, whatever field experience he gained, and whatever books he read or carried with him, a soldier's own innate skill, not to mention such things as luck, weather, the actions of allies, or any other factor or combination of factors, could also influence his military performance or career success. So despite Vegetius's assurance that the experienced and well-trained warrior would "proceed to certain victory,"[66] all the education in the world was only as good as the man receiving it. As the historian Lucien Bély has stated, "the talent of the generals was measured by the use they made of their information,"[67] or, as John Dryden argued at the beginning of this chapter, "the Knowledge of Warfare is thrown away on a General who dares not make use of what he knows."[68] Knowledge of his profession alone, in other words, was not enough to propel a man to fame, fortune, and victory, though it certainly helped.

Notes

1 John Dryden, *Fables Ancient and Modern Translated into Verse from Homer, Ovid, Boccace, & Chaucer, with Original Poems* (London: Printed for Jacob Tonson, 1700), sig. C.

2 J. R. Hale, *Renaissance War Studies* (London: Bloomsbury Publishing Plc, 2003), 226.

3 Keith Thomas, *The Ends of Life: Roads to Fulfilment in Early Modern England* (Oxford: Oxford University Press, 2009), 46–50.

4 I.T., *The A, B, C, of armes, or, An introduction directorie whereby the order of militarie exercises may easily bee vnderstood, and readily practised, where, when, and howsoeuer occasion is offered* (London: Printed by W. Stansby, for Iohn Helmes, 1616), n.p.

5 Robert Ward, *Anima'dversions of vvarre; or, A militarie magazine of the truest rules, and ablest instructions, for the managing of warre Composed, of the most refined discipline, and choice experiments that these late Netherlandish, and Swedish warres have produced* (London: Printed by Iohn Dawson, 1639), 150.

6 For a discussion of the decline of the military page system by the mid-sixteenth century and how this encouraged the use of military literature and academies, see Hale, *Renaissance War Studies*, 230–32. See also Mark Edward Motley, *Becoming a French Aristocrat: The Education of the Court Nobility, 1580–1715* (Princeton: Princeton University Press, 1990), 64–67.

7 The continued importance of household education for inculcating noble values is nicely outlined in Motley, *Becoming*, 18–67.

8 Justus Lipsius, *A Direction for Trauailers,* (London: By R. B[ourne] for Cutbert Burbie, 1592), sig. b4. English modernized. This edition was a free English translation of Lipsius's *Epistola de peregrinatione Italica* by Sir John Stradling. Lipsius also noted approvingly the keen interest in foreign military practices demonstrated by, in Stradling's translation, "the great Alexander: who when any embassadors resorted to kinge Phillip his father from farre countries, and great potentates was woont to demande of them what weapons they vsed in warre: what lawes in peace, how they gouerned their Cities, but especiallie how they ordered their battels" (Sig. b2).

9 Hugo Grotius, *De rebus belgicis, or, The Annals and History of the Low-Countrey-Warrs*, trans. Thomas Manley (London: Printed for Henry Twyford, 1665), 512.

10 Aelianus, *The Tactiks of Ælian Or Art of Embattailing an Army After Ye Grecian Manner Englished & Illustrated with Figures Throughout*, trans. John Bingham (London: For Laurence Lislo, 1616), A2v. See also Geoffrey Parker, "Foreword," in *Exercise of Arms: Warfare in the Netherlands, 1568–1648*, ed. Marco van der Hoeven (Leiden: Brill, 1997), iv.

11 Turenne was at that time merely a young nobleman of fourteen, but later a famous Marshal of France and leading general of both the Thirty Years War (1618–48) and the War of Devolution (1667–68). John A. Lynn, *Giant of the Grand Siècle: The French Army, 1610–1715* (New York: Cambridge University Press, 1997), 515. Lynn cites for this Jean Bérenger, *Turenne* (Paris: Fayard, 1987), 90, but for the original letter see Henri de la Tour d'Auvergne vicomte de Turenne, *Lettres de Turene: extraites des Archives Rohan-Bouillon*, ed. Suzanne d'Huart (Paris: S.E.V.P.E.N., 1971), 170.

12 Roger Burrow Manning, *Swordsmen: The Martial Ethos in the Three Kingdoms* (Oxford: Oxford University Press, 2003), 130. Manning's work also traces the experiences of soldiers from the British Isles apprenticing themselves in continental armies.

13 Lynn, *Giant*, 517. See also Robert Arnauld d'Andilly, *Mémoires*, in *Nouvelle collection des mémoires pour servir à l'histoire de France*, vol. 9, eds. J.F. Michaued and J.J.F. Poujoulat (Paris: Firmin Didot Frères, 1839), 416.

14 François de la Noue, *The politicke and militarie discourses of the Lord de la Nouue Whereunto are adioyned certaine obseruations of the same author, of things happened during the three late ciuill warres of France. With a true declaration of manie particulars touching the same. All faithfully translated out of the French by E.A.* (London: T[homas] C[adman] and E[dward] A[ggas] by Thomas Orwin, 1587 [i.e., 1588]), 181–82.

15 Sir Roger Williams, *A briefe discourse of warre. Written by Sir Roger Williams Knight; with his opinion concerning some parts of the martiall discipline. Newly perused* (London: Thomas Orwin, 1590), 6.

16 Ibid., 26–27.

17 Robert Barret, *The Theorike and Practike of Moderne Warres* (London: Printed by R. Field for William Ponsonby, 1598), 2, 6. See also Clifford J. Rogers, "Tactics and the Face of Battle," in *European Warfare, 1350–1750*, eds. Frank Tallett and David J. B. Trim (Cambridge, UK: Cambridge University Press, 2010), 203–35.

18 Quoted in Sheila Barker, "Cosimo I de Medici and the Renaissance Sciences: 'To Measure and to See,'" in *A Companion to Cosimo I de' Medici*, eds. Alessio Assonitis and Henk Th. Van Veen (Leiden: Brill, 2021), 520–80 (547).

19 Jean Grangnières, comte de Souvigny, *Mémoires du comte de Souvigny, lieutenant général des armées du roi*, vol. 2 (Paris: Librairie Renouard, 1906), 277–78.

20 Ward, *Anima'dversions of warre*, 150. For more on the topic of religious motivations for war, see also Paul Scannell, *Conflict and Soldiers' Literature in Early Modern Europe: The Reality of War* (London: Bloomsbury, 2014), 69–116.

21 Paolo Sarpi, *The free schoole of warre, or, A treatise, whether it be lawfull to beare armes for the seruice of a prince that is of a diuers religion*, trans. William Bedell and Sir Nathaniel Brent (London: John Bill, 1625), sigs. B–Bv.

22 Ibid., sig. Biiiv.

23 Ibid., sig. Fiiiv.

24 There is a large literature on the printing revolution and how it influenced early modern information acquisition and learning in general. See for example Elizabeth L. Eisenstein, *The Printing Revolution in Early Modern Europe*, 2nd ed. (Cambridge, UK: Cambridge University Press, 2005); and Lucien Febvre and Henri-Jean Martin, *The Coming of the Book: The Impact of Printing 1450–1800*, trans. David Gerard (London and New York: Verso, 1976).

25 David R. Lawrence, *The Complete Soldier: Military Books and Military Culture in Early Stuart England, 1603–1645* (Leiden: Brill, 2009), 1. See as well his first chapter, 19–72. Other useful works on this topic include Scannell, *Conflict and Soldiers' Literature*, and Thomas F. Arnold, *The Renaissance at War* (London: Cassell & Co., 2001). For the classic work on military books and learning in Germany, see Max Jähns, *Geschichte der Kriegswissenschaften vornehmlich in Deutschland, vol 1, Altertum, Mittelalter, XV. und XVI. Jahrhundert* (Munich: Oldenbourg, 1889).

26 Sir Clement Edmonds, *Observations, Upon the First Books of Caesars Commentaries, Setting Fourth the Practice of the Art Military, in the Time of the Roman Empire* (London: Printed by Peter Short, 1600), 1.

27 Robert Monro, *Monro his Expedition with the Worthy Scots Regiment Called Mac-Keyes* (London: William Jones, 1637), np.

28 For more on this, see Donald A. Neill, "Ancestral Voices: The Influence of the Ancients on the Military Thought of the Seventeenth and Eighteenth Centuries," *The Journal of Military History* 62, no. 3 (July 1998): 487–520.

29 See Christopher Allmand, *The De Re Militari of Vegetius: The Reception, Transmission and Legacy of a Roman Text in the Middle Ages* (Cambridge, UK: Cambridge University Press, 2011).

30 For Woodcock's discussion of the popularity of Caesar's *Commentaries* and his role as an example of military virtues, see Matthew Woodcock, "'The Breviarie of Soldiers': Julius Caesar's *Commentaries* and the Fashioning of Early Modern Military Identity," in *Early Modern Military Identities: 1560–1639: Reality and Representation,* eds. Matthew Woodcock and Cian O'Mahany (Cambridge, UK: D.S. Brewer, 2019), 56–78. Bingham's quote is in Aelianus, sig. A2v.

31 See Jeanine de Landtsheer, "Justus Lipsius's *De militia Romana*: Polybius Revived or How an Ancient Historian was Turned into a Manual of Early Modern Warfare," in *Recreating Ancient History: Episodes from the Greek and Roman Past in the Arts and Literature of the Early Modern Period,* eds. Karl Enenkel et al. (Leiden: Brill, 2001), 101–22.

32 Niccolò Machiavelli, *The Art of War*, trans. Ellis Farneworth (Cambridge, UK: Da Capo Press, 2001), 5, 93. For Machiavelli and Vegetius, see Allmand, *De Re Militari,* 139–47.

33 François de la Noue, *The Politike and Militarie Discovrses of the lord De La Nowe* (London: For T.C. and E.A. by Thomas Orwin, 1587), in Lawrence, *The Complete Soldier,* 235–37.

34 Gregory Hanlon, *The Twilight of a Military Tradition: Italian Aristocrats and European Conflicts, 1560–1800* (New York: Holmes & Meier, 1998), 343. For more on the complex relationship between warrior masculinity and the new gunpowder weapons, see Patrick Brugh, *Gunpowder, Masculinity, and Warfare in German Texts, 1400–1700* (Rochester, NY: University of Rochester Press, 2019).

35 David J. B. Trim, "Warlike Prowesse and Manly Courage," in *The Chivalric Ethos and the Development of Military Professionalism*, ed. David J. B. Trim (Leiden: Brill, 2003), 30

36 Barret, *Theorike,* 2.

37 Ibid., 13.

38 Thomas, *Ends of Life*, 57.

39 F. González de Leon, "Doctors of the Military Discipline: Technical Expertise and the Paradigm of the Spanish Solider in the Early Modern Period," *Sixteenth Century Journal* 27 (1996): 61–85.

40 For a thorough bibliographic history of military works, see Rainer Leng, *Ars belli: Deutsche taktische und kriegstechnische Bilderhandschriften und Traktate im 15. und 16. Jahrhundert* (Wiesbaden: Reichert Verlag, 2002). An indication of the variety of such texts available in England alone can be seen in M. J. D. Cockle, *A Bibliography of Military Books up to 1642* (London: Simpkin, Marshall, Hamilton, Kent & Co, 1900).

41 David Potter, "Chivalry and Professionalism in the French Armies," in *The Chivalric Ethos,* ed. David J. B. Trim, 174–75.

42 Leonhardt Fronsperger, *Das Kriegsbuch* (Frankfurt am Main: Feyerabend, 1573).

43 Philippe, Duke de Cleves, *Instruction de toutes manières de guerroyer tant par terre que par mer, des choses y servantes* (Paris: G. Morel, 1558).

44 Thomas M. Barker, *The Military Intellectual and Battle: Raimondo Montecuccoli and the Thirty Years War* (Albany: State University of New York Press, 1975), 2–3, 167–68.

45 Barret, *Theorike,* 2.

46 Ibid., 14.

47 Ibid., 13.

48   Blaise de Monluc, *Commentaires et lettres de Blaise de Monluc, maréchal de France,* vol. 1 (Paris: Mme Ve J. Renouard, 1864), 28. My translation. Cf. *The Commentaries of Messire Blaize de Montluc, Mareschal of France,* trans. Charles Cotton (London: Printed by Andrew Clark for Henry Brome, 1674), 2.

49   Gerald W. Kramer and Klaus Leibnitz, "The Firework Book: Gunpowder in Medieval Germany," *The Journal of the Arms & Armour Society* 17.1 (March 2001): 1–88. For more on military literature of the era, see Wolfgang Lefèvre, *Minerva Meets Vulcan: Scientific and Technological Literature—1450–1750* (Cham, Switzerland: Springer, 2021), 71–93. The topic of English military manuals is covered by Steven Walton, "The Art of Gunnery in Renaissance England" (PhD diss., U. Toronto, 2000), 133ff.

50   M. D. Feld, "Middle-Class Society and the Rise of Military Professionalism: The Dutch Army 1589–1609," *Armed Forces & Society* 1, no. 4 (Summer 1975): 419–42.

51   Although the architectural engineering skills and precise mathematical calculations required to design and build the new bastion forts was at first passed along by Italian experts orally and through hand-drawn architectural sketches and scale models, by the mid-sixteenth century, numerous published treatises on the topic and other developments in fortress architecture quickly spread such modern knowledge throughout Europe. See Simon Pepper, "Artisans, Architects and Aristocrats: Professionalism and Renaissance Military Engineering," in *The Chivalric Ethos,* ed. David J. B. Trim, 117–48.

52   Francesco di Giorgio Martini, *Trattato di architettura civile e militare,* ed. Cesare Saluzzo (Torino, Italy: Chirio e Mina, 1841). 249. My translation. See also Horst de la Croix, "Military Architecture and the Radial City Plan in Sixteenth-Century Italy," *The Art Bulletin* 42, no. 4 (1960): 263–90 (269).

53   Williams, *Briefe Discourse,* 34.

54   Lawrence, *The Complete Soldier,* 5–9. See also Sandra L. Powers, "Studying the Art of War—Military Books Known to American Officers and their French Counterparts During the Second Half of the Eighteenth Century," *The Journal of Military History* 70 (2006): 781–814; Hale, *Renaissance War Studies,* 232–34, 429–70.

55   Among the French academies and riding schools that drew on these models were the *Académie des exercises,* established in 1590 at Sedan; two knightly academies in Saumur (one an explicitly Huguenot institution popular with European Protestants, est. circa 1593, the other a possibly older riding academy); and academies at Paris (including the Académie Royal of M. de Pluvinel, est. 1594), Pezenas (1598), and Angers (1600); the Duc de Bouillion's academy at Sedan (1606), plus one at Marseille (1608), Aix-en-Provence (1611), and Besancon (1653). Norbert Conrads, *Ritterakademien der Frühen Neuzeit. Bildung als Standesprivileg im 16. und 17. Jahrhundert* (Göttingen: Vandenhoeck und Ruprecht, 1982), 15–79.

56   According to Keith Thomas, "the unwillingness of the universities [in England] to offer training in riding and fencing led many young gentlemen to go off to the noble academies of France and Italy." Thomas, *Ends of Life,* 50.

57   Thomas Grosser, *Reiseziel Frankreich: Deutsche Reiseliteratur vom Barock bis zur Französischen Revolution* (Wiesbaden: VS Verlag für Sozialwissenschaften Wiesbaden, 1989), 46–47. My translation.

58   For more on this topic, see Francis A. Yates, *The French Academies of the Sixteenth Century* (London: Studies of the Warburg Institute, 1947).

59   For more on the development, role, and significance of the academy for the French nobility, see Motley, *Becoming,* 123–68.

60   Noue, *Politicke,* 73.

61 Conrads, *Ritterakademien,* 105–15, 143–52, 273–322.
62 Werner Hahlweg, *Die Heeresreform der Oranier. Das Kriegsbuch des Grafen Johann von Nassau-Siegen* (Wiesbaden: Selbstverlag der Historischen Kommission, 1973), 568. My translation. For more on the school at Siegen and its curriculum, see Christian Brachthäuser, *"Wie sich ein Fürst zum Krieg soll rüsten": die älteste Militärakademie der Welt: Die Gründung der Ritter-und Kriegsschule in Siegen im Jahr 1616 unter Johann VII. Graf zu Nassau-Siegen* (Groß-Gerau: Ancient Mail Verlag Werner Betz, 2016); and Hubert Zeinar, *Geschichte des österreichischen Generalstabes* (Vienna: Böhlau Verlag, 2006), 62–64.
63 Johann Jacobi von Wallhausen, *Programma Scholae militaris ex veteri veterum Romanorum Instituto ... noviter institutae* (Frankfurt am Main: Paul Jacobi, 1616), 6.
64 Guy Rolands, *The Dynastic State and the Army Under Louis XIV: Royal Service and Private Interest 1661–1701* (Cambridge, UK: Cambridge University Press, 2002), 178–85.
65 Barret, *Theorike,* 8–9.
66 Flavius Vegetius Renatus, *The Military Institutions of the Romans,* trans. John Clark (Harrisburg, Pa.: The Military Service Publishing Company, 1944), 13.
67 Lucien Bély, *Espions et ambassadeurs au temps de Louis XIV* (Paris: Fayard, 1990), 219.
68 Dryden, *Fables,* sig. C.

Further Reading

De Landtsheer, Jeanine. "Justus Lipsius's *De Militia Romana*: Polybius Revived or How an Ancient Historian Was Turned into a Manual of Early Modern Warfare." In *Recreating Ancient History: Episodes from the Greek and Roman Past in the Arts and Literature of the Early Modern Period,* edited by Karl Enenkel et al., 101–22. Leiden: Brill, 2001.

Hale, John Rigby. *Renaissance War Studies.* London: Bloomsbury Publishing Plc., 2003.

Lawrence, David R. *The Complete Soldier: Military Books and Military Culture in Early Stuart England, 1603–1645.* Leiden: Brill, 2009.

Manning, Roger Burrow. *Swordsmen: The Martial Ethos in the Three Kingdoms.* Oxford: Oxford University Press, 2003.

Scannell, Paul. *Conflict and Soldiers' Literature in Early Modern Europe: The Reality of War.* London: Bloomsbury, 2014.

Trim, David J. B., ed. *The Chivalric Ethos and the Development of Military Professionalism.* Leiden: Brill, 2003.

Woodcock, Matthew and Cian O'Mahany, eds. *Early Modern Military Identities: 1560–1639: Reality and Representation,* Cambridge, UK: D.S. Brewer, 2019.

2 Building the Foundations of a Surgical Armory

Johannes Scultetus in Ulm, c.1630–45

Heidi Hausse

In 1639, a garrison soldier in Ulm, in southwest Germany, fell to the floor as if dead after his comrade threw a radish at his right eye.[1] He was brought to the Heilig-Geist-Spital, the imperial city's main hospital, where the physician Johannes Schultheiβ (1595–1645), better known by the Latinized surname Scultetus, was called to examine him. A town native, Scultetus was a doctor of anatomy and surgery educated at the University of Padua by such luminaries of Renaissance anatomy as Girolamo Fabrizi d'Acquapendente (1533–1619) and Adriaan van den Spiegel (1578–1625). In 1625, he took up the position of town physician in Ulm, where he practiced for twenty years.[2] The radish-tossing incident was but one of a wide variety of cases Scultetus treated during his career. In 1633, for example, he inserted a metal tube into the chest of a German horseman, injured in a duel, to drain blood from the thoracic cavity. A year later, he used external medicaments to treat a woman from nearby Bermaringen with cancer of the breast. All the while he took notes on his patients, eventually drawing out individual cases to edit and organize into a manuscript for publication. On a merchant treated for a putrefying shin bone in 1634, Scultetus reflected that the patient could walk without crutches "until the year 1645, in which I write this."[3]

The career and writings of Johannes Scultetus offer a compelling illustration of how different bodies of knowledge came together in the construction of a learned surgical treatise. In his *Armamentarium Chirurgicum* (1655), or "surgical armory," Scultetus gathered information from university lectures, Latin and vernacular treatises, and hands-on work with patients, using the act of writing to allow these different sources to inform and instruct one another. As the vast majority of surgeons in the Holy Roman Empire trained by apprenticeship, Scultetus presents a rare form of surgical practitioner and author in the German context, and his example holds implications for how learned and craft worlds of medicine intersected in German-speaking lands and outside of them. A member of the *Collegium medicum* (College of Physicians) in Ulm, he was a *medicus* and performed surgery himself only in extraordinary situations. His impact on surgery in Europe instead came from his widely circulated treatise, divided into two parts. Part I, famed for its detailed illustrations of surgical instruments, in

DOI: 10.4324/9781003157700-4

many ways follows the learned example of Girolamo Fabrizi d'Acquapendente. Part II, however, is a collection of case histories, or *observationes*, that draw readers into the stormy years of unrest in and around Ulm that created the extraordinary cases—the injuries caused by duels, marauding soldiers, skirmishes, and open battle—which required Scultetus to employ in practice some of his treatise's most well-known instruments. Sixty-one of the 100 *observationes* date between 1630 and 1645, the last fifteen years of Scultetus's career.[4] This chapter uses these case histories, published in the 1666 German and Latin editions, as well as archival records and chronicles from Ulm, to explore a particularly complex example of the many-layered ways of learning and doing surgery in a time of war.

Historians often treat learned medicine, surgery, and military surgery as separate categories in early modern scholarship. Yet, all three uniquely overlapped in the latter part of Scultetus's career because he was a doctor of medicine and surgery in a German town during the Thirty Years War (1618–48). His case histories shed light on practicing and writing about surgery in a moment of remarkable disruption. Although Ulm managed to avoid battle, siege, and sack, its burghers could not help but feel the effects of the conflict.[5] 1635, for example, a devastating year when thousands perished in a pestilence that swept through the city, followed the arrival of large numbers of refugees fleeing the destructive aftermath of the Battle of Nördlingen.[6] In his chronicle of Ulm, Joseph Furttenbach estimated a staggering death toll of over 14,000 within the city, 4,000 of whom were burghers and their families, the rest a combination of rural folk and outsiders.[7] Moreover, not only did burghers pay contributions (war taxes) to appease armies, but at times the city also allowed armies to winter in its territory.[8] The *observationes* reflect this influence: the majority date to 1630–45, and of those nearly thirty percent focus on violent injuries caused by soldiers. This does not mean that roughly a third of Scultetus's day-to-day practice consisted of treating military injuries. Rather, it suggests the sustained violence of the Thirty Years War created opportunities for Scultetus to gain hands-on experience performing operations.

It was also during these years that Scultetus began to write *Armamentarium Chirurgicum*. His work attempts to bring together the genre of learned surgical textbooks, steeped in references to ancient and contemporary authorities, with the new genre of *observationes*, which prized empiricism and firsthand experience. Although they have generally received little attention, the *observationes* of Part II provide a glimpse into how empirical knowledge contributed to Part I. The instruments in Part I were a mixture of objects owned and used by Scultetus and objects he had never used himself; of designs previously appearing in other publications and ones of his own invention; of recently devised tools and ones dating back to antiquity. Part II reveals which instruments Scultetus *did* have firsthand experience using and suggests how this experience influenced his presentation and discussion of them in Part I.

The following chapter examines Scultetus as a complex example of "doing" surgery—of thinking about, practicing, and communicating surgical techniques. I first explore the publication of *Armamentarium Chirurgicum* and the epistemological emphasis its *observationes* suggest. Scultetus's educational background informed his writing, but his treatise was not simply a reflection of the relationship between abstract theory and hands-on practice. It was also itself a tool for processing Scultetus's surgical knowledge and experience. I then outline the broad characteristics of the *observationes* from 1630 to 1645 as encounters that Scultetus chose to record, think with, and share, highlighting elements of the social, political, and professional contexts in which Scultetus worked and which in turn shaped his practice and writing. Finally, I use a close reading of two case histories to consider the connections between Parts I and II. The cross-references embedded in the *observationes* reveal cognitive links in which the author drew from a firsthand encounter to write about surgical knowledge and practice more generally. Scultetus's *observationes* served as the practical foundation of his learned surgical armory.

Armamentarium Chirurgicum

Armamentarium Chirurgicum, first published in Ulm in 1655, presented a learned, instrument-centered surgical guide.[9] The first half of the work ("Part I") featured a series of forty-three plates illustrating instruments with explanatory text describing how they were used, the wounds or diseases they were used to treat, and brief anecdotes. Part I follows the model of Fabrizi's *Operationes* (1619) and includes several instruments of Fabrizi's own design.[10] The second half of the work ("Part II") consisted of 100 case histories (*observationes*) intended to further elucidate the information presented in Part I. This contributed to the growing medical genre of *observationes* while continuing to emphasize the use of instruments, connecting firsthand observation to more general instruction through comprehensive cross-referencing between Parts I and II. The forty-three plates of the first edition expanded to fifty-six in the 1666 revised edition, where plates 26–56 elaborated on the use of the instruments presented in plates 1–24.[11] A thoroughly reorganized Part II still consisted of 100 *observationes*, most carried over and some newly added. In the same year and from the same publisher came its German translation, *Wund-Artzneyisches Zeug-Hauß*.

Both the 1655 edition and its expanded versions appeared posthumously since Johannes Scultetus had died unexpectedly in the winter of 1645, at the age of fifty.[12] Yet, fourteen new *observationes* appeared in the Latin and German editions of 1666. To explain this peculiarity, and to consider the case histories of Part II as evidence of Scultetus's experience, the publication history of his manuscript warrants unspooling. In particular, one should note that the publication of *Armamentarium Chirurgicum* (1655) and the subsequent editions of 1666 all sprang from a communal network of medical practitioners in southwest

Germany and a posthumous collaboration between Scultetus and his nephew, Johannes Scultetus the Younger.[13]

The dedication of *Armamentarium Chirurgicum*, composed by the nephew, offers glimpses into the manuscript's creation, transmission, and publication. Scultetus the Younger reflects on his uncle: "every day he acquired new *observationes*, which he wrote on paper with the idea and resolve that, after revising them, they might eventually be made public."[14] The nephew explains that it was his duty to carry out the wishes of the deceased, whose career he had modeled by attending the University of Padua and obtaining a position as town physician in Ulm in 1653.[15] He also states that he had received many letters from friends requesting that he publish his uncle's work. Scultetus's original manuscript has not survived. While its state at the time of his death cannot be determined, it has been suggested it was close to completion.[16] Indeed, the nephew reveals that Scultetus had advanced far enough in the project to have recorded his plans to dedicate the work to the city fathers of Ulm.[17] The author's preface also appears to have been written by Scultetus, who describes his manuscript and surgical practice as following in the footsteps of Fabrizi.[18] The nephew edited and published the work in 1655, declaring on the title page that it was an "*opus posthumum.*"

Soon after the publication of the first edition, Scultetus the Younger embarked on an extensive expansion and revision, which he completed in manuscript form in 1662. He then began to translate the new manuscript into German, and, on his deathbed in 1663, he asked his long-time friend Amadeus Megerlin to finish it.[19] Megerlin, a physician working in the duchy of Württemberg, did not possess a humanist surgical education and did not appear to share the ambition of Scultetus or his nephew to contribute to a corpus of learned surgical writings. On the whole, his translation faithfully followed the expanded Latin edition.[20] The first edition of *Armamentarium Chirurgicum* and its expanded versions of 1666, then, were published posthumously on two levels—the first edition after Scultetus's death, and the expanded versions after his nephew's.

A discussion of Scultetus's *observationes* requires a measured view of the treatise as edited and shaped by Scultetus *and* his nephew. There are places in which Scultetus the Younger inserts examples of his own experience without specifying that these were not his uncle's. Plate 42 of the 1655 edition, for instance, includes a figure of a two-headed infant, accompanied by a firsthand account of its dissection in the duchy of Württemberg in 1651, when Scultetus the Younger was practicing there.[21] At times he also updates his uncle's *observationes*. For example, in the 1655 edition, Observation 47 discusses Melchior Frick, an Ulm fuller stabbed in 1645, and concludes simply that after successful treatment the patient experienced no more pain.[22] In the expanded 1666 editions, Frick's stabbing appears as Observation 58 and ends with an additional phrase noting that after his cure he lived in good health until at least 1655—a full decade after Scultetus's death.[23]

Understanding the different filters through which the *observationes* come complicates rather than diminishes their usefulness as evidence. The vast majority of source material for both the first and expanded editions derives from writings Scultetus left. Despite its retrospective addendum, the case history of Melchior Frick, for example, is still firmly rooted in Scultetus's original source material. Supporting evidence appears in town council notes, the *Ratsprotokoll*. The *observatio* dates Frick's treatment to August 1645; in October the *Ratsprotokoll* records a complaint that Scultetus ("Herr D. Schultheißen") charged Melchior Frick ("Melchior Frickhenes") too high a fee for his treatment.[24] In addition, while most of the fourteen new *observationes* of the 1666 editions are undated, those with dates consistently contain details suggesting that they come from Scultetus's practice.[25] Keeping in mind the role of his nephew in their publication, the *observationes* of 1630–45 remain valuable sources for exploring Scultetus's approach to writing, thinking about, and practicing surgery.

Part II of *Armamentarium Chirurgicum* is entitled "Armamentarii chirurgici pars secunda. Sive observationes chirurgicae." The expanded 1666 editions reorganize the *observationes* to loosely follow a head-to-toe ordering of cases. Most focus on a single patient's case and include dates, the patient's name and occupation, the cause of disease or injury, and the course of the affliction and treatment. At times a follow-up report indicates the status of the patient several years later, if known. For the most part, these conform to the characteristics typical of the growing genre of medical writing that, as we see in Scultetus's publications, provided readers with collections of case histories, or *observationes*. Such works could be slender volumes or massive tomes with detailed illustrations, exemplified by the six *Centuriae* (treatises containing 100 *observationes* each) published by Wilhelm Fabry von Hilden (1560–1634).[26]

This genre, which emerged in the late sixteenth century, was deeply influenced by a renaissance humanistic medicine that cultivated new habits of taking notes.[27] As Gianna Pomata and others have shown, note-taking in medical training and record-keeping of patient cases in one's daily practice became widespread among physicians.[28] Practice records could take many forms, from alphabetical commonplacing to chronological journals.[29] Aspiring authors then used such notes as source material for publications. Volker Hess and Sabine Schlegelmilch have recently observed that subsequent edits, from corrected misspellings to insertions of learned references, can allow one to "even track the journey of such notes right up to their publication."[30]

The *observationes* of *Armamentarium Chirurgicum* were presumably culled in a similar fashion from notes Scultetus kept from his student days to his death. As with his handwritten book of medical recipes, referenced in later sources, such records from his daily practice do not survive.[31] However, we can place Scultetus in a wider context of note-taking habits among medical practitioners in this period. Scultetus was a unique figure in Ulm because he attended the University of Padua and obtained a degree as a doctor of medicine and surgery.

University-educated surgeons were a rarity in the Holy Roman Empire, where most practicing surgeons trained within the master-apprentice system. Padua, a premier institution of Renaissance anatomy, followed the humanist medical tradition that promoted note-taking.[32] In his treatise, Scultetus cites lectures and refers to cases of his mentors, Adriaan van den Spiegel and Girolamo Fabrizi d'Acquapendente, that he witnessed at university. The earliest *observationes* in Part II date to his student days in 1622.[33] In Observation 48, for instance, he describes dissecting the corpse of a French nobleman and includes the subsequent lecture that "Herr D. Spiegel," or "Excellentissimus Spigelius," gave on the case.[34] This suggests these early *observationes* were drawn from notes preserved for over two decades.

The new genre likely appealed to Scultetus in several ways. In a broad sense, there was the stupendous success of the *Centuriae* of Fabry von Hilden, whom he frequently cited and at times criticized in his manuscript. Domenico Bertolini Meli has referred to Fabry as one of the founding authors of surgical observations, which he identifies as a subgenre of *observationes* distinct from medical observations.[35] Nearer to home was the example set by Gregor Horst, Scultetus's elder colleague in Ulm, who published a set of medical observations in 1625, the same year Scultetus joined the *Collegium medicum*. Further sets of medical observations, by Horst and Johann Peter Lotichius, were published in Ulm by 1644.[36] By preparing his own manuscript, Scultetus was planning to take part in this trend, and his focus on surgery set him apart in Ulm.

The *observationes* do not mirror Scultetus's everyday practice, for they are too highly filtered. Rather, they are evidence of Scultetus thinking through and presenting his surgical practice, with hints at the wider surroundings in which it took place. Pomata, after all, has called *observationes* an epistemic genre because "they provided a framework for gathering, describing and organizing the raw materials of experience."[37] The rise of this genre was tied to a growing esteem for firsthand observation.[38] *Armamentarium Chirurgicum* straddles two worlds with its two parts: Part I in a learned tradition prioritizing doctrine over experience, and Part II in a growing tradition prioritizing experience over doctrine, and the particular over the universal. The relationship between doctrine and experience in Scultetus's writing was complex. Part II introduces the *observationes* to the reader as supplemental to Part I, as its full title declares that the case histories "affirm and explain all that was briefly mentioned in the first part."[39] Yet, as the following sections show, the *observationes* reveal that much of Scultetus's knowledge of instruments came from patient cases, and these inform—and at times decisively mold—the general discussions of Part I.

The *Observationes*

Armamentarium Chirurgicum emphasizes the role of surgical instruments. Yet the practice of surgery in early modern Europe involved much more. There were different operative techniques and broader conceptions of health, disease, and the human body at play, as well as the social, political, and professional contexts within which practitioners worked. As a tool for processing Scultetus's surgical knowledge and experience, the *observationes* of 1630–45 are practical encounters that he chose to record, think with, and share.[40] These encounters are filtered through Scultetus's thoughts and his goals for publication, but they are also shaped by the wider environment in which he practiced. In those years, the effects of the ongoing military conflict were interwoven with the day-to-day cases of a town physician.

In Ulm, Scultetus was a physician—a *medicus*—who performed surgery only on extraordinary occasions. A member of the *Collegium medicum*, he received a salary as a town physician in addition to the fees he charged.[41] Patients could seek him out, and municipal authorities sent him to treat specific cases.[42] His everyday practice applied knowledge of the body's inner workings, particularly of the humors, to treat patients by monitoring diets, administering purgatives or laxatives, ordering bloodletting, and prescribing medicine. Fifty-six *observationes* (nearly ninety-two percent) of those dated 1630–45 are presented as Scultetus's firsthand experience.[43] Approximately twenty-seven percent of these involve Scultetus consulting on a case, overseeing a surgical case without operating himself, or treating a patient strictly through the noninvasive measures of a *medicus*. As was typical in the Holy Roman Empire, he worked in an urban center that regulated which kinds of practitioners could provide which medical services, and what they could charge. Surgery was normally the purview of hands-on practitioners trained by apprenticeship rather than at university. Their shifting titles—barbers, barber-surgeons, surgeons—often indicated which major and minor procedures they could perform. As in other cities, surgeons in Ulm could also call in physicians such as Scultetus for more serious cases.[44]

When Scultetus undertook a surgical case, occasionally it was because an injury was so severe he was immediately called, sometimes even to the scene of the incident.[45] Usually, however, it was after another practitioner's initial treatment failed, rendering the patient's condition more serious. Half of the *observationes* describe the failures of barbers, barber-surgeons, surgeons, and itinerant specialists as a prelude to Scultetus's involvement.[46] In the context of the early modern medical marketplace, the actual number of patients from the case histories who first attempted treatment elsewhere was probably higher.[47] Some seventy-three percent of his firsthand *observationes* involve Scultetus performing invasive manual operations. The text often clarifies when he advised or oversaw the handiwork of others and when he operated himself. For example, in the case of a soldier with a gunshot wound in his groin, Scultetus had the patient's blood let "by the barber-surgeon."[48] By contrast, the use of the first person makes clear that Scultetus held forceps to remove an obstruction from the opening of the patient's urethra: "I seized the said piece of crust with the surgical tweezers and pulled it out."[49]

The very title of *Armamentarium Chirurgicum*—or the "surgical armory"—conceptually situates surgical instruments in a militaristic setting, with operations as battles to be waged with skill and precision. The *observationes* suggest that the Thirty Years War provided ample opportunity for Scultetus to perform surgery. Nearly thirty percent of case histories from 1630 to 1645 involve injuries caused by soldiers. Not only was military violence responsible for a significant proportion of the injuries Scultetus treated (Tables 2.1–2.3), but it also disproportionately required the riskiest operations of Part II. The gravest of them were related to head wounds. Soldiers cause patient injury in forty-seven percent of case histories about head wounds. The most invasive and dangerous procedures to treat such injuries—drilling or cutting into the skull—also occur disproportionately in these *observationes*. Over fifty-five percent of the cranial operations Scultetus singled out for publication came from a violent episode with a soldier.

That Scultetus did in fact perform such procedures is supported by traces left in the archival record. Notes from the town council discuss a complaint that he charged a local shoemaker an exorbitant sum of thirty Reichtaler "for trepanning."[50] The circumstances, patient's name and occupation, and dates match Scultetus's *observatio* about Hans Jacob Hechingen.[51] In the *observatio*, the shoemaker's treatment takes place in October and November of 1638; the *Ratsprotokoll* discussion of Scultetus's fee occurs in January and February of 1639.[52] The cause of the shoemaker's injury is unclear—the town council ordered Scultetus to see the patient, and he found him in bed with a fractured skull.[53] The *observatio* makes no mention of a complaint about the cost of Scultetus's services, nor that the Bürgermeister rebuked him for charging too much.[54] The shoemaker's case reminds us that the case histories present information selectively.

Table 2.1 Observationes: Patients, c.1630–45[a]

Categories of patients	56 observationes *Scultetus treated*	61 observationes *total*
	(percent)	(percent)
Soldier or officer	23.21	22.95
Civilian	75.00	75.40
Sheep dissection	1.78	1.63
Male (soldier and civilian)	76.78	77.04
Female	19.64	19.67
Male and female	1.78	1.63

[a] This includes two case histories (Obs. 13, 16) with treatment lasting from 1629 to 1630.

The above provides statistics about patients appearing in *observationes* dated c.1630–45 in Johannes Scultetus, *Armamentarium Chirurgicum* (Frankfurt, 1666), and *Wund-Artzneyisches Zeug-Hauß* (Frankfurt, 1666). The data is divided into two parts: percentages among the total number of case histories (sixty-one *observationes*) and the number of case histories presented as Scultetus's firsthand experience (fifty-six *observationes*). Percentages reflect numbers of *observationes*, not individual patients.

Table 2.2 Observationes: Civilians as patients, c.1630–45[a]

Case information	46 observationes of civilians	56 observationes Scultetus treated	61 observationes total
	(percent)	(percent)	(percent)
Cause of affliction			
Injury	47.82	33.92	36.06
Human violence	28.26	17.85	21.31
Inflicted by soldier	8.69	5.35	6.55
Inflicted by civilian	13.04	8.92	9.83
Attacker unidentified	6.52	3.57	4.91
Likely soldier	4.35	3.57	3.27
Likely civilian	2.17	0	1.64
Accident	19.56	16.07	14.75
Illness/disease	41.30	32.14	31.14
Original cause unidentified	10.86	8.92	7.19
Social status of patient			
Upper class	19.56	16.07	14.75
Nobility	10.86	8.92	8.19
Patrician or town elite	8.69	7.14	6.55
Ulm town official (a bailiff)	2.17	1.78	1.63
Ulm town citizen	4.34	3.57	3.27
Ulm craftsman/family member	17.39	12.5	13.11
Peasant farmer/family member	17.39	8.29	13.11
Other occupation (e.g., innkeeper)	13.04	10.71	9.83
Relative of soldier	4.34	3.57	3.27
Multiple patients of different statuses	2.17	1.78	1.63
Unspecified	19.56	16.07	14.75
Geographical origins of patient			
Town of Ulm	41.30	32.14	31.14
Environs of Ulm (Ulm territory and neighboring lands)	43.47	32.14	32.78
Multiple patients with different origins	2.17	1.78	1.63
Unspecified	13.04	8.92	9.83

[a] This includes two case histories (Obs. 13, 16) with treatment lasting from 1629 to 1630.

The above provides statistics about civilian patients in *observationes* dated 1630–45 in Johannes Scultetus, *Armamentarium Chirurgicum* (Frankfurt, 1666), and *Wund-Artzneyisches Zeug-Hauß* (Frankfurt, 1666). The data is divided into three parts: percentages among the total number of case histories (sixty-one *observationes*), the number of case histories presented as Scultetus's firsthand experience (fifty-six *observationes*), and the total number of case histories with civilian patients (forty-six *observationes*). Percentages reflect numbers of *observationes*, not individual patients.

The majority of Scultetus's *observationes*—some seventy-five percent—focus on civilian patients like the shoemaker Hans Jacob (Tables 2.1 and 2.2). These suggest the broader social fabric of Scultetus's practice. He records a wide social spectrum among his civilian patients, including town elites, craftsmen, and peasant farmers. The inclusion of a bargeman and a boatman's son

Table 2.3 Observationes: Soldiers as patients, 1630–45

Case information	*14* observationes of soldiers	*56* observationes Scultetus treated	*61* observationes total
	(percent)	(percent)	(percent)
Cause of affliction			
Injury	85.71	19.64	19.67
Military combat (e.g. battle)	35.71	7.14	8.19
Unofficial violence (e.g. duel)	42.85	10.71	9.83
Unspecified	7.14	1.78	1.63
Illness/Disease	14.28	3.57	3.27
Social status of patient			
Common soldier	42.85	10.71	9.83
Officer or noble	57.14	14.28	13.11
Political affiliation of patient			
Imperial forces	28.57	5.35	6.55
Ulm forces	21.42	5.35	4.91
Swedish forces	21.42	5.35	4.91
German army allied with Swedes	7.14	1.78	1.63
Unspecified	14.28	3.57	3.27
Geographical origins of patient			
Ulm and its territory	21.42	5.35	4.91
Outside Ulm	35.71	7.14	8.19
Holy Roman Empire	21.42	3.57	4.91
Sweden	14.28	3.57	3.27
Unspecified	42.85	10.71	9.83

The above provides statistics about soldiers as patients in *observationes* dated 1630–45 in Johannes Scultetus, *Armamentarium Chirurgicum* (Frankfurt, 1666), and *Wund-Artzneyisches Zeug-Hauß* (Frankfurt, 1666). The data is divided into three parts: percentages among the total number of case histories (sixty-one *observationes*), the number of case histories presented as Scultetus's firsthand experience (fifty-six *observationes*), and the total number of case histories with soldiers as patients (fourteen *observationes*). Percentages reflect numbers of *observationes*, not individual patients.

is indicative of the importance of the Danube and its connected waterways to Ulm's economy.[55] When geographic origins are indicated, these *observationes* present local cases: the patients come from Ulm, its hinterlands, or its neighbors.

Nearly half of the *observationes* featuring civilian patients involve traumatic injuries. Those caused by accidents run from the mundane, such as a man falling from his horse, to the bizarre, including a baker who sticks himself with a knife at a wedding.[56] Injuries caused by violent means in particular highlight moments of strained social relationships within communities, as well as the dangers posed by itinerant soldiers. Of thirteen *observationes* identifying human violence as the cause of injury, approximately forty-six percent are caused by other civilians, thirty-one percent by soldiers, and twenty-three percent by unidentified attackers. Motivations for violence are rarely given. Scultetus does not comment on why a gardener stabbed a fuller, or what led a tailor's apprentice to attack his master.[57] Soldiers designated "imperial" or "Bavarian" lurk in the text as outside threats to sedentary dwellers of town and countryside. This reflected Ulm's confessional composition and its stance in the ongoing war. Ulm, a Protestant imperial city, maintained neutrality through the Thirty Years War, while the Catholic duchy of Bavaria held some form of alliance with the emperor through virtually the entire conflict.[58] In the case of Maria Lutzin, for example, Scultetus writes vaguely that Bavarian soldiers treated her "very badly" (*sehr übel*; *miserrimè*), resulting in a head contusion.[59]

Scultetus rarely expounds on the ongoing war directly, yet the *observationes* comment on the ethical or—more often—unethical behavior of soldiers to a far greater extent than civilians.[60] This might take the form of a general remark: once healed, a soldier might go on to fight with honor.[61] However, the larger pattern in the text conveys moral disapprobation of soldiers' conduct. This might emerge in a detail outside the strict scope of the case at hand. In one instance, bystanders steal a diamond ring under the pretext of helping an injured officer.[62] Most frequently noted, however, is the negative role of alcohol. Scultetus remarks that a lieutenant injured in a duel on the evening of St. John's, a significant festival, was "quite drunk" (*wol bezecht*; *benè potus*) when he fought.[63] The inclusion of the lieutenant's inebriated state implies that the fight broke out while celebrating—a drunken sword fight rather than a sober ritual of honor. Consumption of alcohol also led to relapse and even death.[64] Perhaps the most wrenching case history involving military violence against civilians is that of the farmer Balthasar Steiger and his wife, who were shot on the road several times in January 1644 by a soldier who was "well and truly drunk" (*vollen und wolbezechten*; *bene poto & temulento*).[65] Local chronicles show that this took place while Bavarian and imperial forces were holding their winter quarters in Ulm territory.[66] Indeed, Balthasar had to hurry home in the middle of his treatment to see to the soldiers billeting in his house.[67] Here and elsewhere Scultetus does not openly criticize soldiers' behavior, but he does include enough contextual information for the reader to infer it is immoral.

The *observationes* can be revealing in other ways. Scattered throughout are details about his sensory experience treating patients. Treatises often offered information with which surgeons could compare and interpret sensory data, and Scultetus was thorough in his visual descriptions of patients' conditions and the use of his hands to treat them. On several occasions, for instance, he removes the thin layer of connective tissue protecting the surface of bones by scraping it back with his fingernails.[68] Yet, the exercise of writing *observationes*, a genre that invited authors to dwell on individual experiences rather than general concepts, sometimes led him to include more dramatic details. At times the reader can hear what is happening. In 1640, Scultetus must halt operating on the putrefying shin bone of an eleven-year-old boy because of the patient's "violent wailing and bellowing."[69] In a case from 1634, the reader can *smell* the sickbed: Scultetus writes that while treating the wounded jaw of a Swedish officer, the fetid odor became so terrible he could hardly remain in the room.[70] This is a rare admission of a moment so physically revolting to him that he struggled to stay on site to treat a patient. Such instances hint at aspects of a patient encounter that vividly stuck with him, and that imprinted on his memory so powerfully that he did not disentangle them from his recounting.

The case histories give a sense of the different locations of Scultetus's practice, as well as gesture to a network of practitioners with whom he worked. *Observationes* take place in Ulm's "Spital" and pest houses, as well as in patients' or surgeons' homes, and inns or taverns.[71] Scultetus even performs the dissection of a sheep's head in his own home.[72] Many cases refer to the nameless assistants who were undoubtedly present in most major surgical operations.[73] More significantly, in thirty percent of Scultetus's firsthand *observationes*, he characterizes himself as treating patients alongside or in coordination with other practitioners. These include barber-surgeons, surgeons, three physicians, and even a field surgeon. Among his colleagues in the *Collegium medicum*, the physician Johann Regulus Villinger recurs most frequently.[74] Scultetus also had a collaborative relationship with guild surgeons, with Georg Riedlein and Hans Georg Baulern in particular appearing in multiple *observationes*.[75] While Scultetus critiques nameless empirics and inexperienced barbers, he refers to the named surgeons with respect. Riedlein, for example, is a "well-experienced surgeon."[76] This points to how Scultetus navigated his unique circumstances in Ulm as a doctor of medicine and surgery: he collaborated with and expressed professional esteem for local guild surgeons while maintaining his superior authority over them.

The *observationes* are a tailored collection of experiences that Scultetus processed through the act of writing. He reflected on these encounters as important moments in which he was able to perform surgery, use surgical instruments, and with his own hands work with and learn from the living human body. The landscape in which he worked influenced not only the opportunities for such encounters but also the ways in which they unfolded.

General and Particular Forms of Knowledge

The title page and headings of *Armamentarium Chirurgicum* divide the work into two parts: the elaborately illustrated plates with annotations of the first part and the *observationes* of the second part. These two halves are distinct, yet to shift between them is less jarring than one might expect. If read from cover to cover, the treatise moves through different modes of thinking about surgery in three stages. The first presents instruments as solitary objects (Part I, plates 1–24); the second illustrates their use, with disembodied hands operating on disembodied heads, legs, arms, and so forth (Part I, plates 26–56); and the third recounts specific patient cases requiring surgical treatments from beginning to end in 100 *observationes* (Part II).[77] This movement in the published treatise, from the general to the specific, is in fact a reversal of the pattern of thought that went into its original creation. A close reading of two case histories shows that the embedded cross-references in Part II to plates from Part I reveal crucial cognitive links in the way Scultetus thought about, wrote about, and practiced surgery.

The 1635 case of Michael Schneider, reported in Observation 12, gives a detailed account of a head injury. Schneider's wife requested Scultetus's aid six weeks after her husband, an Ulm soldier on guard in nearby Elchingen, suffered two head wounds from an imperial soldier's sword. A barber-surgeon's initial treatment failed, and Schneider experienced unbearable headaches, followed by frenzy and convulsions of the entire body, and then apoplexy. Scultetus examined the unconscious patient and found his wounds complex: a depression and fissure of the skull, and injury to the outermost membrane of the brain, the dura mater. He discovered the damage first by reading the outward signs with his eyes. The black, stinking matter issuing from the patient's nostrils and mouth indicated that the apoplexy was caused by injury to the skull and the inner parts. Scultetus investigated further by using his fingers to assess the wound on the top of the patient's head, which had healed over solid. There he felt a very large pit, indicating a depression in the skull. He declared the patient in grave danger and prepared to operate. Over the next week, he performed several delicate procedures. After reopening the flesh with a scalpel, he trepanned the skull by drilling several holes around the edges of the depression. Then he sawed between the holes to detach and remove the depressed portion completely, applied pressure to the brain to allow the corrupt matter to escape, and finally left a small ulcer in the flesh for matter to continue to flow. After several months of good health, the patient neglected to keep the ulcer open, and, according to Scultetus, subsequently died as a result.[78]

As a stand-alone narrative, Schneider's case reveals much about Scultetus as a surgical practitioner and author. Reading external symptoms as signs of internal problems, as Scultetus did with the dark matter running from Schneider's nose and mouth, was fundamental to Hippocratic-Galenic medicine.[79] In the *observatio*, physical touch was crucial—only by exploring the injury with his fingers could Scultetus diagnose the fractured depression and decide to operate. During surgery, his hands were instruments. He used his fingernails to remove splintered bits of the skull after opening the flesh to reveal the fracture. Drawing from both contemporary and ancient authors for recipes, Scultetus treated the patient as a *medicus* and learned surgeon. He used a cloth with a salve from Girolamo Fabrizi d'Acquapendente to lay over the injured cerebral membrane and placed a cataplasm—containing a soft, warm mixture of herbs and other ingredients—from Hippocrates on top. The same was true of his dressings and instruments. He cut the flesh over the fracture with a basic scalpel discussed in

Hippocratic texts and by the first-century Roman medical writer Aulus Cornelius Celsus, used a trepan designed by Fabrizi to drill holes, and dressed the entire wound with a technique attributed to the second-century Greek physician Galen. The narrative identifies instruments with care, noting names, the order in which they were used, and their purpose. Subtle but important was the inclusion of witnessing, a typical feature of the genre. Scultetus notes that Schneider's wife was still living (presumably in 1645, when Scultetus was working on the manuscript) and refers to his discussions with Schneider's family during treatment.[80]

Embedded cross-references connect the operations performed on Schneider to the instruments and general instructions found elsewhere in the treatise. Eight instruments are designated with parenthetical references to figures in the illustrated plates of Part I. Two, in particular, show cognitive connections Scultetus made between general and particular forms of surgical knowledge.[81] The first is a small saw with a cogwheel, one of the few instruments Scultetus presents as his own design. In the *observatio*, Scultetus uses it to cut between the holes made by the trepan in Schneider's skull. In parentheses appears "Tab. XXVIII. Fig. VIII." (i.e. plate 28, fig. 8), which directs the reader to a plate in Part I with several disembodied heads, each serenely looking downward as disembodied hands perform various operations on a portion of the skull where the hair has been shaved off and the flesh cut open (see my Figure 2.1).[82] The text accompanying fig. 8, which displays two disembodied hands using the cogwheel saw, states that Scultetus already described the instrument in plate 5.[83] This turns the reader's attention to an earlier plate depicting the saw and its components as objects floating in empty space (see my Figure 2.2). Here the explanatory text describes only the components of the saw; one must see the *Armamentarium*'s plate 28, fig. 8, to learn its "rechter Gebrauch," or proper use.[84] From Schneider's specific case, the cross-references lead the reader first to a general depiction of the saw's use, then to an isolated image of the instrument, and back again.

Parenthetical references do not always appear in three-link chains of *observatio*—illustration of instrument use—illustration of a solitary object. Two such parentheticals are more than enough to suggest a cognitive link, as a second example from Schneider's case demonstrates. In the *observatio*, Scultetus refers to multiple kinds of forceps used to remove corrupt matter from between the skull and the dura mater. Here one parenthetical reference to plate 29, fig. 4, takes the reader to a plate of disembodied heads undergoing operations.[85] These include the use of "parrot-beak forceps" (*Papagey-schnabel*; *rostro psittaceo*), another Scultetus design. At this point, the treatise does not provide an embedded reference to the beak-like instrument's first illustration as a solitary object in an earlier plate, but a curious reader could find this by checking its name in an index of instruments.[86] The pattern of movement from particular to general remains.

The links between the two halves of *Armamentarium Chirurgicum* are particularly significant for understanding Scultetus's relationship with the instruments he presents in Part I. As Bertolini Meli has pointed out in reference to

Figure 2.1 Johannes Scultetus, *Wund-Artzneyisches Zeug-Hauß* (Frankfurt am Main, 1666), plate 28. Photo: Universitätsbibliothek Heidelberg/doi: 10.11588/diglit.21962/Tabula XXVIII.

Fabry von Hilden's *Centuriae*, cross-referencing with objects connects texts with materials in powerful ways indicative of the role of materiality in the thinking and practice of authors.[87] While Fabry tied his text to objects in a physical museum, Scultetus connected his to the surgical instruments he used on patients. Scultetus himself explained that he did not have firsthand experience with every

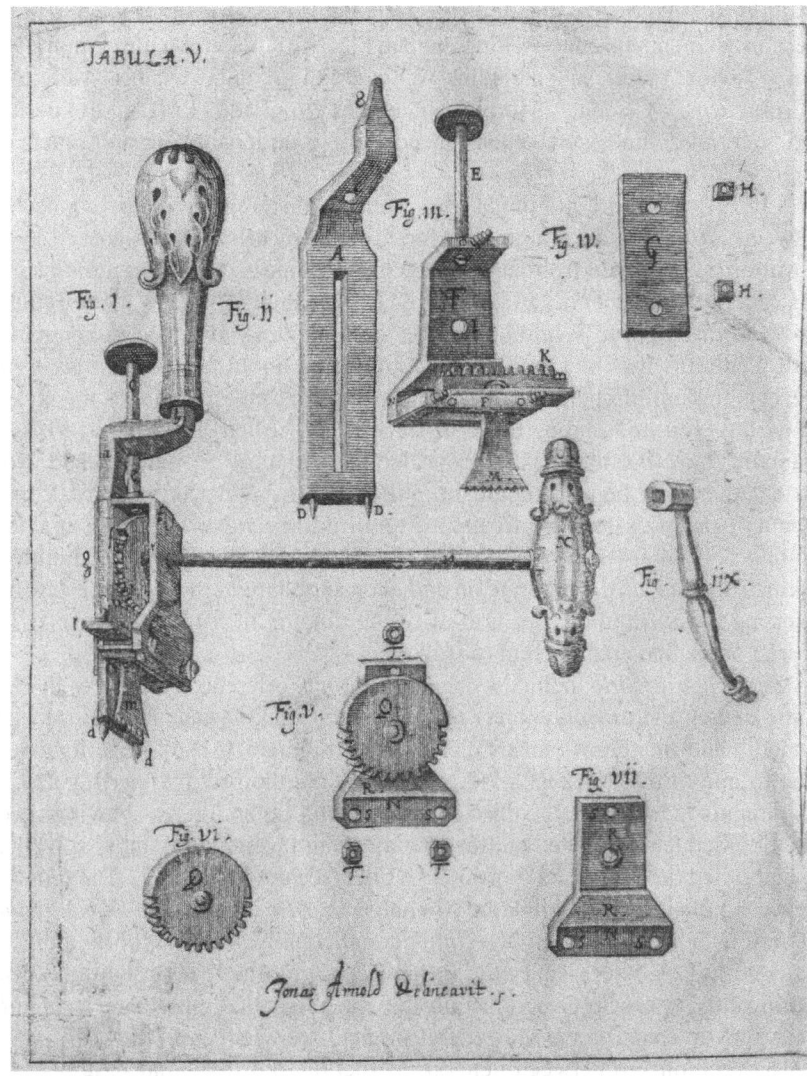

Figure 2.2 Johannes Scultetus, *Wund-Artzneyisches Zeug-Hauß* (Frankfurt am Main, 1666), plate 5. Photo: Universitätsbibliothek Heidelberg/doi: 10.11588/ diglit.21962/Tabula V.

instrument illustrated in Part I. In his discussion of eye-surgery instruments in plate 31, for example, he clarifies that he possessed the instruments needed to operate on a cataract and had even watched others perform the procedure, but had never done so himself.[88] The references embedded in the *observationes* of Part II highlight the instruments that Scultetus *used*, including both recent inventions and tools from antiquity. As Scultetus processed his encounters with these instruments through writing, he engaged with material culture in multiple ways. Textual practices—including writing, editing, and revising—were material practices, as Hannah Murphy has recently argued.[89] Firsthand experience and material culture were crucial to Scultetus's understanding and practice of surgery.

A brief look at an additional *observatio* further reinforces the connections between Parts II and I. Augustin Mertz, the Ulm merchant discussed in Observation 83, suffered from an ulcer on his right shin in the summer of 1634.[90] His case history recounts treatment over more than 200 days, with Scultetus involved throughout. While he was the only *medicus* involved in Schneider's case, Scultetus was one of three who conferred about Mertz. The *observatio* presents the initial operation as a demonstration: Scultetus makes the incision down the shin and shows the foul tibia to his colleagues. He reports, with dramatic flair, that the incision revealed a cartilage-like material had grown so that the bone beneath could be pulled out from it "as a sword from its sheath."[91] Mertz's lengthy treatment required several procedures, including drilling several holes into the shin bone. The *observatio* repeatedly highlights the instruments used to operate on and even supplement the natural body, from inserting a wax candle into a hollow of the tibia to fitting the patient's leg with an iron brace that enabled him to walk.

Mertz's *observatio* includes nine parenthetical references to figures in Part I. Some depict instruments, others techniques for making incisions and dressing wounds, and one shows an extracted bone fragment. All of these figures belong to one of three plates: 2, 12, and 54. While the first two portray assorted instruments in isolation, the third contains a total of ten figures showing several disembodied legs stretched out for operations and wound dressings, as well as a disembodied (and diseased) portion of a tibia (see my Figure 2.3). The cross-references to plate 54 firmly link the two halves of *Armamentarium Chirurgicum*.[92] At various points, Mertz's *observatio* parenthetically cites half of the figures in plate 54: figs. 3–5 (dressing), fig. 6 (incision), and fig. 9 (diseased tibia). More significantly, these figures illustrate the role Scultetus's firsthand encounters could play in constructing the general surgical knowledge of Part I.

The relationship between general instruction and particular case history is so entangled in plate 54 that it is impossible to separate them. Figs. 3–5 provide general instructions for dressing the lower leg, but figs. 6–10 explicitly demonstrate lower-leg operations using Scultetus's experience with Mertz. The text for fig. 6, which displays a shin with an open incision, explains how Scultetus healed Mertz's tibia, and uses a parenthetical aside to indicate that the case is also the subject of Observation 83.[93] The subsequent figures provide an abbreviated account of Mertz's surgical operations. The cognitive links emerge through the juxtaposition of the general and the particular among all ten images on the plate. When Scultetus presents his *general* instructions for bandaging

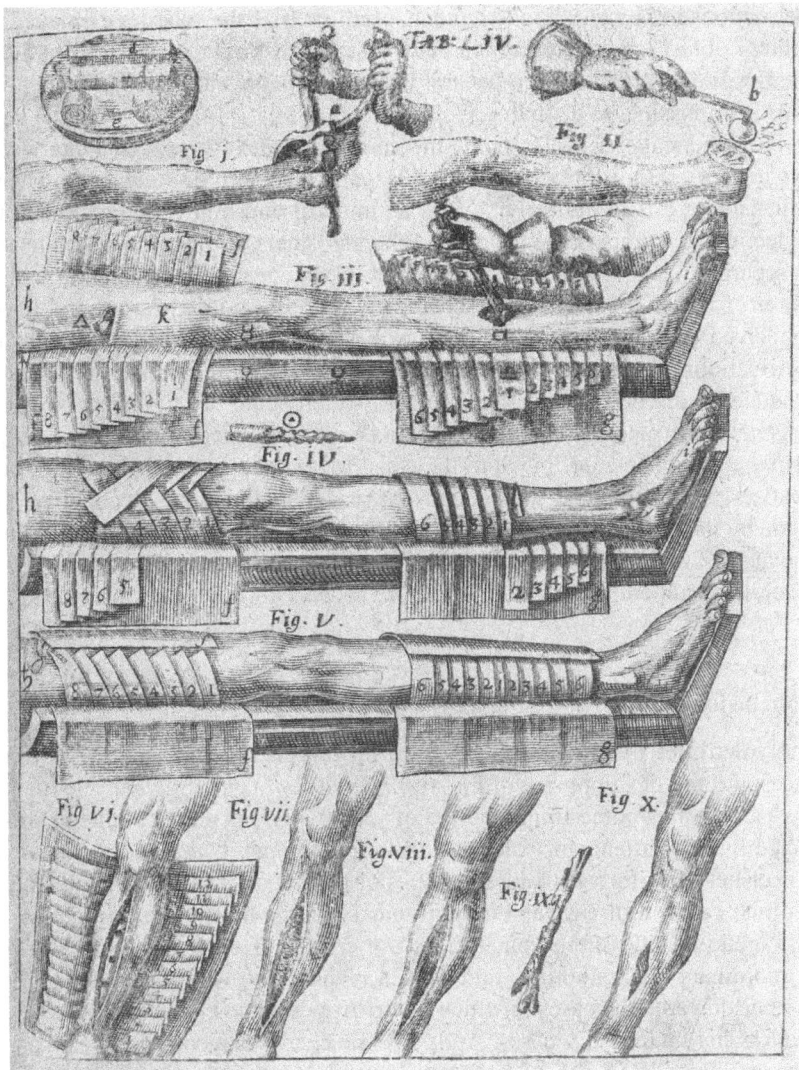

Figure 2.3 Johannes Scultetus, *Wund-Artzneyisches Zeug-Hauß* (Frankfurt am Main, 1666), plate 54. Photo: Universitätsbibliothek Heidelberg/doi: 10.11588/ diglit.21962/Tabula LIV.

a lower leg, he thinks of Mertz's *particular* case and uses it to develop a discussion about the operations required for a lower leg with a putrefying tibia. That Mertz's *observatio* in Part II does not cite every figure illustrating Mertz's case in plate 54 is particularly telling. Scultetus treated Mertz, used that experience to explain how to operate on fractured shin bones, and drew from those instructions to present the instruments used, such as the scissors displayed as a solitary object labeled "fig. 4" in plate 12 and cross-referenced to plate 54 and Mertz's *observatio*.[94] Where general instruction begins and firsthand encounter ends is impossible to distinguish.

Parenthetical cross-referencing in *Armamentarium Chirurgicum* interweaves two genres—surgical textbook and *observationes*—into a single work. The significance here is twofold. First, it is an unusual and innovative approach to a collection of surgical case histories. Whereas others published such collections as separate works, Scultetus crafted his to be in constant conversation with his plates of instruments and their accompanying descriptions. Parts I and II were conceived of and meant to be read as a whole. Second, it offers insight into the relationships among different kinds of knowledge Scultetus drew from to produce his learned manuscript. The cross-referencing ties firsthand experience to general instructions about instruments and techniques, suggesting how particular experience contributed to the construction of general information in the treatise. The textual links included for the reader are thus signs of cognitive links made by the author. They show that Scultetus thought about and drew from his encounters with patients, recorded in the *observationes* of Part II, to discuss several instruments and techniques more broadly through the illustrated plates of Part I.

Conclusion

The volatilities of war played a significant role in shaping Scultetus's experiences and the development of his *Armamentarium Chirurgicum*. The majority of case histories come from a period of years when his practice in Ulm was affected by the violence of soldiers attacking civilians, Ulm guards wounded in skirmishes, and foreign soldiers requiring treatment. These were interspersed with the cases of disease and injury typical of a town physician's practice. The sustained violence of the Thirty Years War created an environment in which the extraordinary was quotidian, and what was quotidian was extraordinary. From this period came the cases in which Scultetus performed the most invasive procedures his treatise discusses. And, of course, it was also when he wrote the manuscript itself.

Scultetus may have only operated on patients in extraordinary cases, but over the course of his career, he thought about, prepared for, performed, and wrote about surgery, bringing his experiential and textual knowledge to bear on the subject. His practice and writings of 1630–45 exemplify one of the many-layered ways to "do" surgery in early modern Europe—which in his case meant in a time of war. The *observationes* of Part II reveal patient cases that helped to build the practical foundation of Part I of *Armamentarium Chirurgicum*. The interlinking of Parts I and II through cross-referencing reflects the utility of Scultetus's firsthand experiences for thinking about how to do surgery, what objects are needed for surgery, and how to communicate these things to others. This approach is also evident in the treatise's three indices of instruments, *observationes*, and "noteworthy things," found in both the original and expanded editions. The instruments correspond to Part I, the *observationes* to Part II, and "noteworthy things" can refer to items in either part. The indices allow readers to move effortlessly between the two halves of the treatise. When Scultetus refers to an instrument like surgical tweezers (*Korn-Zanglein*; *volsellam*) in an *observatio* without a parenthetical reference to Part I, the reader can consult the register of instruments to find a corresponding illustration and explanation. Scultetus, in other words, envisioned his readers utilizing the two parts together. In the same way, the different modes of thinking about surgery presented in the treatise were interlinked and essential for Scultetus himself, as a practitioner and as an author.

Notes

1 Johannes Scultetus, *Wund-Artzneyisches Zeug-Hauß* (Frankfurt am Main, 1666), 2:57. [Henceforth: *Zeug-Hauß*.]
2 Johann Dieterich Leopold, *Memoria Physicorum Ulmanorum seu Biographia Medicorum ordinariorum*, Stadtarchiv Ulm, H Leopold 2, 129. [Henceforth: *Memoria*.]
3 *Zeug-Hauß*, 2:100, 121, 124, 198: "biß auff das Jahr 1645. in welchem ich dieses schreibe." All translations are mine.
4 Of the remaining *observationes*, sixteen date from 1622 to 1629, nineteen are undated, and four give general treatment methods only.
5 Hans Heberle and Gerd Zillhardt, *Der Dreissigjährige Krieg in zeitgenössischer Darstellung: Hans Heberles "Zeytregister" (1618–1672), Aufzeichnungen aus dem Ulmer Territorium: ein Beitrag zu Geschichtsschreibung und Geschichtsverständnis der Unterschichten* (Stuttgart: Kohlhammer, 1975), 14–40. [Henceforth: *Zeytregister*.]
6 *Zeytregister*, 25–29; Anneliese Seiz et al., *Johannes Scultetus und sein Werk* (Ulm: Merckle KG, 1974), 21; Annemarie Kinzelbach, *Gesundbleiben, Krankwerden, Armsein in der frühneuzeitlichen Gesellschaft: Gesunde und Kranke in den Reichsstädten Überlingen und Ulm, 1500–1700* (Stuttgart: F. Steiner, 1995), 209–11.
7 Joseph Furttenbach, *Cronica oder historische Beschreibung waß sich in der löbl. Reichsstatt Ulm...Der Ander Teil.* Stadtarchiv Ulm, H Furttenbach 2, 19. [Henceforth: *Cronica*.]
8 *Cronica*, 120; *Zeytregister*, 178, 191.
9 I cite the reproduction printed a year later, which duplicates all text and images: Johannes Scultetus, *Cheiroplotheke, seu...Armamentarium Chirurgicum* (Hagae-Comitum, 1656). [Henceforth: *AC-1656*.]
10 E.g., Hieronymus Fabricius ab Aquapendente, *Operationes Chirurgicae. In duas Partes divisa* (Venice, 1619), 1:5, 7; *Zeug-Hauß*, 1:4, 7.
11 Johannes Scultetus, *Armamentarium Chirurgicum* (Frankfurt am Main, 1666). [Henceforth: *AC*-1666.]
12 *Cronica*, 139; *Memoria*, 130.
13 On Scultetus the Younger: *Memoria*, 174–77; Seiz et al., *Scultetus*, 25–29.
14 *AC*-1656, sig.)(3v: "cotidie novas observationes didicit, quas foliis chartaceis inscripsit, ea opinione & eo animo, ut aliquando revisae in vulgus exirent."

15   *Memoria*, 175–76.
16   Seiz et al., *Scultetus*, 20.
17   *AC*-1656, sig.)(3v.
18   Ibid., sig.)(5v.
19   *Zeug-Hauß*, 1:1.
20   On Megerlin's translation: Seiz et al., *Scultetus*, 30, 71–78.
21   *AC*-1656, 172; *Memoria*, 176.
22   *AC*-1656, 258.
23   *AC*-1666, 2:86; *Zeug-Hauß*, 2:142.
24   Stadtarchiv Ulm, A 3530, Bd. 95, RP (1645), 454v, 465v.
25   The exception is a report from Scultetus's colleague and friend, Johann Georg Gockeln. *AC*-1666, 2:87; *Zeug-Hauß*, 2:144.
26   Janus Abraham à Gehema, *Zwantzig sonderbahre Chirurgische Observationes* (Frankfurt am Main, 1690); Wilhelm Fabricius Hildanus, *Observationem & Curationum Chirurgicarum Centuriae* (Basel, 1606).
27   Gianna Pomata, "Sharing Cases: The *Observationes* in Early Modern Medicine," *Early Science and Medicine* 15:3 (2010): 210–11, 215.
28   Pomata, "Sharing Cases," 221; Volker Hess and Sabine Schlegelmilch, "*Cornucopia officinae medicae*: Medical Practice Records and Their Origins," in *Medical Practice, 1600–1900. Physicians and Their Patients*, ed. Martin Dinges et al. (Leiden: Brill Rodopi, 2016), 20–21.
29   Pomata, "Sharing Cases," 221; Hess and Schlegelmilch, "Medical Practice Records," 27–30; Michael Stolberg, "Medical Note-Taking in the Sixteenth and Seventeenth Centuries," in *Forgetting Machines: Knowledge Management Evolution in Early Modern Europe*, ed. Alberto Cevolini (Boston: Brill, 2016), 249–60.
30   Hess and Schlegelmilch, "Medical Practice Records," 26.
31   The German naturalist and biographer Johann Dietrich Leopold (1702–36) listed three handwritten works by Scultetus in his collection, now lost: a collection of recipes, an epitome of surgery, and a work suggestively titled *Volumen variorum consultationum medicarum Patavii calamo currento exceptarum*. *Memoria*, 131–32.
32   On Padua: Cynthia Klestinec, *Theaters of Anatomy. Students, Teachers, and Traditions of Dissection in Renaissance Venice* (Baltimore: Johns Hopkins University Press, 2012).
33   *AC*-1666, 2:79, 87; *Zeug-Hauß*, 2:130, 143.
34   *AC*-1666, 2:61; *Zeug-Hauß*, 2:102–103.
35   Domenico Bertolini Meli, "'Ex Museolo Nostro Machaonico': Collecting, Publishing, and Visualization in Fabricius Hildanus," *Journal of the History of Medicine and Allied Sciences* 72, no. 1 (2017): 105.
36   Gregor Horst, *Observationum medicinalium singularium libri quatuor priores* (Ulm, 1625), and *Observationum medicinalium singularium libri quatuor posteriores* (Ulm, 1628–31); Johann Peter Lotichius, *Consiliorum et observationum medicinalium libri VI* (Ulm, 1644).
37   Pomata, "Sharing Cases," 197.
38   On medical observations and the rise of empiricism: Gianna Pomata, "*Praxis Historialis*: The Uses of *Historia* in Early Modern Medicine," in *Historia: Empiricism and Erudition in Early Modern Europe*, eds. Gianna Pomata and Nancy G. Siraisi (Cambridge, MA: MIT Press, 2005), 105–46; and J. Andrew Mendelsohn, "The World on a Page: Making a General Observation in the Eighteenth Century," in *Histories of Scientific Observation*, eds. Lorraine Daston and Elizabeth Lunbeck (Chicago: Chicago University Press, 2011), 396–420.
39   *AC*-1666, 2:1; *Zeug-Hauß*, 2:1.

40 Unless otherwise stated, this section discusses the sixty-one *observationes* of 1630–45 from Part II of the 1666 Latin and German editions.
41 Seiz et al., *Scultetus*, 15–18; Ruth Schilling et al., "Stadtarzt oder Arzt in der Stadt? Drei Ärzte der Frühen Neuzeit und ihr Verständnis des städtischen Amtes," *Medizinhistorisches Journal* 46 (2011): 99–133; Hannah Murphy, *A New Order of Medicine: The Rise of Physicians in Reformation Nuremberg* (Pittsburgh: University of Pittsburgh Press, 2019), 6–10, 51–54.
42 E.g., *AC*-1666, 2:59; *Zeug-Hauß*, 2:98.
43 The remaining five are second-hand reports of his colleagues.
44 Robert Jütte, *Ärzte, Heiler und Patienten: Medizinischer Alltag in der frühen Neuzeit* (Munich: Artemis & Winkler, 1991), 19–32; Mary Lindemann, *Medicine and Society in Early Modern Europe*, 2nd ed. (Cambridge, UK: Cambridge University Press, 2010), 128–36; Claudia Stein, *Negotiating the French Pox in Early Modern Germany* (Farnham, Surrey, UK: Ashgate, 2009), 154–56; Murphy, *New Order*, 94; Seiz et al., *Scultetus*, 20.
45 *AC*-1666, 2:51; *Zeug-Hauß*, 2:85.
46 E.g., *AC*-1666, 2:78; *Zeug-Hauß*, 2:128.
47 Lindemann, *Medicine and Society*, 238–46; Jütte, *Ärzte*, 17–19.
48 *Zeug-Hauß*, 2:159: "durch den Barbier." Cf. *AC*-1666, 2:95.
49 *Zeug-Hauß*, 2:160: "ich besagtes Ruffen-Stücklein mit dem Korn-Zänglein ergriffen und…herauβ gezogen habe." Cf. *AC*-1666, 2:95.
50 Stadtarchiv Ulm, A 3530, Bd. 89, RP (1639), 91v: "für daβ treponiren." [Henceforth: RP-1639.]
51 *AC*-1666, 2:5–7; *Zeug-Hauß*, 2:7–9.
52 RP-1639, 56v–57r, 91v–92r.
53 *AC*-1666, 2:5; *Zeug-Hauß*, 2:7.
54 RP-1639, 92r.
55 *AC*-1666, 2:7, 28; *Zeug-Hauß*, 2:10, 47.
56 *AC*-1666, 2:35, 70; *Zeug-Hauß*, 2:59, 117.
57 *AC*-1666, 2:85, 133; *Zeug-Hauß*, 2:140, 222.
58 Peter H. Wilson, *The Thirty Years War: Europe's Tragedy* (Cambridge, UK: Belknap Press, 2009), 295, 726–27.
59 *Zeug-Hauß*, 2:48; *AC*-1666, 2:28.
60 Of three notable exceptions, two come from Johann Georg Gockeln. E.g., *Zeug-Hauß*, 2:26–29.
61 *AC*-1666, 2:11; *Zeug-Hauß*, 2:17.
62 *AC*-1666, 2:74; *Zeug-Hauß*, 2:122.
63 *Zeug-Hauß*, 2:117; *AC*-1666, 2:71.
64 *AC*-1666, 2:21, 95; *Zeug-Hauß*, 2:34, 160.
65 *AC*-1666, 2:98; *Zeug-Hauß*, 2:164.
66 *Zeytregister*, 195, 197; *Cronica*, 120.
67 *AC*-1666, 2:100; *Zeug-Hauß*, 2:167.
68 *AC*-1666, 2:10, 112; *Zeug-Hauß*, 2:15, 188.
69 *Zeug-Hauß*, 2:188: "hefftigen klagen und heulens." Cf. *AC*-1666, 2:112.
70 *AC*-1666, 2:46; *Zeug-Hauß*, 2:76.
71 E.g., *AC*-1666, 2:17, 54, 93, 135; *Zeug-Hauß*, 2:35, 89, 155, 226. On medical institutions in Ulm: Kinzelbach, *Gesundbleiben, Krankwerden, Armsein*, 323–52; Hans-Joachim Winckelmann et al., *Medizinhistorischer Streifzug durch Ulm* (Ulm: Jan Thorbecke Verlag, 2016), 65–72.
72 *AC*-1666, 2:19; *Zeug-Hauß*, 2:30.
73 E.g., *Zeug-Hauß*, 2:193; *AC*-1666, 2:116.

74 E.g., *AC*-1666, 2:135; *Zeug-Hauß*, 2:226.
75 *AC*-1666, 2:19; *Zeug-Hauß*, 2:31; Annemarie Kinzelbach, *Chirurgen und Chirurgie-Praktiken: Wundärzte als Reichsstadtbürger 16. bis 18. Jahrhundert* (Mainz: Donata Kinzelbach, 2016), 7.
76 *Zeug-Hauß*, 2:42: "wolerfahrnen Wund-Artzt." Cf. *AC*-1666, 2:25.
77 *AC*-1666; *Zeug-Hauß*. Likewise, in the original 1655 edition, plates 1–17 present isolated objects and plates 21–43 (with the exception of plate 22) show objects in use.
78 *AC*-1666, 2:20–21; *Zeug-Hauß*, 2:32–35.
79 Pomata, "Sharing Cases," 216; Claudia Stein, "The Meaning of Signs: Diagnosing the French Pox in Early Modern Augsburg," *Bulletin of the History of Medicine* 80:4 (2006): 617–48.
80 *AC*-1666, 1:5; 2:20–21; *Zeug-Hauß*, 1:6; 2:32–34.
81 While the following draws from the 1666 Latin and German editions and uses their numbering of plates, the cross-referencing strategy discussed is also consistent with the 1655 edition.
82 *AC*-1666, 1:Tabula XXVIII; 2:20; *Zeug-Hauß*, 1:Tabula XXVIII; 2:33.
83 *AC*-1666, 1:40; *Zeug-Hauß*, 1:58.
84 *Zeug-Hauß*, 1:Tabula V, 15; cf. *AC*-1666, 1:Tabula V, 11.
85 *AC*-1666, 1:Tabula XXIX; 2:21; *Zeug-Hauß*, 1:Tabula XXIX; 2:33.
86 *AC*-1666, 1:Tabula IV, fig. 2; *Zeug-Hauß*, 1:Tabula IV, fig. 2.
87 Bertolini Meli, "Hildanus," 106.
88 *Zeug-Hauß*, 1:95; *AC*-1666, 1:62.
89 Murphy, *New Order*, 16.
90 *AC*-1666, 2:115–19; *Zeug-Hauß*, 2:192–98.
91 *AC*-1666, 2:116; *Zeug-Hauß*, 2:193.
92 The figures of plate 54 in the 1666 editions correspond to plates 27–28 of the 1655 edition. Some reproduce original figures (1–2, 6–10), some combine multiple original images into one figure (3), and some are new images that elaborate on the stages of a technique originally illustrated in a single figure (4–5). Although the order and numbering of figures differ in the 1655 and 1666 editions, the chain of reasoning argued here remains the same.
93 *Zeug-Hauß*, 1:210; *AC*-1666, 1:125.
94 *AC*-1666, 1:19; 2:116, 119; *Zeug-Hauß*, 1:26;.2:194, 198.

Further Reading

Bertolini Meli, Domenico. "'Ex Museolo Nostro Machaonico': Collecting, Publishing, and Visualization in Fabricius Hildanus." *Journal of the History of Medicine and Allied Sciences* 72, no. 1 (January 2017): 98–116.

Hausse, Heidi. "Bones of Contention: The Decision to Amputate in Early Modern Germany." *The Sixteenth Century Journal* 47, no. 2 (2016): 327–50.

Kinzelbach, Annemarie. "Erudite and Honoured Artisans? Performers of Body Care and Surgery in Early Modern German Towns." *Social History of Medicine* 27, no. 4 (2014): 668–88.

Leong, Elaine. "Learning Medicine by the Book: Reading and Writing Surgical Manuals in Early Modern London." *BJHS Themes* 5 (2020): 93–110.

Nutton, Vivian. "Pieter Van Foreest: The Physician as Writer on Surgery." *Journal of The History of Medicine and Allied Sciences* 72, no. 1 (January 2017): 87–97.

3   Sighs of War and Peace

Feeling Prayer through Song in Lutheran Germany during the Thirty Years War[1]

Thomas Marks

In February 1631, Protestant leaders of the Holy Roman Empire gathered in Leipzig at the request of Saxony's elector, Johann Georg I, to discuss strategies for upholding the imperial constitution, a set of agreements that they felt on the verge of dissolution due to the increasing aggression of Emperor Ferdinand II and the ongoing difficulties of the Thirty Years War.[2] Among the many distinguished guests in attendance at this meeting, which came to be known as the Leipzig Convention or Conference, was Johann Casimir (1564–1633), the duke of Saxe-Coburg.[3] During the week that Casimir and the Protestant leaders convened, citizens in the duke's capital city of Coburg marked the gathering at Leipzig with their own unique observance. On the Sunday of the Feast of the Chair of Saint Peter (*Cathedra Petri*), a newly written composition for four voices—penned by the *Kapellmeister* at Coburg, Melchior Franck (c. 1579–1639)—was sung in honor of the assembled leaders in Saxony. Franck titled the short composition *Hertzlicher Seufftzer der Christlichen Kirchen in Deutschland* [Heartfelt Sigh of the Christian Church in Germany]; today, only the tenor part of this work survives.[4] The text of the vocal composition, a paraphrase of Psalm 122 likely penned by the composer himself, describes the gathering of the rulers of Israel's tribes and concludes with a prayer for peace—an apt text to mark the Convention at Leipzig.

The title of Franck's work, with its references to both the (Protestant) church's "heartfelt sighs" and the warring state of Germany, suggests direct relationships between music, emotion, and the Thirty Years War.[5] Of further significance is the fact that Franck's sigh-composition is not unique, but rather one of numerous musical compositions that juxtapose the contemporary experience of the Thirty Years War with the expressive sigh.[6] Collectively, these sigh-compositions that address the experience of the war—which are eclectic in style and substance—are the subject of this essay. What is a musical sigh-composition, more generally? Out of what context did they arise, and what might a consideration of these compositions reveal about the emotionality of music and warfare in early modern Germany?

DOI: 10.4324/9781003157700-5

To engage with these questions, I rely throughout this essay on approaches developed in recent decades by historians of emotion, who recognize first and foremost that emotions are not ahistorical categories but are rather experienced and expressed according to the particular historical milieu in which they occur.[7] With this in mind, this essay opens with a careful consideration of the meanings ascribed to the sigh in early modern Germany. Because sigh-compositions emerged out of specifically Lutheran devotional contexts, I focus especially on Lutherans' construal of the sigh as a sacred gesture—one that was closely associated with the practice of prayer. Indeed, the proliferation of this musical genre almost exclusively within Lutheranism relied in part on the coterminous development of literary sigh-prayers in private devotional manuals; while contemporary Catholic vocal compositions often alluded to the sigh in their texts, such compositions were rarely titled musical "sighs"—a genre classification—in the same way that Lutheran counterparts were.[8] From my consideration of this emotional history, I argue in this essay that these wartime sigh-compositions offered Lutherans a set of prayerful, emotive scripts that, when performed, might afford words, sounds, and feelings to their otherwise indescribable experiences of war.

Experiencing the Sigh

The sigh held a variety of meanings for early modern Lutherans, but by far one of its most widely understood functions was as a kind of affective prayer that existed within the heart and was entirely nonverbal.[9] When unable to speak prayer—either because of some physical impediment or a more momentary cognitive lapse—the Lutheran might instead sigh from her heart and, by this gesture, communicate to God all that which could not be articulated in language. Martin Luther emphasized the centrality of the sigh as a form of prayer in his *Tischreden* [Table Talk], writing that though Christians do not continuously pray with words, "their hearts pray continually, sleeping and waking; for the sigh of a true Christian is a prayer."[10] Drawing from Matthew 6:7—in which Christ admonishes his followers to avoid long, loquacious prayers—Luther likewise wrote in the preface to his 1523 *Betbüchlein* [Little Prayer-Book] that "a good prayer does not lie in many words, as Christ says in Matthew 6, but in many and frequent heartfelt sighs to God, which should be unceasing."[11]

Luther's construal of the sigh as a form of prayer was not idiosyncratic, nor was it entirely novel in the sixteenth and seventeenth centuries; instead, it reflects the emotional gesture's long history within Christianity, the origins of which lie in Scripture itself. The apostle Paul confirmed the efficacy of the sigh as a form of prayer in Romans 8:26, which details the emotional gesture's capacity to convey the Christian's prayer to God through the intercessory role of the Holy Spirit: "Likewise, the Spirit helps our weakness. Then we know not how we should pray to our benefit; but the Spirit itself pleads for us best with unspeakable sighs."[12] Writers in the early church, following Paul, likewise confirmed the functionality of the sigh as a form of prayer. In a letter addressed to the Roman noblewoman Proba, for example, Augustine of Hippo—whose writings on the sigh were informative sources for early modern Lutherans—expressed a general skepticism toward verbose, spoken prayers, favoring instead engagement with God through tears and sighs.

> Prayer is to be free of much speaking, but not of much entreaty, if the fervor and attention persist. To speak much in prayer is to transact a necessary piece of business with unnecessary words, but to entreat much of Him whom we entreat is to knock by a long-continued and devout uplifting of the heart. In general, this business is transacted more by sighs than by speech, more by tears than by utterance.[13]

Augustine's understanding of the sigh challenges the logocentric superiority of spoken prayer, encouraging instead an affective mode of communication and spiritual engagement with God. Despite his recommendation, he is simultaneously aware of the ways in which language might nonetheless be necessary in order first to move Christians to feel and experience these affective gestures. Continuing in the same letter on prayer to Proba, Augustine conceded:

> Words, then, are necessary *for us so that we may be roused* and may take note of what we are asking, but we are not to believe that the Lord has need of them, either to be informed or to be influenced. Therefore, when we say "Hallowed be thy name," we rouse *ourselves* to desire that His Name, which is always holy, should be held holy among men also, that is, that it be not dishonored, something which benefits men, but not God.[14]

For Augustine, God does not attend to the words of our mouths, but rather the affective stirrings that words effectuate in us. Language is only useful in prayer, in other words, to the extent that it rouses Christians to desire God and long for him with tears and sighs.

As Paul articulated in Romans 8:26, the capacity for the Holy Spirit to communicate a Christian's prayers to God through nonverbal heart-sighs occurred especially during times in which "we know not how we should pray." For early modern Lutherans, experiences of intense physical or spiritual hardship—what they called *Anfechtungen,* or "spiritual assaults"—were frequently considered occasions where human thought and language faltered, thus necessitating the onset of the sigh.[15] In the experience of *Anfechtung,* hardships are heaped upon the Lutheran to such an extent that the afflicted comes to understand the world as a locus of inherent suffering. The Christian in *Anfechtung* turns her thoughts away from the world and relies more exclusively on God, fortifying her faith and ultimately—through suffering—advancing in a surefooted path toward salvation.[16] For the preacher and theologian Martin Moller (1547–1606), the biblical matriarch Hannah offered an especially potent example of how the sigh might function as a prayer when the Christian lost the capacity to speak under the weight of *Anfechtung*. Drawing from 1 Samuel 1:9–13, which recounts the way in which the barren Hannah prayed silently to God to grant her a child, Moller wrote in the preface of his 1603 *Thesaurus Precationum: Andächtige Gebet/und tröstliche Seufftzer* [Treasury of Prayer: Devotional Prayers and Comforting Sighs] that it was not the rhetorical sophistication of Hannah's prayer that eventually led her to conceive a child, but rather the inner sighs of her heart—the affective, non-linguistic product of her anguish and suffering.[17]

> Hannah knows well that God does not attend to the great, grand words of the mouth, but rather to the afflicted, groaning, faithful heart. It is therefore right to note the scripture through which her prayer is conveyed. She cried, she prayed for a long time, she spoke in her heart, her voice was heard by no one, her lips alone moved. From this it can be seen that she was able to come slowly to proper prayer through suffering and sadness, [that she] sighed at one time, cried at another, and at the third time spoke alone to herself until she had rightly poured out her heart before God.[18]

As Moller's example of Hannah suggests, the sigh was at once a gesture of bodily and spiritual suffering—an expression that emerged at the utter collapse of human agency and the dissolution of language—and also simultaneously a salvific gesture that opened the possibility of deliverance at the height of suffering. As Dennis Ngien recognizes, the total collapse of language at extreme moments of suffering "is not the end but the beginning of hope, as words are replaced by sigh, which will surely be heard by God."[19] Even as the Christian's ability to speak is rendered impotent in suffering, it is precisely this impotence that gives way to a more affective and effective form of prayer that sustains the Christian through her tribulation.

The sigh and language maintained a complex relationship in early modern Germany—one that was at times paradoxical if not outright contradictory. The vocal gesture was, by definition, a non-linguistic form of prayer that existed within the heart and was beyond the scope of words. And yet, beginning especially in the first half of the seventeenth century, Lutherans published prayer-books whose contents—literary and linguistic—were called sigh-prayers by their authors. Volumes such as the aforementioned *Thesaurus Precationum: Andächtige Gebet/und tröstliche Seufftzer* (1608) by Martin Moller; the 1616

Andächtige Kirch Seufftzer [Devotional Church-Sighs] by Johann Heermann; and the anonymous *Suspiria Hominis Moribundi Extrema, Das ist: Hertzen-seufftzerlein eines sterbenden Christen* [*Suspiria Hominis Moribundi Extrema,* That is: Heart-sighs of a Dying Christian] (1616) provided literate audiences with myriad prayer-texts that were named after a decidedly non-linguistic form of prayer.[20] But this seeming contradiction necessitates contextualization. On the one hand, these prayer-books actually fulfilled the ontological functionality of the sigh. Sigh-prayers emerged when the Christian was beset with such affliction that she was unable to speak her own words in prayer. But compilations of literary sigh-prayers actively filled in these linguistic voids, replacing the Christian's loss of words with the devout words of others. On these occasions, the sigh-prayer continued to function as it ought: the Christian was able to pray through her loss by means of the provided text of the sigh-prayer. On the other hand, though, these literary sighs evoked the Augustinian interdependence between language and feeling in the act of prayer. Words, while of no use to God, were nonetheless the devices by which humans were moved to feel in their prayers. Lutheran sigh-prayers published in devotional literature provided these affective scripts, the texts of which prompted Christians to experience salvific tears and sighs—the truly efficacious form of prayer to which God attentively listened.

As particular communities within the Empire experienced the hardships that often accompanied the Thirty Years War—especially famine, plague, and military pillaging—devotional authors began concertedly to tailor their sigh-prayers to address these increasingly persistent tribulations.[21] One of the most popular of these wartime collections of sigh-prayers was the posthumously published *Ernewerte Hertzen-Seufftzer* [Renewed Heart-Sighs] (1633) by Josua Stegmann (1588–1632).[22] The centrality of the Thirty Years War as the manual's primary framing device is immediately apparent in the book's two engraved title pages, illustrated below in Figures 3.1 and 3.2. The top portion of Figure 3.1 features a scene from Noah and the Great Flood, detailed in Genesis 8:15–21. Foregrounded in the image is Noah and his family, who burn offerings to God on an altar as a gift of thanks for the family's safe deliverance through the flood. In the background, Noah's ark rests atop Mount Ararat and a rainbow—a sign of God's covenant not to punish the Earth again for humanity's sins—stretches across the sky. Figure 3.2 depicts a contemporary and grisley scene of war. Marauding soldiers set fire to a village, townspeople beg for their lives, and the bodies of presumably innocent victims lie strewn on the ground. In the top left corner of the scene, one woman reaches out of a window, gesturing toward an angel in the sky who brandishes a sword; behind her, a pillar of smoke rises heavenward.

The two scenes invite early modern readers to draw direct comparisons between themselves and their biblical predecessors. Like Noah, who suffered and endured God's wrath by persisting through the Great Flood, those in early modern Germany similarly persist through God's anger, enacted in the form of war

Figure 3.1 Detail of right-facing title-page engraving from Josua Stegmann's *Ernewerte Hertzen-Seufftzer* (Lüneberg, 1633). Staats- und Stadtbibliothek Augsburg, Th Pr 2515.

for the sins of Germany's people.[23] The depiction of smoke in both Figures 3.1 and 3.2 gestures toward the contemporary understanding that the materiality of the sigh was in some ways akin to the rising clouds of smoke that resulted from burning incense. The Lutheran preacher Johann Arndt (1555–1621), for example, described the relationship between sighs and smoke in the second volume of his *Wahren Christentumb*: "As soon as one places frankincense, myrrh, and other herbs in the fire," Arndt writes, "so raises up a little cloud of smoke and gives off a pleasant smell, which cannot be sent without fire. As long as the fire of the Holy Spirit touches our heart, and that it not be hindered, so then the smell of a little sigh and prayer ascends."[24] Just as Noah sent clouds of burning smoke

Figure 3.2 Detail of left-facing title-page engraving from Josua Stegmann's *Ernewerte Hertzen-Seufftzer* (Lüneberg, 1633). Staats- und Stadtbibliothek Augsburg, Th Pr 2515.

to heaven as a sign of thanks for his safe deliverance from the Great Flood, so too does the woman who hangs from the window in Figure 3.2 send her own clouds of smokey sighs to heaven during her time of suffering and *Anfechtung*.

Stegmann included metered poems, prose texts, and songs in this collection of sigh-prayers that address, among other topics, the experience of war, its accompanying hardships, and the desire for peace.[25] The author's original contributions to the manual demonstrate his adroit rhetorical skills, especially in the way he renders the contemporary experience of war in highly moving and emotional language. In his "Gebet zu Kriegszeiten" [Prayer for Times of War], for example, the author begins by highlighting the ways in which only just recently the German people lived with relative peace and security, but now with the quick and sudden onset of war, all manner of daily life was fundamentally disrupted.[26] The prayer concludes with a moving appeal for God's merciful intercession:

> Look [God] at our great suffering of heart, our heartfelt sadness, our sad lamenting, our lamentable misery, our miserable adversity, our adverse grief and our grief-ridden adversity. We plead, we search, we knock, we lament, we howl, we cry before your countenance, we earnestly call [out], we imploringly call, we unceasingly call. Hear us, hear us, hear us; help us, help us, help us; save us, save us, save us. Hear our humble, our fearful, our imploring prayer, our whimpering and our lamentation. We ask you—we ask you imploringly—do not abandon us.[27]

Singing Sighs in the Thirty Years War

It was out of the rich practice of Lutheran prayer that musical sighs such as the *Hertzlicher Seufftzer der Christlichen Kirchen in Deutschland* by Melchior Franck, discussed in this essay's introduction, emerged. Prior to the beginning of the war in 1618, only one sigh-composition was published in Lutheran Germany.[28] Beginning around 1625, though, Lutheran composers set sigh-prayers to music that commented more specifically on the increasing hostilities within the Empire and the suffering it heaped on inhabitants. The first of these compositions to make explicit reference to the war was Andreas Rauch's (1592–1656) *Thymiaterium musicale, Das ist: Musicalisches Rauchfäßlein mit dem edlesten vnnd Gott dem HERRN angenemsten Rauchwerck/ das ist/sehnlichen Seufftzen vnnd Gebetlein der betrangten Christlichen Kirchen [Thymiaterium musicale,* That is: Musical Censer with the Precious and Pleasing Incense to God the Lord, That Is, Longing Sighs and Little-Prayers of the Hard-Pressed Christian Church] (1625).[29] The publication includes twenty-five motets for four to eight voices with basso continuo, all of which feature an eclectic array of texts extracted from the Bible (especially the Old Testament and the four Gospels), contemporary hymns, and devotional prayers.[30]

Rauch, a Lutheran, lived and worked as an organist in the small city of Hernals, which is today an incorporated neighborhood of Vienna. Though located in what was a Catholic state, Hernals had become a haven for Austrians who wished to practice Lutheranism. When Emperor Ferdinand II strode into confessional politics during the Thirty Years War, enacting reforms to re-Catholicize those lands that had gone the way of Protestantism since the signing of the Peace of Augsburg in 1555, Hernals (among other locations, especially in Austria and Bohemia) became a targeted place of interest. In 1625, Lutherans were faced with the decision either to abandon their faith and convert to

Catholicism or leave their homeland and migrate to nearby Protestant territories.[31] Many in Hernals, including Andreas Rauch, chose the latter option and entered into exile.[32]

Ferdinand II's disruptive reforms must have been palpable when Rauch penned the dedication of his musical sighs on the feast day of the Saints Philip and James (May 1) in 1625. In this paratext, the composer compared the visible presence of the Protestant church to the fluctuating light of the moon—sometimes it shines brightly for all to see while other times it is obscured behind clouds. For the past sixteen years, the Protestant church in Hernals flourished openly: Lutherans celebrated the sacraments, attended church services, and observed the public preaching of ministers. At the time of the dedication's writing, however, God had scattered the city's Lutherans to foreign territories, placing them in exile from the community that they called home. As a response to this event, Rauch published his collection of musical sighs to offer comfort to all those who were stricken with "adversity, fear, affliction, and persecution;" the vernacular German texts featured in the publication's motets further ensured that the work might "be understood by the learned and unlearned, now and then, here and in other places, ignited through heart and senses, and through this might be made a sweet smell to the LORD, if he would shine his holy Word and light longer to us and let cease his anger upon us."[33] Rauch's attention to "the sweet smell" that these sigh-compositions produce when performed—along with their ability to "ignite" the heart—evoke the image of smoke and incense often associated with the substance and materiality of the sigh in early modern Lutheranism. Like the prayer manual by Josua Stegmann, the composer likewise metaphorically compares his musical compositions to prayerful burnt offerings that, when sung, might carry the Lutheran's concerns heavenward to God. For Rauch, the performance of his collection's sonorous prayers was as pleasing to God's nose as it was to God's ear.[34]

While composers such as Rauch culled the texts of their wartime sigh-compositions from eclectic scriptural and liturgical sources, others turned to contemporary devotional prayer manuals for inspiration. One especially popular source was the aforementioned *Ernewerte Hertzen-Seufftzer* by Josua Stegmann. For example, in the *Musica oratoria et laudatoria* [Music of Prayer and Praise] (1630)—a collection of twenty-nine occasional works for three to six voices—the composer Johann Dilliger (1593–1647) set ten of Stegamnn's sigh-prayers to music.[35] The third piece in the collection, a three-voiced *Lamenta & suspiria super miserrimo & tantum non deploratissimo statu Germaniae* [Lament and Sigh Over the Most Miserable and the So Hopeless Condition of Germany], was composed for the birthday of Georg Mundig von Rodach, a notable jurist employed in Coburg, Dilliger's city of residence. The music appears in score format with each of the three voices—generically labeled voice I, II, and III—published on a single folio. Example 3.1 is Dilliger's strophic

Example 3.1 Johann Dilliger, *Lamenta & suspiria super miserrimo & tantum non deplo-ratissimo statu Germaniae* (1630), mm. 1–23. Landesbibliothek Coburg. Mus Mo B 32.

composition—shown with original clefs and meter—with the first of the song's sixteen verses printed below each voice.

Because this piece was published in 1630, the composer must have extracted the music's text from one of the earlier editions of Josua Stegmann's sigh-prayer manual, likely the now-lost 1628 edition, *Suspiria Temporum* [Sighs of These Times].[36] The song's text appears in Stegmann's 1633 edition under the title "Reimen Klage vber die Verwüstung des Vaterlandes" [Rhymed Lament for the Desolation of the Fatherland].[37] As its title suggests, the poem is a wrenching account of the various tribulations that Germany faced during the war. In verse one, the narrative voice speaks from the "I" perspective, beginning the lament by describing an image of Germany drowning in a sea of tears and unrest.

1.

Ich klag dich O Deutschland/in
Vnfriedsmeer ersäufft/

Dein Augn vnd dein Gesicht mit
Thränen überhäufft/

Der Pfeil deins Hertzenleyds thut meine
Seele kräncken/

Wenn ich an dein Noht vnd Elend
thu gedencken.

1.

I lament you, O Germany,
drowned in a sea of unrest,

Your eyes and your face deluged
with tears.

The dart of your heart-pain sickens
my soul

When I think on your hardship and
suffering.

Subsequent verses recognize God as the ultimate source of protection from, and intervening force within, the war—it is God, not man, who has the capacity to protect the innocent, grant wisdom to Germany's leaders, and end the carnage afflicting the Empire.

Because of the strophic structure of the composition, Dilliger is unable to set individual words to specific musical gestures; instead, he establishes a generally somber mood with a concise musical structure that, when repeatedly sung to the changing text, colors each of the poem's verses with severity. This is especially apparent at the end of the piece's first section in mm. 5–8, where the middle voice clashes continuously in a chain of syncopated suspensions against the bass line: here, dissonant intervals sound on beats one and three as the second voice moves counter to the other two, resolving downward by step on beats two and four into a consonance against the first voice and bass. The effect produces a series of jarring tensions that immediately release and resolve as the voices descend to the song's formal division at the repeat sign. This dissonant, descending passage immediately undoes itself in the following measures (mm. 9–11) in which all voices reverse direction to counteract the weighty decent that has just occurred—here, the first and second voices surge upward in syncopation against the bass. At the conclusion of the piece, though, Dilliger's music again descends toward the song's final cadence: all voices move downward in parallel motion in mm. 16–23 (with the exception of the middle voice's octave displacement in the final two measures). Both the first and second sections of the composition, then, end with descending gestures that pull against temporary moments of uplift—when coupled with the piece's frequent dissonances and overall minor tonality, the sinking gestures add continual weight and gravity appropriate to and reflective of the severity of Stegmann's original prayer for war-torn Germany.

While Dilliger set Stegmann's metric poetry to music, Johann Hildebrand (1614–84)—organist and composer from Eilenburg—set prose prayers from the *Ernewerte Hertzen-Seufftzer* in his own collection of musical sighs, the *Krieges-Angst-Seufftzer* [Fearful Sighs of War] (1645).[38] Hildebrand's collection is divided into two parts: the first of these features seven short works for solo voice and basso continuo, while the second part (a *Zugabe,* or appendix) contains strophic pieces for four parts. The texts the composer set to music in the collection's first half were extracted from both the Old Testament (specifically, the books of Job, Ezekiel, and Jeremiah) and, as my own research has uncovered, Josua Stegmann's *Ernewerte Hertzen-Seufftzer.* Musicologists have previously assumed the non-scriptural texts that appear in the first part were likely penned by the composer himself; Hildebrand, though a musician and organist, exercised his poetic proclivities in such publications as his *Johann Hildebrands In deutsche Reime übersetzter Jesus Syrach* (1662), a rhymed paraphrase of the entire Book of Sirach.[39] On closer examination, though, it is evident that Hildebrand selected multiple prayers from Stegmann's devotional manual, synthesizing the texts together for the collection's third, fifth, and sixth sighs.

In Hildebrand's fifth sigh, the composer combined extracted texts from two separate prose prayers from Stegmann's manual, respectively labeled "Gebet zu Kriegszeiten" [Prayer for times of war] and "Gebet zu Kriegszeiten/ein anders/vmb den lieben Frieden" [Prayer for Times of War: another one, for dear peace].[40] Hildebrand faithfully adapts the first of Stegmann's prayers—the "Gebet zu Kriegeszeiten"—into his composition, only slightly altering the text by removing or substituting an occasional word. But the second prayer, "vmb den lieben Frieden," was more liberally adapted. In Stegman's original prayer for peace, the theologian elaborates on the destructive forces of war, pleading to God for the abatement of his wrath and the cessation of an all-consuming fire of destruction. In his adaptation of this text, Hildebrand removes these references, focusing instead on the instillation of peace within the Empire, the protection of the church, and the prospect of a safe space in which to live out one's life. Especially important is the composer's own original addition to the end of the text, in which he includes a final petition for God to provide a safe space to peacefully die ("in Friede selig sterben mögen"), a plea that appears nowhere in Stegmann's original. In Table 3.1, Hildebrand's rendering of these two texts appears in the left column and an English translation is provided on the right; the first words of the second prayer for peace, "vmb den lieben Frieden," are emphasized in bold to highlight the point at which the composer joins these two separate prayers together. Formally, the composition is organized around these two conjoined prayers. Hildebrand divides the music into two parts that are separated by a repeat at the exact point where Stegmann's "Gebet zu Kriegeszeiten" ends and the prayer "vmb den lieben Frieden" begins.

Table 3.1 Text and translation of "Ach Gott! Wir habens nicht gewust" from Johann Hildebrand's *Krieges-Angst-Seufftzer* (1645)

Ach Gott! wir habens nicht gewust/was Krieg vor eine Plage ist/nun erfahren wir es leyder allzusehr/das Krieg eine Plage über alle Plagen ist/den da gehet Gut weg/da gehet Muth weg/da gehet Blut weg/da gehet alles weg/da muß man sein Brot mit Sorge[n] im Elende essen/da muß man seine Wasser mit beben trincken/da höret man nichts als auff allen Strassen/Weh! Ach! wie sind wir so verderbet! O du Gott des Friedes/gönne uns doch wieder deinen Him[m]lischen Frieden/laß Kirchen und Schulen nit zerstöret/laß den Gottesdiesnt und gute Ordnung nicht vertilget werden/hilff uns mit deinen ausgestreckten Arm/beschere uns ein Ortlein/da wir bleiben/ein Hüttlein/darinne wir uns auffhalten/ein Räumlein/da wir sicher seyn/und deinen Namen dienen kön[n] - en/daß wir in Friede deine[n] Tempel besuche[n]/in Friede dich loben und preisen/in Friede selig sterben mögen.	Ah God! We did not know what a plague is war, but now we experience it unfortunately all too much—that war is a plague above all plagues in which all good goes away, all courage goes away, all blood goes away, everything goes away. One has to eat one's bread with worry and in suffering; one must drink one's water with trembling. One hears nothing on the streets but "Woe! Ah! We are so totally ruined!" O, God of peace, grant us again your heavenly peace. Let not church and school be destroyed. Let not your church services and good order be exterminated. Help us with your outstretched arm. Bestow us a little place where we can remain—a little house, in which we can stay; a little room, where we can be safe and serve your name— that we may visit your temple in peace, laud and praise you in peace, die a holy death in peace.

Unlike Dilliger's strophic musical sigh, Hildebrand's fifth sigh-composition is through-composed and features a monodic musical texture—a solo alto voice is supported by the basso continuo, which provides harmonic support to the vocal line and facilitates a declamatory style resembling an impassioned speech. Indeed, as Stefan Hanheide has noted, the composer consciously drew in this collection from musical rhetorical lament-topoi especially characteristic of contemporary Italian music.[41] This highly expressive musical vocabulary is

Example 3.2 Johann Hildebrand, "Der V. Krieges-Angst-Seufftzer," *Krieges-Angst-Seufftzer* (1645), mm. 1–11. Leipziger Städtische Bibliotheken–Musikbibliothek II.4.57.

immediately apparent in the sigh's opening measures, shown above with the original clefs in Example 3.2. The initial "Ach Gott" [Ah God!] begins with abrupt chromatic and harmonic alterations in which the alto, singing a C-natural on the word "Ach," rises only by half step to C-sharp on "Gott." Harmonically, the basso continuo supports these initial pitches with an A minor chord that is followed immediately by a shift in quality to A major, a chordal progression highly atypical of other contemporary pieces that would have surely produced a striking effect on the listener. Hildebrand sets the subsequent, repeated expressions of "Ach Gott" to equally jarring melodic gestures—in m. 2, the alto voice performs an awkward downward leap (the interval of a major 6th) before rising again through stepwise motion to G-sharp in m. 3. The repeated references to the plague of war that follow in subsequent measures add additional urgency to the prayer's arresting opening gestures. In mm. 6–8, the alto sings the text "was Krieg vor eine Plage" two times in a tense, rising sequence that culminates near the top of the singer's range on the pitch F in m. 8. Collectively, these striking musical gestures that open Hildebrand's sigh-composition were likely intended to capture the attention of not only listeners and performers but also God, to whom the suppliant prayer is addressed.

On the one hand, the highly expressive character of Hildebrand's sigh might reflect a preexisting, impassioned performance practice of prayer—an imitation of the ways in which the composer's contemporaries delivered their sighs during a time of great hardship in the Thirty Years War. On the other hand, the composer's musical scripts might have equally served as pedagogical models that instructed Lutherans about the ideal way to pray a sigh. With its inherent ambiguity, paradoxical ontology, and lack of semantic meaning, the sigh was an open site that necessitated interpretation. Literary sigh-prayers, published in devotional manuals such as Stegmann's *Ernewerte Hertzen-Seufftzer,* added some clarity to the emotional gesture by assigning it linguistic meaning. But musical works such as Dilliger's *Lamenta & suspiria* and Hildebrand's *Krieges-Angst-Seufftzer* clarified for the Lutheran the appropriate and emphatic methods by which to deliver these wartime texts. The music's frequently somber harmonies, repeated words, abrupt shifts in mood and tone, and full exploration of the voice's expressive range further encouraged the Lutheran to *feel* the sentiments they expressed. As contemporary sources suggested, the efficacy of sigh-prayers resided not in their rhetorical sophistication, but in the affective conditions they cultivated within the Christian's heart. Musical sighs during the Thirty Years War provided Lutheran audiences with an additional level of affective engagement: when sung, these pieces might move the Christian to engage deeply with her prayer with the ultimate aim of increasing its efficacy. For contemporaries affected by the hardships of the Thirty Years War, the consequences of these works' performances, if done with sincere feeling, would have profound effects on the state of the Empire. God sent the war as a form of punishment for Germany's sins; simultaneously, God had the ability to end it and restore peace. By singing their sighs and emotionally engaging with God through them, pious Christians might appeal to God's divine mercy which, when exercised, could reinstate order during what was an immensely chaotic and difficult time.

Notes

1   Some of the research conducted for this essay was supported with funding from the *Deutscher akademischer Austauchdienst* (DAAD).
2   Peter Wilson, *The Thirty Years War: Europe's Tragedy* (Cambridge, MA: The Belknap Press of Harvard, 2009), 465–67.
3   Casimir's name, for example, was published on a list of Protestant leaders in attendance at the Leipzig conference in 1631. See *Außschreiben Deß Durchleuchtigsten/ Johann Georgen...An die Sämptlichen Evangelischen Stände* (1631), Aiij^v. VD17 14:004469D. In this essay, I have provided VD16 and VD17 numbers for sources that are digitized and available online through these respective catalogues.
4   Melchior Franck, *Hertzlicher Seufftzer der Christlichen Kirchen in Deutschland* (Coburg, 1631). A digitized copy of the tenor part is available via the Staatsbibliothek zu Berlin: http://resolver.staatsbibliothek-berlin.de/SBB0001A81100000000
5   For studies of musical life in Lutheran Germany during the Thirty Years War, see especially Mary Frandsen, "Music in a Time of War: The Efforts of Saxon Prince

Johann Georg II to Establish a Musical Ensemble, 1637–1650," *Schütz-Jahrbuch* 30 (2008): 33–68; Stephen Rose, "Music Printing in Leipzig during the Thirty Years' War," *Notes* 61, no. 2. (December 2004): 323–49; Derek Stauff, "Schütz's *Saul, Saul, was verfolgst du mich?* and the Politics of the Thirty Years War," *Journal of the American Musicological Society* 69, no. 2 (2016): 355–408. Most recently, several essays were published in a special edition of the *Journal for Seventeenth Century Music* that address musical life during the Thirty Years War. See Lois Rosow and Victor Coelho, eds. *Journal of Seventeenth Century Music* 26, no. 1 (2020).

6   Additional examples include Melchior Franck, *Suspirium Germaniae publicum* (Coburg, 1628); Johann Erasmus Kindermann, *Musicalische FriedensSeuffizer* (Nuremberg, 1642); and Georg Neumark, *Hertzliches Friedens-Seuffizen* (Königsberg, 1645).

7   For recent overviews on methodologies and approaches to the history of emotions, see Jan Plamper, *The History of Emotions: An Introduction,* trans. Keith Tribe (New York: Oxford University Press, 2015); Rob Boddice, *The History of Emotions* (Manchester: Manchester University Press, 2018). Recently, musicologists including Bettina Varwig have incorporated work from the history of emotions into seventeenth-century music studies. See Bettina Varwig, "Music in the Thirty Years War: Towards an Emotional History of Listening," *Journal of Seventeenth Century Music* 26, no. 1 (2020). https://doi.org/10.17863/CAM.57733

8   In my research of seventeenth-century sigh-compositions, I have found only two Catholic composers who wrote and published musical sighs: Augustin Grieninger, *Suspiria Mariana* (Augsburg, 1672), and Johannes Khuen, *Cor Contritvm et humiliatum: Engelfrewd oder Bußseuffizer* (Munich, 1640).

9   Kristiina Savin, "Sighs of Desire: Passionate Breathing in Medieval and Early Modern Literature," in *Pangs of Love and Longing: Configurations of Desire in Premodern Literature,* ed. Anders Cullhed et al. (Newcastle-upon-Tyne: Cambridge Scholars Publishing, 2013), 163–66

10   Martin Luther, *The Table Talk of Martin Luther,* trans. William Hazlitt, ed. Thomas S. Kepler (Mineola, NY: Dover Publications, 2005), 118–19.

11   "Den[n] es ligt nicht an vil worten ein güt gepet/ wie Christus sagt Mat. 6. Sondern an vil vn[d] oft hertzlich seufftze[n] zü got/ wölchs solt wol on vnderlaß sein." Martin Luther, *Ein Betbüchlin und leßbuchlin* (Augsburg: Steiner, 1523), preface unpaginated. VD16 L 4089. All German translations are my own unless otherwise noted. In all transcriptions of original German, the superscript E has been replaced with an umlaut.

12   "Desselbigen gleichen auch der Geist hilfft vnserer schwachheit auff. Dan[n] wir wissen nit wz wir beten sollen/ wie sichs gebüret/ sondern der Geist selbst vertritt vns gewaltiglich/ mit vnaußsprechlichen seufftzen." Rom. 8:26. *BIBLIA: Das ist: Die gantze Heilige Schrifft Teutsch: D. Mart. Luther* (Straßburg, 1621). VD17 32:697976V.

13   Augustine of Hippo, *Letters (83–130),* trans. Sister Wilfrid Parsons, S. N. D. (Washington D. C.: Catholic University of America Press, 1953), 391.

14   Ibid., 391–92. Emphasis is my own.

15   Ronald Rittgers translates *Anfechtung* as "spiritual assaults" as opposed to the more traditional "temptation." See Ronald Rittgers, *The Reformation of Suffering: Pastoral Theology and Lay Piety in Late Medieval and Early Modern Germany* (New York: Oxford University Press, 2012), 93.

16   Paul Bühler, *Die Anfechtung bei Martin Luther* (Zürich: Zwingli-Verlag, 1942), 207.

17   "After they had eaten and drunk at Shiloh, Hannah rose and presented herself before the Lord. Now Eli the priest was sitting on the seat beside the doorpost of the temple of the Lord. She was deeply distressed and prayed to the Lord, and wept bitterly.

She made this vow: 'O Lord of hosts, if only you will look on the misery of your servant, and remember me, and not forget your servant, but will give to your servant a male child, then I will set him before you as a Nazirite until the day of his death. He shall drink neither wine nor intoxicants, and no razor shall touch his head.' As she continued praying before the Lord, Eli observed her mouth. Hannah was praying silently; only her lips moved, but her voice was not heard; therefore Eli thought she was drunk." 1 Sam 1:9–13. Biblical quotations in this essay are taken from *The New Oxford Annotated Bible with Apocrypha: New Revised Standard Version*, ed. Michael D. Coogan et al. (Oxford: Oxford University Press, 2010).

18 "Hanna weiß wol/ das Gott nicht sihet auff grosse breyte Wort des Mundes/ sondern auff ein zuschlagenes/ zuknirrschtes/ rechtgleubigs Hertze. Sind derhalben die Wort wol zumercken/ dadurch jhr Gebet commendiret wird: Sie habe geweinet/ sie habe lang gebetet/ sie habe in jrem Hertzen geredet/ jre Stim[m]e habe man nicht gehöret/ allein jre Lippen haben sich gereget. Darauß den zu sehen: Das sie für Jam[m]er vnd Trawrigkeit langsam zum rechten Gebet komen können/ habe einmal geseufftzt/ das ander mal geweinet/ das dritte mal bey sich selbst geredet/ biß sie jhr Hertz für Gott wol außgeschüttet hat." Martin Moller, *Thesaurus Precationum* (Görlitz: Rhambaw, 1608), Bij^R–Biij^R.

19 Dennis Ngien, *Fruit of the Soul: Luther on the Lament Psalms* (Minneapolis: Augsburg Fortress Publishers, 2015), 195.

20 Johann Heermann, *Andächtige Kirch Seufftzer, oder Evangelische Schließ-Glöcklein* (Leipzig, 1616); *Suspiria Hominis Moribundi Extrema, Das ist: Hertzenseufftzerlein eines Sterbenden Christen* (Dresden, 1616). The explosion of sigh-prayer literature coincided with the budding *Frömmigkeitsbewegung* (Pietistic movement)—an impulse within Lutheran orthodoxy that sought to move away from doctrinal arguments and address more attentively the individual, pietistic needs of worshiping Christians. Martin Brecht, "Das Aufkommen der neuen Frömmigkeitsbewegung in Deutschland," in *Der Pietismus vom siebzehnten bis zum frühen achtzehnten Jahrhundert*, ed. Martin Brecht (Göttingen: Vandenhoeck & Ruprecht, 1993), 113–18. Patrice Veit, for instance, notes that the proliferation of prayer-manuals throughout the late sixteenth and seventeenth centuries was intended in part to establish the Lutheran home as an "Ort des Gebets gegenüber der Kirche aufzuwerten [place of prayer compared to the church]" and that devotional authors such as Johann Arndt, Johann Habermann, and Martin Moller helped to initiate with their devotional literature "ein Prozeß der Verhäuslichung des religiösen Lebens [a process of the domestication of religious life]." Patrice Veit, "Die Hausandacht im deutschen Luthertum: Anweisungen und Praktiken," in *Gebetsliteratur der frühen Neuzeit als Hausfrömmigkeit,* eds. Ferdinand van Ingen and Cornelia Niekus Moore (Wiesbaden: Harrassowitz Verlag, 2001), 194–95. On the use of prayer-books in early German Lutheranism, see, in the same volume, Johannes Wallmann, "Zwischen Herzensgebet und Gebetbuch. Zur protestantischen deutschen Gebetsliteratur im 17. Jahrhundert," in van Ingen and Moore, *Gebetsliteratur der frühen Neuzeit*, 13–46. The widespread popularity of prayer-manuals such as Johann Arndt's *Paradiesgärtlein* put religious piety quite literally into the hands of Lutherans who could perform prayer-practices in their own homes, and Arndt's manual subsequently and broadly influenced Lutheran poetry, music, and prayer-literature throughout the seventeenth century. Jeung Keun Park, *Johann Arndts Paradiesgärtlein: eine Untersuchung zu Entstehung, Quellen, Rezeption und Wirkung* (Göttingen: Vandenhoeck & Ruprecht, 2018), 197. While numerous authors have noted the importance of prayer-literature in seventeenth-century Lutheranism, few have devoted particular attention to the sigh-prayer despite its popularity and widespread usage. For my own work with

this literary genre of prayer and its relation to Lutheran musical culture, see Chapter 3 of my dissertation, Thomas Marks, "Sighs of the German People: An Emotional History of Musical Sigh-Compositions during the Thirty Years War (1618–1648)" (PhD diss., Graduate Center of the City University of New York, 2019), 79–136.

21 David Lederer reminds us that suffering the indirect effects of the war was a highly localized phenomenon and "varied greatly from region to region, from town to town, indeed from village to village." David Lederer, "The Myth of the All-Destructive War: Afterthoughts on German Suffering, 1618–1648," *German History* 29, no. 3 (2011): 386–87.

22 Josua Stegmann, *Ernewerte Hertzen-Seufftzer* (Lüneburg, 1633). A digitized copy of this source is available through the Staats- und Stadtbibliothek Augsburg at the permalink: https://sbaoz2.bib-bvb.de/00/bvnr/BV010809436 For an account of the complex publication history of Stegmann's manual, see Marian Szyrocki, "'Himmel Steigente HertzensSeufftzer' von Andreas Gryphius," *Daphnis* 1, no.1 (1972): 42–43.

23 On the Thirty Years War as a form of punishment for sin within the Empire, see Anton Schindling, "Das Strafgericht Gottes. Kriegserfahrungen und Religion im Heiligen Römischen Reich Deutscher Nation im Zeitalter des Dreissigjährigen Krieges. Erfahrungsgeschichte und Konfessionalisierung," in *Das Strafgericht Gottes: Kriegserfahrungen und Religion im Heiligen Römischen Reich Deutscher Nation im Zeitalter des Dreißigjährigen Krieges*, 2nd ed., eds. Matthias Asche and Anton Schindling, 44–47 (Münster: Aschendorff Verlag, 2001).

24 "So bald man ein Weyrauch/ Mirrhen/ vnnd ander Kreutlein ins Fewr legt/ so steiget ein Reuchlein/ auff/ vnd gibt einen lieblichen Geruch/ welches ohn Fewer nicht geschicht: Also so bald das Fewer des heiligen Geistes vnser Hertz berühret/ vnnd daß er nicht verhindert wird/ so bald steiget ein Geruch eines seufftzerleins vnd deß Gebets auff." Johann Arndt, *Das ander Buch Vom wahren Christentumb* (Magdeburg, 1610), 401. VD17 1:039798P.

25 While many of the texts included in this manual of sigh-prayers were written by Stegmann himself, a number of the songs included in the manual—such as the hymn *Wie Schön leuchtet der Morgenstern* by Philipp Nicolai—were penned by his contemporaries. For a list of hymn-texts that were included in the collection of sigh-prayers, but which were not written by Stegmann, see Albert Fischer, "Josua Stegmanns Lieder," *Blätter für Hymnologie* 11 (1888): 162–69.

26 Stegmann, *Ernewerte Hertzen-Seufftzer*, 348–54.

27 "Sihe doch an vnser grosses Hertzenleid/ vnser hertzliches Trawren/ vnser trawriges Klagen/ vnser klägliches Jammern/ vnser jämmerliches Elend/ vnser elendes Betrübniß vnd betrübtes Elend/ Wir bitten/ wir suchen/ wir klopffen an/ wir klagen/ wir heulen/ wir weinen vor deinem Angesicht/ wir ruffen ernstlich/ wir ruffen inständiglich/ wir ruffen vnnachlässig/ höre doch/ höre doch/ höre doch/ hilff doch/ hilff doch/ hilff doch/ rett doch/ rett doch/ rett doch/ höre vnser demütiges/ vnser ängstigliches/ vnser flehentliches Gebet/ vnser Winseln vnd Wehklagen/ Wir bitten dich/ wir bitten dich inständiglich verlaß vns nit." Ibid., 352.

28 Melchior Franck, *Suspiria Musica* (Coburg, 1612).

29 Andreas Rauch, *Thymiaterium musicale. Das ist: Musicalisches Rauchfäßlein mit dem edlesten und Gott dem Herrn angenemsten Rauchwerck, das ist sehnlichen Seufftzen und Gebetlein der betrangten Christlichen Kirchen* (Nuremberg, 1625).

30 No complete critical edition of Rauch's work has been published. For an overview of the collection, see especially Andrea Solya, "Thymiaterium Musicale (1625) by Andreas Rauch," (PhD diss., University of Illinois at Urbana-Champaign, 2010), 7–21.

31 For a brief history of Protestantism in Hernals and its re-Catholicization during the Thirty Years War, see Rudolf Leeb, "'Europa niemals kannte ein größere Kommun....' Die evangelischen Pfarrzentren für Wien außerhalb der Stadtmauern in der Reformationszeit," in *Brennen für den Glauben: Wien nach Luther*, eds. Rudolf Leeb, Walter Öhlinger, and Karl Vocelka (Vienna: Residenz Verlag, 2017), 192–97.

32 Rauch ultimately migrated to nearby Ödenburg (today Sopron), Hungary, where he lived the rest of his days as an exile. Andrew Weaver, "The Materiality of Musical Diplomacy in Early Modern Europe: Representation and Negotiation in Andreas Rauch's *Currus triumphalis musicus* (1648)," *Journal of Musicology* 35, no. 4 (2018): 463.

33 "noth/ angst/ trübsal vnd verfolgung…von gelehrt vnd vngelehrt verstanden/ hin vnnd wider/ hie vnd anderer Ort/ durch Hertz vnnd Sinn angezündet/ auch also dem HERRN ein süsser Geruch/ ob er vnns noch lenger sein heilig Wort vnd Liecht leuchten/ vn[d] sein Zorn über vnns fallen lassen wolte/ dardurch gemacht werden möchte." Andreas Rauch, *Thymiaterium musicale*, tenor partbook, Xiij[R]. A facsimile reproduction of the work's title-page and dedication are reprinted in Josef Pausz, *Andreas Rauch: Ein evangelischer Musiker—1592 bis 1656* (Vienna: Evangelischer Pressverband in Österreich, 1992), 55–62.

34 Andrea Solya has suggested that "the title of the collections [sic] cleverly named after the composer. Rauch means 'smoke' in English and so as in Latin." Andrea Solya, "Thymiaterium Musicale (1625) by Andreas Rauch," 4. While Rauch's title does evoke the composer's family name, it also concurrently gestures to the smokey materiality of the sigh-prayer, thus evoking multiple levels of meaning.

35 Each of the twenty-nine songs included in Dilliger's collection were composed for and dedicated to the composer's acquaintances in honor of special occasions such as birthdays, funerals, and new-year celebrations. Johann Dilliger, *MUSICA ORATORIA ET LAUDATORIA Oder Bet vnd LobMusica* (Coburg, 1630). Landesbibliothek Coburg. Mus Mo B 32.

36 In his piece *Horribile spectaculum horum TEMPORUM*, also published in the collection *Musica oratoria et laudatoria*, Dilliger notes that the text for the composition is "zu finden in den *suspirijs temporum* H. D. Josuae Stegmanni [found in the *suspirijs temporum* of Mr. Dr. Josua Stegmann]." Ibid., F[R].

37 Stegmann, *Ernewerte Hertzen-Seufftzer*, 361.

38 Johann Hildebrand, *Krieges=Angst=Seufftzer* (Leipzig, 1645). Leipziger Städitsche Bibliothek–Musikbibliothek. II.4.57. A digitized copy of the work is available through the Sächsische Landesbibliothek—Staats- und Universitätsbibliothek at the permalink: http://digital.slub-dresden.de/id45616152X. In this chapter's Example 3.2, the original basso continuo figures ♭5 and slash-6 have been replaced with 5 and ♯6, respectively, to reflect more standard notational practices. Little is known about the life of the composer outside of his long, thirty-seven-year tenure as an organist for the St. Nicholas Church in Eilenburg. For a brief biography of the composer, see Stefan Hanheide's introductory remarks in Johann Hildebrand, *Krieges-Angst-Seufftzer: Sieben Monodien und sechs Choralsätze mit Basso continuo (1645)*, ed. Stefan Hanheide (Osnabrück: Electronic Publishing Osnabrück, 2014), VII-IX.

39 Johann Hildebrand, *Johann Hildebrands In deutsche Reime übersetzter Jesus Syrach* (Halle, 1662). Numbers three, five, and six of the collection were published with no text attribution. Stefan Hanheide writes in the preface to his critical edition of the collection that "bei den dichterischen Ambitionen des Komponisten ist es naheliegend, dass die Texte von ihm selbst stamen, wenngleich er als Autor nicht eigens genannt ist [it is probable according to the poetic ambitions of the composer that the texts

come from him even though he is not named as the author]." Hildebrand, *Krieges-Angst-Seufftzer*, IX.

40 Stegmann, *Ernewerte Hertzen-Seufftzer*, 348–49 and 372–73.
41 Stefan Hanheide, "Die *Krieges-Angst-Seufftzer* von Johann Hildebrand aus dem Jahre 1645: Lamento-Stil im Dienste der Friedenssehnsucht," *Kirchenmusikalisches Jahrbuch* 90 (2006): 19–32. Hildebrand's incorporation of Italian musical topoi in the collection is consistent with seventeenth-century German musical style. On the preference for Italianate music at the Dresden court in the seventeenth century, for example, see Mary E. Frandsen, *Crossing Confessional Boundaries: The Patronage of Italian Sacred Music in Seventeenth-Century Dresden* (New York: Oxford University Press, 2006).

Further Reading

Boddice, Rob. *The History of Emotions*. Manchester: Manchester University Press, 2018.
Hanheide, Stefan. "Musical Lamentations of the Thirty Years' War." In *1648: War and Peace in Europe*, edited by Klaus Bussmann and Heinz Schilling, translated by David Allison, 439–47. Münster: Westfälisches Landesmuseum, 1998.
Marks, Thomas. "Singing Repentance in Lutheran Germany during the Thirty Years War (1618–1648)." *Music & Letters* 103, no. 2 (2022): 226–63.
Savin, Kristiina. "Sighs of Desire: Passionate Breathing in Medieval and Early Modern Literature." In *Pangs of Love and Longing: Configurations of Desire in Premodern Literature*, edited by Anders Cullhed et al., 157–75. Newcastle-upon-Tyne: Cambridge Scholars Publishing, 2013.
Varwig, Bettina. "Music in the Thirty Years War: Towards an Emotional History of Listening." *Journal of Seventeenth-Century Music* 26, no. 1 (2020).

4 The Fate of Jesuit Art and Architecture in Germany during the Thirty Years War

Jeffrey Chipps Smith

In 1643, the Jesuit neo-Latin poet Jacob Balde wrote a discourse on the dreadful state of Germany due to the war that by then had raged for thirty-five years (Figure 4.1).[1] He had experienced the war directly though from the relative safety of the heavily fortified town of Ingolstadt. He identified the standing woman in Wolfgang Kilian's engraving that graced the discourse as the Virgin of Germany. She wears a veil in mourning for the Holy Roman Empire, whose double-headed eagle symbol decorates the cloth. Like the Virgin Mary of Seven Sorrows, Germany's breast is pierced by a sword. Her banderole reads, *Threni Germaniae* (laments of Germany). A city burns right beneath the stormy, lightning-filled sky. A lute and fiddle hang from willow trees on the opposite bank of the Rhine River. The inscription below the print references Psalm 136 (137):2 about the woes of the Israelites during their captivity in Babylon. The psalm's first two verses read, "By the rivers of Babylon, there we sat down, yea, we wept, when we remembered Zion. We hanged our harps upon the willows in the midst thereof." Balde's Virgin of Germany weeps over the silencing of music (and all of the arts) in her war-ravaged land.

This essay addresses the often highly destructive impact of the Thirty Years War (1618–48) on the Society of Jesus and its artistic heritage. The topic is vast so I shall focus on 1631–35. These were the most devastating years of the conflict, when war, famine, plague, and death, the four horsemen of the Apocalypse, afflicted much of Germany. The military intervention of the staunch Lutheran King Gustavus Adolphus of Sweden (Gustav Adolf, r. 1611–32) and his armies starting in 1630 altered the war's dynamic, moving it into an even more destructive phase that wreaked economic and psychological havoc across much of the German lands. In the winter of 1634–35, for example, during the imperial army's siege of Augsburg, then occupied by Swedish troops, the town's population plummeted from a pre-war estimate of 45,000 to 16,400, largely due to famine and disease.[2] I wish to consider how the themes of devastation, curiosity, and construction help us to comprehend the experiences of residents in several south German towns during this unsettled age.

DOI: 10.4324/9781003157700-6

In falicib, in medio eius fufpendimus
organa nostra. Pf. 136.

Figure 4.1 Wolfgang Kilian, *Laments of Germany*, in Jacob Balde, S.J., *Sylvarum Libri VII* (Munich, 1643), book 4, opposite p. 82, engraving.

The Society of Jesus, officially founded in 1540, quickly became the most dynamic arm of the Roman Church striving for the renewal and expansion of Catholicism in Europe and beyond, especially in the aftermath of the Council of Trent. The Society thus played a much more active role than older religious orders in strengthening the Catholic faith and contesting Protestantism across the Holy Roman Empire. Between the 1570s and 1648, the Jesuits also expanded dramatically in their three German provinces—Upper Germany, Upper Rhine, and Lower Rhine. By 1640, they were well established in at least forty towns, where they built schools and constructed new or renovated existing churches.[3] The Jesuits' successes, their militant defense of the Roman Catholic Church, and their close access to the ruling elite made them the ideal target for their Protestant adversaries as well as some older jealous Catholic orders, notably the Dominicans.[4]

For example, the Jesuits were blamed following the disastrous victory of the imperial general Johann Tserclaes, count of Tilly, and his armies at Magdeburg on 20 May 1631.[5] During the siege, fire whipped by strong winds burned most of the city to the ground and up to 20,000 people died. In one related contemporary broadsheet, Gustavus Adolphus, king of Sweden, is cast in the role of physician (Figure 4.2).[6] He stands behind a Jesuit priest who sits on the papal

Figure 4.2 Anonymous, *Gustavus Adolphus Washing a Jesuit Priest's Hair with Magdeburg Lye*, c. 1632, broadsheet.

cathedra, which is marked with the coat of arms of the current pope, Urban VIII (r. 1623–44). Gustavus Adolphus cleanses the priest's hair using "Magdeburg lye." Lye, a major ingredient in soap and cleaners, is made by allowing water to wash through wood ash. Here the soap, dripped from the suspended bucket, is fashioned from the hot ashes of the burnt city. As conveyed by the text appearing beside the Swedish king, Gustavus Adolphus assumes that once he has fully cleansed the Jesuit, the man's true identity as the antichrist will be revealed.[7] The Catholic priest cries out, "O heiß, ist genug, ist g[e]nug" ("O hot, [it] is enough, [it] is enough"). The third protagonist is the Lutheran pastor standing behind Gustavus Adolphus. He asks the Swedish king to use his hand (his military power) to drive the Jesuits from Germany.[8] At the bottom of the broadsheet is a further dialogue between the priest and the pastor, who again urges the expulsion of the Jesuits.

With their victory over the imperial army at the battle of Breitenfeld by Leipzig on 17 September 1631, Gustavus Adolphus and his allies moved steadily into southern Germany. These Protestant forces occupied Würzburg on October 15, followed in 1632 by their control of Bamberg in February; Augsburg in April; and Neuburg, Landshut, and Munich in May. The Protestant majority in bi-confessional Augsburg hailed the Swedish king as a liberator. Jacob and Raphael Custos's engraving entitled *PATRIAE LIBERATORI À DEO ... MISSO. AUGUSTANI*, for example, portrays Gustavus Adolphus's arrival in the Augsburg Weinmarkt (an important public square) on 24 April 1632, amid a welcoming throng of citizens and a phalanx of soldiers, as the king addresses them from the window of a patrician house.[9] After a Lutheran service in the church of St. Anna, the king then removed the Catholic city council and appointed a Protestant one. The people's response to Gustavus Adolphus's arrival highlights the existing confessional tensions during this period. Just a year earlier about 8,000 Protestant residents had left the town for religious reasons.[10]

For our purposes, the turmoil of these years helps to explain the hostility directed at the Society of Jesus.[11] In the aftermath of Emperor Ferdinand II's Edict of Restitution of 6 March 1629, which sought to restore the territorial *status quo* reached at the Peace of Augsburg in 1555, Catholic authorities in Augsburg moved quickly to close Protestant churches and suppress their public and private services.[12] For instance, the church of St. Anna changed confessional hands multiple times during this period. It was originally a Carmelite monastery and a favorite burial site for wealthy patricians. Later it became Augsburg's first Lutheran church. On 6 November 1630, Bishop Heinrich von Knöringen transferred St. Anna's church, its library, and its school to the Jesuits.[13] By October 1631, the Jesuits were offering classes there. When Gustavus Adolphus arrived a year later, he and his officials returned the church and school to Lutheran use. Jesuit rector Konrad Reiching feared Protestant mob violence, especially as rumors circulated that members of the Society had stored weapons in graves, or they planned to set off explosives during a Protestant service in the Annakirche.[14] The complex was controlled again by the Jesuits from 1635 until 1648 when, as part of the Treaty of Westphalia, it reverted permanently to the Lutherans.

On his arrival at Augsburg, Gustavus Adolphus demanded the local members of the Society pay 3,000 gulden as a war contribution. Unable to raise this sum, they signed a promissory note requiring them to deliver 260 gulden per month, which they still increasingly found difficult to fulfill. Liturgical textiles were then seized and on 9 January 1633 their sacristy was plundered in lieu of payment by Axel Oxenstierna, the Swedish *Rikskansler* (Lord High Chancellor), who commanded the Swedish forces following Gustavus Adolphus's death. Even though the Jesuits, like other Catholic clergy in Augsburg, were still officially under Swedish royal protection, their reality proved harsher. Rector Reiching and a few other Jesuits were arrested and later released only to be incarcerated yet again. They were also required to quarter ten officers, their servants, and horses at their college near the cathedral. Later 120 foot soldiers, along with their accompanying women and children, were housed there.

Oxenstierna also demanded that Augsburg's Catholic clergy swear an oath of loyalty to the Swedish crown. On 19 May 1633, Augsburg's remaining Jesuits, monks, and Catholic clergy who refused to take this oath were banished.[15] They left the city at midday. A Protestant broadsheet, dated 1633, portrays a long procession of the expelled Catholics, many holding crosses, monstrances, statues, liturgical objects, and banners, departing Augsburg on their way to Landsberg or Munich (Figure 4.3).[16] Rector Reiching and fourteen Jesuits left that day, and both the Jesuits and the Capuchins, identifiable by their attire, are shown prominently among the dispossessed. The print's title calls this a miracle, one that healed the Augsburg Confession, which had suffered harm. Gustavus Adolphus

Figure 4.3 Anonymous, *The Expulsion of the Jesuits and Catholic Clergy from Augsburg*, 19 May 1633, broadsheet.

is heralded for restoring Protestant rights and the statement of faith first articulated in the Augsburg Confession of 1530 and codified in the Peace of Augsburg of 1555. The light of God's eye is shown shining over the town. By contrast, darkness, the absence of God's radiance, blankets the departing Catholics since the orb of the earth blocks God's light from reaching them. This detail symbolically alludes to the solar eclipse experienced in Augsburg on 8 April 1633. In April 1632, Gustavus Adolphus rescinded the imperial Edict of Restitution, which had restricted the religious and political freedoms of Augsburg's Protestant residents.[17] As the pendulum of war swung in the other direction, however, imperial troops besieged and in 1635 retook Augsburg, and Catholic services were restored. Five Jesuits, exiled for two years, returned on Palm Sunday, 1 April 1635.[18] They discovered their college had been ransacked by the Swedes and by local residents. They reopened their school in May, though plague limited their teaching. The following year fourteen Jesuits were again working in the city.

The sheer number of anti-Jesuit prints and broadsheets, especially those dating between 1618 and 1648, shows how often they were a target of Protestant wrath.[19] In 1631–32, Jesuits and other Catholic clergy were forced to flee as Swedish troops conquered or threatened many of the towns along the Main and Rhine Rivers.[20] Another frequently copied broadsheet from 1632 depicts a Jesuit forced to flee from Würzburg (Figure 4.4).[21] He carries a large knapsack over his shoulder. A divine hand guides a mouse who gnaws through the seven layers of a priest's knapsack, spilling indulgences, astrology and other false books, musical instruments, liturgical objects, a cardinal's hat and two

Figure 4.4 Anonymous, *The Jesuits in Exile*, 1632, broadsheet.

bishop's miters, and other Catholic symbols. The mouse is Gustavus Adolphus, who was sometimes referred to as the *Wasser Mauß* (water mouse), presumably because he crossed the Baltic Sea to become an agent or tool of God.[22] In the background are the towns of Würzburg, Frankfurt, Mainz, Koblenz, Cologne, and, at the far right, as the text informs us, the Jesuit's destination—the shrine of St. Raspinus. St. Raspinus (Saint Prison) refers to the correctional workhouse or prison in Amsterdam.[23] There the prisoners were put to various manual tasks including rasping wood. As the Jesuits are chased from town to town, their only hope, according to the print, is confessional rehabilitation at the rasphouse. Some of the later editions of this broadsheet assert the Jesuits were found too weak to work and were directed instead to Purgatory.[24]

As this and other broadsheets suggest, the Jesuits forced to flee Protestant advances often sought refuge in more secure Catholic towns within the Empire and soon also in other Catholic countries. Athanasius Kircher, the famous Jesuit scholar, was a second-year student at the Jesuit college in Paderborn in January 1622 when Christian von Braunschweig, the Protestant Administrator of Halberstadt, occupied the town and took up residence in the Jesuit school. Some of Christian's soldiers then paraded through the streets dressed in the hats and cloaks of the Jesuits. As he recounted years later, Kircher was forced to flee Protestant troops.[25] He almost died of cold and exposure as he fell through the ice while crossing the Rhine River on his way to Cologne. Nine years later, as a professor at the college in Würzburg, he had to escape once again from another advancing Protestant army, since shortly after Gustavus Adolphus entered Würzburg on 15 October 1631, Kircher was among those who, under the leadership of the Jesuit rector Peter Facies, left town.[26] Their church and college were plundered. Kircher then moved to Avignon and, in 1633, to Rome, where he doubtlessly felt safer.

Let us now consider the fates of the Jesuit communities in Bamberg and Eichstätt. The military control of Bamberg changed hands thirteen times between 1632 and 1643.[27] Most of the Jesuits, some dressed as peasants, fled before the initial arrival of the Swedish troops, which occurred on 11 February 1632, when, after a three-day siege, the town fell to Swedish General Gustav Horn and Lieutenant Wildenstein. One eyewitness to this siege was Maria Anna Junius, a nun in the Convent of the Holy Sepulcher, just outside Bamberg. Even though she began writing the account of her experiences of the war only in 1633, she included earlier memories. She recalled that when the Swedes first occupied Bamberg, "Just as we were in the greatest fear and terror a Jesuit came to us, saying: 'Good virgins, it is true, the enemy is here; stay in your convent; no harm will come to you but we [Jesuits] will be shown no mercy. ... I must make my escape quickly. Many hundred thousand good nights, dear virgins, I must be away, they are chasing after me. My name is Dominicus.'"[28]

On their arrival, the Swedish military leaders ordered their soldiers not to harm St. Martin's, the Jesuits' church, which formerly had been the Carmelite cloister of St. Maria.[29] Still, some soldiers under Lieutenant Wildenstein were quartered in the college and looting occurred, especially later, when cavalry under Johann Bülow were stationed in the college and choir stalls and other furniture were chopped up and used as firewood.[30] While most of the paintings in the church were spared, relics were removed, images of saints and crucifixes were destroyed, and the great organ was badly damaged while the smaller organ was smashed.[31] Horses were stabled in the church, sacristy, great aula (or hall), refectory, bakery, and cloister. General Tilly and his imperial army liberated Bamberg on 10 March 1632, and a day later the Jesuit college rector Jodok Döring returned from Forchheim, where he had taken refuge. In a letter dated March 12, Döring describes how a few hundred horses had used their buildings as a stall. All was "voll von mist" or full of shit.[32] The Jesuit chronicler, likely citing Döring, remarked that throughout the whole college "alles glich einem Augiasstall," a reference to the Augean stable that Hercules labored to clean.[33] Despite cleaning, the smell of dung lingered for months within the church. The altars were dirty, all of the bed sheets and pillows were looted, and most of the college's records and account books were lost. Most likely the latter were burned to warm the soldiers.

Duke Bernhard of Saxe-Weimar, one of the prominent Protestant military leaders following the death of Gustavus Adolphus at the Battle of Lützen on 6 November 1632, reconquered Bamberg on 8 February 1633. As was customary, he demanded the city contribute 200,000 talers to support the war effort. He eventually settled for 12,000. The Jesuits here and in other towns were forced to contribute toward this sum goldsmith works, liturgical vessels, and other forms of portable wealth, normally destined to be melted down. The Swedes plundered the college yet again on 11 April 1634. Only in the years following the Treaty of Westphalia in 1648 could the Society confidently repair their church and college. A new high altar was installed about a decade later. The Jesuits' church in Bamberg, which previously served as a monastery, was finally replaced with a grander new building in 1686–96 (Figure 4.5).

Johann Christoph von Westerstetten, bishop of Eichstätt (r. 1612–36), invited the Jesuits to town in 1614.[34] They initially used the small St. Catherine's church but two years later they obtained Rome's approval to erect a new church dedicated to the *Schutzengel* or Guardian Angel. The building was constructed between January 1617 and 1620 on a site adjoining the city's eastern wall. Inspired by the Society's Studienkirche Mariä Himmelfahrt (1610–17) in Dillingen, the new building was constructed as a wall pillar church with a barrel vault. It had a wide aisle-less nave, three chapels spaced between the

Figure 4.5 St. Martin's Church, Bamberg, 1686–96.

wall pillars on both the north and south sides, and a choir terminating with a rounded apse. A tall bell tower stood adjacent to the south side of the apse. Wolfgang Kilian's engraving is the only view of the original church before its destruction in 1634 (Figure 4.6).

At the end of April 1633, the Protestant general Bernhard of Saxe-Weimar seized Eichstätt and demanded the town pay an 18,000 gulden "contribution." The Jesuits were unable to pay their share of 3,000 gulden, though they provided animals and large supplies of flour and food. Imperial troops freed Eichstätt that October, but by December the Swedes had again retaken the town. They burned fifty houses and lit other fires as a penalty for the town's failure to pay protection money. They stole more cattle and practiced jumping their horses inside the Jesuit refectory. At the beginning of February 1634, Bernhard sent Johann of Hesse-Darmstadt with 4,000 soldiers to destroy the town. After plundering Eichstätt once again, soldiers put wood, straw, and gunpowder in a storage space in the Jesuit church. The next day, February 12, they started fires here and across town. The conflagration obliterated six churches and 444 houses in Eichstätt. Only 127 useable houses remained, and the Jesuits' complex was severely damaged. There was not a single habitable room in the college. Fire claimed the vaults and most of the walls of the nave of their church. The choir vault and tower survived but the half-cupola of the apse at the eastern end crashed down into the garden. Fire and smoke ruined all of the church's interior decoration, including the choir stalls and furniture.

Figure 4.6 Wolfgang Kilian, *View of the Jesuit College and Schutzengelkirche in Eichstätt*, detail, c. 1627, engraving.

Losses included three major altarpieces painted by Johann Matthias Kager of Augsburg.[35]

Kager's high altar, dedicated like the church to the guardian angel, had been consecrated on 30 August 1620, while his side altars representing the Nativity and Assumption of the Virgin had been consecrated the next day.[36] A large preparatory drawing, now in Windsor Castle, likely records the high altar's original composition (Figure 4.7).[37] The design features the heavenly ranks of angels. God the Father and Christ sit enthroned on a cloud. Each rests a hand on the world orb positioned between them while the dove of the Holy Spirit hovers above. Below, the archangels Gabriel with a lily, Michael with a sword, and Raphael with a walking staff, backed by a host of music-making angels, stand ready to aid Christians. At the bottom, a small group of angels peer down, presumably toward Earth, as if searching for those in need. There were additional side altars, dedicated on 14 May 1622, to the newly canonized Jesuit saints Ignatius of Loyola and Francis Xavier. The Hessian troops burned all of these altars and their decorations.

Figure 4.7 Johann Mathias Kager, *Holy Trinity in Glory Adored by Three Archangels*, Design for the High Altarpiece of the Jesuit Church in Eichstätt, c. 1620, drawing. Windsor Castle, Royal Collection Trust.

Repairs began only in 1638 once the military situation was more stable (Figure 4.8). The choir was covered in 1639 and the nave, provisionally, the next year. The stone floor throughout the interior had to be replaced. The church was re-consecrated only in 1656. Repairs to the tower and façade were carried out in 1660 and the nave was finally vaulted in 1661. The form of the church, which still exists, largely matches the original building. That includes the broad west façade with paired Tuscan pilasters flanking a central portal topped by an axial window and, above the entablature, a circular window set into a tall curved pediment. The new interior decorations date from 1717.[38] The adjoining college was slowly rebuilt starting in 1637.[39] The Aula Mariana, its largest room, was

Figure 4.8 Façade of the Schutzengelkirche, Eichstätt, 1617–20 and 1638–61 (reconstruction).

available for use by the students and members of the Marian sodality in 1658.[40] The situation in Eichstätt thus exemplifies how long it took the Jesuit community to recover from the tumult of the 1630s.

The Swedish king Gustavus Adolphus was a curious traveler who often instructed his officers and soldiers not to damage Catholic churches. Sentries were frequently posted by church entrances to prevent looting, the worst of which then occurred after his death in November 1632. As an example of the king's attitude toward the preservation of sacred spaces and objects, let us briefly consider his visits to Neuburg an der Donau and Munich. His troops, led by a Captain

Binheim, entered Neuburg on 18 April 1632, just days after General Tilly was mortally wounded nearby at the battle of Rain am Lech.[41] Between April 7 and 15, that is just days before the arrival of the Protestant forces, Neuburg's Jesuits had moved most of their portable treasures to nearby Ingolstadt.[42] Their caution was well-founded, but the occupying Protestant garrison was well behaved.

A few weeks later, on May 3, Gustavus Adolphus and several Protestant nobles, including Count Palatine Frederick V (the Winter King) and Duke August von Sulzbach, entered Neuburg. Friedrich Hundpiss, rector of the Jesuit college, greeted them.[43] As they passed into the Hofkirche Unserer Lieben Frau (Court Church of Our Dear Lady), Gustavus Adolphus questioned the rector about the liturgy of the Catholic mass (Figure 4.9). The king inquired why the celebrant's words, quietly recited in Latin, could not be heard or understood by the congregants. The rector responded that the Canon of the Mass instructs that the words should be said in a soft voice. Furthermore, the Jesuit said the Eucharist liturgy was not intended to instruct the faithful. That was the purpose of preaching, catechism, and other opportunities. Gustavus Adolphus continued to the eastern end of the church where he admired the *Last Judgment* high altar (1617) and side altars of the *Nativity* and *Pentecost* (1619), which Peter Paul Rubens had painted at the request of Count Wolfgang Wilhelm.[44] Interestingly, the Jesuits had made no effort to remove Rubens's painting, unlike in Freising cathedral,

Figure 4.9 Interior view, Hofkirche Unserer Lieben Frau, Neuburg, 1607–18.

where Rubens's high altar painting was rolled up and transported for safety to Austria.[45] The king commented on the paintings' beauty but asked whether the Jesuits prayed to the paintings. Rector Hundpiss explained about the Catholic veneration of saints. Their discussion, while touring the college, also included a debate about whether General Tilly, who died on April 30, would go to heaven or hell. After surveying the entire building, Gustavus Adolphus praised the construction of the Jesuits' complex though he criticized its cost and then asked whether St. Peter would want such a lavish temple. The rector replied that Peter already had a grander church, the Basilica in Rome where he is buried. Their recorded conversation ended with the king exclaiming that he had no quarrel with the Jesuits and their college but with the Holy Roman Emperor and the elector of Bavaria.

Unfortunately, Bernhard of Saxe-Weimar did not share Gustavus Adolphus's constraint or theological curiosity. When his troops captured Neuburg yet again in October 1633, they stabled the horses of seventy knights in the Jesuit college, emptied the food stores, and caused other damage. Most of the Jesuits had fled but two, including Rector Hundpiss, were captured, imprisoned for five months, and later ransomed.[46]

Munich, the capital of Maximilian I, elector of Bavaria, had long been a prize sought by Gustavus Adolphus.[47] On the night of 16–17 April 1632, his troops arrived outside the walls. Munich surrendered without a fight. As a contemporary broadsheet documents, Gustavus Adolphus made his formal entry a month later on May 17, when he received the keys to the city (Figure 4.10).[48] He resisted calls to destroy Munich in retaliation for the imperial army's destruction of Magdeburg in the previous year. Instead, he levied a steep war tax on

Figure 4.10 Anonymous, *King Gustavus Adolphus's Entry into Munich on 17 [16] May 1632*, broadsheet.

the city and took forty-two leading citizens, including six Jesuits, as hostages.[49] Soon after they were finally released on 3 April 1635, Lucas Kilian published an elaborately engraved broadsheet showing the freedmen kneeling in prayer before the skyline of Munich and beneath an apparition of the Virgin and Christ Child, whom they credited for protecting them.[50] Angels beside Mary and Christ hold scrolls listing the names of the former hostages. The lengthy inscription at the bottom of the print describes the men's three-year-long incarceration.

On May 20, the Swedish king, along with Count Palatine Frederick and other high-ranking men, rode to St. Michael's, the Society's church and college (Figure 4.11). Erected between 1583 and 1597, thanks to the lavish funding supplied by Duke Wilhelm V of Bavaria, St. Michael's was not only the first great Jesuit church in northern Europe, it was the first major Catholic church erected in Germany since the advent of the Protestant Reformation.[51] Statues of Wilhelm, his father Albrecht V, other Wittelsbach ancestors, and allies adorn the façade, as does Hubert Gerhard's monumental bronze of *St. Michael Vanquishing Lucifer*. The façade also offers a militant statement about the Bavarian

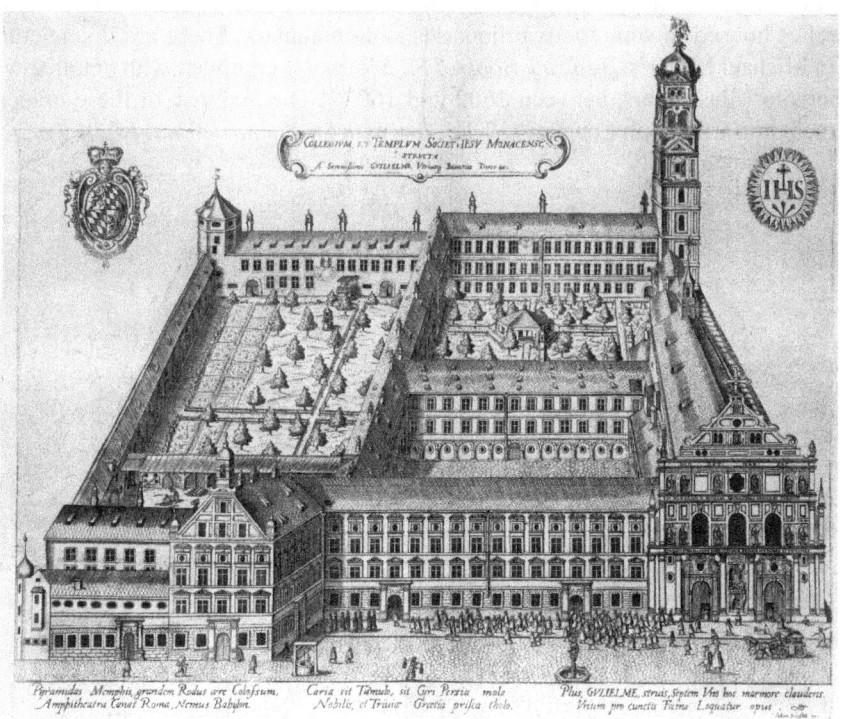

Figure 4.11 Johann Smissek, *St. Michael's Church and the Jesuit College in Munich*, c. 1644–55, etching.

ducal support for the Jesuits and the resurgent post-Tridentine Roman Catholic Church.[52] On that same day of Gustavus Adolphus's arrival, the Jesuit community, consisting of twenty-two priests, one master, and seventeen lay brothers, decided to remain in Munich rather than go into exile.[53] As mentioned, six Jesuits were soon after included among the hostages held as security that Munich would pay its war contribution.

As Gustavus Adolphus entered St. Michael's and walked toward the choir, he removed his hat. The rector gave the king a tour of the church (Figure 4.12). Just as he did in Neuburg, Gustavus Adolphus quizzed the Jesuit about the Catholic mass and other religious questions. The church with its sumptuous decoration was not damaged. The Jesuits were even told they could continue preaching in their church and helping the sick in the city. There is no evidence that troops and horses were quartered in the college.

Nevertheless, the Jesuits were ordered to surrender all objects made of silver as their contribution to the city's ransom. Here and in other towns the Jesuits and other Catholic clergy were forced to sacrifice liturgical silver, reliquaries, and other valuable devotional objects to fund the armies of the Swedes and their allies. Under Duke Wilhelm, St. Michael's had been richly endowed with holy relics housed in sumptuous reliquaries and containers. These are documented in Michael Miller's *Treasury Book of St. Michael's*, compiled with detailed watercolor illustrations between 1602 and 1607.[54] The majority of these objects, including Christoph Lencker's *Reliquary with St. Michael Vanquishing Lucifer*,

Figure 4.12 Interior of St. Michael's Church, Munich, 1583–97.

Figure 4.13 Christoph Lencker, *Reliquary of St. Michael Vanquishing Lucifer*, 1596 as recorded in Michael Miller, *Treasury Book of the Church of St. Michael's in Munich*, vol. 1, fol. 7, c. 1602/05, illuminated drawing with mixed techniques, St. Michael's Church, Munich.

completed in 1596, are lost (Figure 4.13).[55] Lorenz Seelig notes that while some silver decorations on the reliquaries were removed to help pay for Swedish "fire protection," most of the items in the treasury were not disturbed.[56] Far greater losses occurred in 1773 following the suppression of the Society of Jesus.[57]

At the invitation of Elector Maximilian I of Bavaria, the Jesuits opened a new college in Landshut in 1629.[58] At this moment, Catholic forces were winning the war and Gustavus Adolphus had not yet crossed the Baltic Sea. The foundation stone for the new church of St. Ignatius, designed by the Jesuit architect Johann

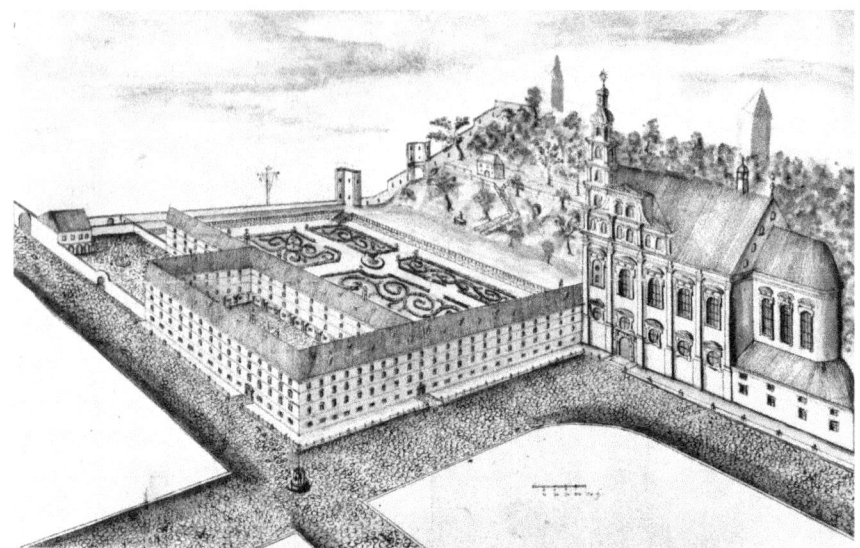

Figure 4.14 Anonymous, *View of St. Ignatius and the Jesuit College in Landshut*, c. 1665, engraving.

Holl, was placed on 31 July 1631 (Figure 4.14). This was the only major church begun during this period. Work had barely progressed before the Swedish army began their successful siege of Landshut on 8 May 1632. Swedish officers then took up residence in the Jesuit college and Gustavus Adolphus demanded Landshut pay a war contribution of 100,000 talers. The Swedish army left Landshut, only to recapture the town again on 22 July 1634. Bernhard of Saxe-Weimar and General Gustav Horn made harsh new financial demands on the Jesuits and local residents. Rector Ulrich Speer tried to raise the 3,000 gulden war tax (*Lösegeld*), but a contemporary report dated August 1 stated that soldiers smashed the college's four exterior doors and then broke or stole everything they could, so that there was not a piece of bread left in the college.[59] The soldiers engaged more violently with the citizens than they had when King Gustavus Adolphus was still alive. At least one Jesuit was shot and killed and another wounded.

Construction on the church in Landshut resumed in April 1637, following the departure of the Swedish troops and their allies from the region. The building was consecrated on 25 November 1640; however, the permanent vaults and most of the interior decoration took the next three decades to complete (Figure 4.15). The first significant adornment was the life-size bronze *Crucifix* (1643), designed and modeled by Hans Joachim Krum and cast by Bernhard Ernst. Until it was moved to the nave in 1958, this statue was mounted on an 8.5-meter tall cross that stood at the entrance to the choir. Eight wooden statues of the founders of different Catholic religious orders adorning niches lining

Figure 4.15 Johannes Holl, St. Ignatius Church, Landshut, interior.

the nave also were added in 1643.[60] Six painted altarpieces, including two by Joachim von Sandrart, were placed in the side chapels between 1640 and 1644. Johann Christoph Storer's high altar, depicting *Christ's Appearance to St. Ignatius at La Storta*, was erected only in 1662 as Bavaria began to recover from the war. Likewise, the school reopened in 1635 but the college complex was not finished until 1665.

The Jesuits' experiences discussed here were typical of their communities as well as those of other Catholic clergy across Germany. The terror of war with all of its uncertainties halted virtually all of the Society's building campaigns and major artistic projects. A few important towns, notably Cologne, were not threatened. There the Jesuits erected the church of Mariä Himmelfahrt (Assumption of Mary) between 1618 and 1629.[61] The lavish decorative program was completed in the 1630s. In general, however, war gave the Virgin of Germany, seen at the beginning of this essay in Wolfgang Kilian's engraving, ample justification for her sorrow. The Society of Jesus witnessed their colleges and churches being occupied, plundered, and often severely damaged, especially in the aftermath of King Gustavus Adolphus's death in 1632. Thereafter, Rikskansler Axel Oxenstierna and other Protestant military leaders were far less restrained than the king had been, especially as the war dragged on. Portable wealth, especially in the form of precious metal, such as liturgical silver, was seized and melted down to support the Swedish war effort. Catholic paintings and sculptures were sometimes attacked as readily available and defenseless non-human targets, whether for iconoclastic reasons or simply amid the raw fury of war. The years from 1631 to 1635, when the war most deeply affected Franconia, Bavaria, and Swabia, saw the Jesuits, their students, and their patrons simply trying to survive its horrors. In some towns the Jesuits and their students fled or were exiled; in others they remained, perhaps due to local circumstances where they did not feel as threatened or when some level of accommodation allowing them to protect their property was reached. Many Jesuit communities attempted to continue their teaching and preaching missions when possible. The restoration of their churches and the renewal of the artistic decorations had to wait until the decades following the signing of the Treaty of Westphalia in 1648. They responded slowly at first and then more energetically during the last quarter of the seventeenth century, as economic and political conditions improved. Between 1648 and the Society's suppression in 1773, the Jesuits in their three German provinces erected twenty-four new or greatly renovated churches, including St. Martin's in Bamberg.[62] A harder matter to gauge is whether the Society of Jesus post-1648 ever recaptured the same missionary or triumphant zeal that characterized their educational, pastoral, and artistic efforts within the German lands during the decades around 1600. The scars of their wartime experiences during the 1630s and 1640s lingered.

Notes

1   Jakob Balde, *Sylvarum Libri VII* (Munich: Cornelius Leysser, 1643), book 4–engraving, opposite p. 82, opens the lament. Jeffrey Chipps Smith, *Sensuous Worship: Jesuits and the Art of the Early Catholic Reformation in Germany* (Princeton: Princeton University Press, 2004), fig. 188.

2   Peter H. Wilson, *The Thirty Years War. Europe's Tragedy* (Cambridge, MA: Harvard University Press, 2009), 779–821 ("The Human and Material Cost"), here 791.

3   Smith, *Sensuous Worship*, 4–6, fig. 2. An earlier version of this paper was delivered at a conference on the Thirty Years War held at the National Library of Russia in St. Petersburg on 30 October 2018. I wish to thank Dr. Zinaida Laurie and Prof. Andrey Yu. Prokopiev of Saint-Petersburg State University for the invitation.

4 Robert Bireley, *The Jesuits and the Thirty Years War: Kings, Courts, and Confessors* (Cambridge, UK: Cambridge University Press, 2003). Also Bernhard Duhr, *Hundert Jesuitenfabeln. Volksausgabe der Jesuitenfabeln* (Freiburg im Breisgau: Herdersche Verlagshandlung, 1913).

5 Wolfgang Harms, ed., *Deutsche illustrierte Flugblätter des 16. und 17. Jahrhunderts*, vol. 1.2, *Die Sammlungen des Herzog August Bibliothek in Wolfenbüttel* (Munich: Kraus, 1980–89), no. II, 29; Jeffrey Chipps Smith, "The Destruction of Magdeburg in 1631: The Art of a Disastrous Victory," in *Disaster, Death and the Emotions in the Shadow of the Apocalypse*, eds. Jenny Spinks and Charles Zika (London: Palgrave, 2016), 249–73.

6 John Michael Paas, *The German Political Broadsheet 1600–1700*, 9 vols. (Wiesbaden: Otto Harrassowitz, 1985–2007), vol. 5, no. P-1454; Michael Niemetz, *Antijesuitische Bildpublizistik in der Frühen Neuzeit* (Regensburg: Schnell & Steiner, 2008), 131–32, fig. 133; also see fig. 134 for a different composition with the same title.

7 "Nun Zwag ich dir mit dieser laugen. / Dein Heyligs haupt und klare augen. / Damit du auch sein sauber bist. / Wan du wirst werden der Antechrist." (Now I force you to leach with this. Your holy head and clear eyes. So that you are clean too. When you will become the Antichrist.)

8 "Lieber Schwed Zwag mit einer Hand. Als Esawitee auß Teutschlandt." There were various derogatory plays on the German word *Jesuiter* (Jesuit), such as *Jesuwidder* (a *Widder* is a goat ram), *Jesuwider* (working against Jesus); *Esauiter* (a power usurper, a reference to Esau and Jacob–Genesis 25: 21–16). Paas, *Broadsheet*, vol. 1, 74–75.

9 Beate Rattay, *Illustrierte Flugblätter aus den Jahrhunderten der Reformation und der Glaubenskämpfe*, intro. and ed. Wolfgang Harms, exh. cat. (Coburg: Kunstsammlung der Veste Coburg, 1983), 184–85, no. 89.

10 Harms, *Flugblätter*, vol. 1.3, 460.

11 On Augsburg's Jesuits during these years, see Placidius Braun, *Geschichte des Kollegiums der Jesuiten in Augsburg* (Munich: Jakob Giel, 1822), 53–64; Bernhard Duhr, *Geschichte der Jesuiten in den Ländern deutscher Zunge*, 4 vols. (Freiburg and Regensburg: Herder, 1907–28), vol. 2.1, 416–20.

12 Duhr, *Geschichte*, vol. 2.1, 226–27.

13 Braun, *Augsburg*, 50–52; Duhr, *Geschichte*, vol. 2.1, 227–28.

14 Braun, *Augsburg*, 53–59 for what follows.

15 Braun, *Augsburg*, 60–61. Braun reports that many Catholics came to the Jesuits' church to pray, receive communion, and be comforted on the 18th and morning of the 19th.

16 Niemetz, *Bildpublizistik*, 144–45, fig. 158; also see fig. 157–*Der [Aug]spurgisch ver[lauf]*, 1632. Niemetz notes the print's date of "9. 19. May" refers to the refusal of the Protestant authorities in Augsburg to adopt the Gregorian calendar.

17 Another etched broadsheet dated 1632 depicts the gigantic seven-headed apocalyptic beast and the ram wearing a Jesuit biretta vomiting Jesuits and Catholic monks over Augsburg. This references the Edicts of Restitution as well as the departure of about 8,000 Protestants from Augsburg in 1631 in the aftermath of the closing of Protestant churches and schools. See Harms, *Flugblätter*, vol. 1.2, 460–61, no. II.265.

18 Braun, *Augsburg*, 62–64.

19 Niemetz, *Bildpublizistik* provides a good survey of the polemical anti-Jesuit prints most of which date to the first half of the seventeenth century.

20 On the flight of Jesuits and some of their students to the Society's safer communities, see Duhr, *Geschichte*, vol. 2.1, 444–51. The rector of the Jesuit college in Innsbruck

sought financial help from Rome since during the years 1632 to 1635 his community housed between fifty and ninety exiles.

21 Harms, *Flugblätter*, vol. 1.2, no. II.294; Paas, *Broadsheet*, vol. 6, no. P-1781 also P-1780 and, with different letterpress P-1785; Niemetz, *Bildpublizistik*, 128–29, fig. 123.

22 Harms, *Flugblätter*, vol. 1.2, 514–15, no. II.294; Niemitz, *Bildpublizistik*, 128.

23 An earlier print about the expulsion of Jesuits from Bohemia and Hungary in 1618–19 shows a procession of Jesuits on foot and eight Jesuits, each named, in a carriage on the "Via ad S. Raspinūm," whose gate and town (Amsterdam) loom in the distance. Paas, *Broadsheet*, vol. 2, nos. P-467 to P-471; Bireley, *Jesuits*, 130. On the corrections house in Amsterdam, see Pieter Spierenburg, *The Prison Experience: Disciplinary Institutions and their Inmates in Early Modern Europe* (Amsterdam: Amsterdam University Press, 1991), 96–98.

24 Harms, *Flugblätter*, vol. 1.2, no. II. 294.

25 Duhr, *Geschichte*, vol. 2.1, 398–401; and more broadly on the war emergency, see his 392–451.

26 Duhr, *Geschichte*, vol. 2.1, 406–7.

27 Heinrich Weber, "Bamberg im Dreissigjährigen Krieg. Nach einer gleichzeitigen Chronik bearbeitet," *Bericht des Historischen Vereins für die Pflege der Geschichte des ehemaligen Fürstbistums Bamberg* 48 (1886), 1–132, esp. 5–57 for the years 1631–33, as documented in a Jesuit chronicle; Duhr, *Geschichte*, vol. 2.1, 406–09; Tilmann Breuer and Reinhard Gutbier, eds., *Stadt Bamberg. Innere Inselstade* (Munich: Deutscher Kunstverlag, 1990), esp. 57 and 60; Geoff Mortimer, *Eyewitness Accounts of the Thirty Years War, 1618–48* (Baskingstoke: Palgrave, 2002), 100; Smith, *Sensuous Worship*, 203.

28 Mortimer, *Eyewitness*, 100.

29 On the history of the church, see Breuer and Gutbier, *Bamberg*, 48–145. The monastery was transferred to the Jesuits on May 16, 1611.

30 Weber, *Bamberg*, 18 and 23.

31 Weber, *Bamberg*, 18–19.

32 Duhr, *Geschichte*, vol. 2.1, 408.

33 Weber, *Bamberg*, 22.

34 Duhr, *Geschichte*, vol. 2.1, 436–38; Smith, *Sensuous Worship*, 112–14; Claudia Grund, "Templum Honoris. Zur Baugeschichte von Kirche und Kolleg der Jesuiten zu Eichstätt im 17. und frühen 18. Jahrhundert," in *Die Schutzengelkirche in Eichstätt*, eds. Sibylle Appuhn-Radtke, Julius Oswald, and Claudia Wiener (Regensburg: Schnell & Steiner, 2011), 196–217; Julius Oswald, "Episcopale et Academicum Gymnasium Societatis Jesu Eustettense. Geschiche der Jesuiten in Eichstätt," in Appuhn-Radtke, *Die Schutzengelkirche*, 55–71, esp. 62–64.

35 Tilman Falk, "Johann Matthias Kager und seine Altarbilder für die Schutzengelkirche (1620)," in Appuhn-Radtke, *Die Schutzengelkirche*, 248–61.

36 Grund, "Templum Honoris," 203. Kager's design for the Assumption of the Virgin altarpiece is recorded in two copies after his lost picture. Falk, "Kager," 259–60, figs. 108–9.

37 Pen, grey ink, and watercolour | 40.7 x 24.5 cm (sheet of paper), Royal Collection Trust, Windsor Castle, RCIN 912963; Falk, "Kager," 256 and 258, fig. 101.

38 Grund, "Templum Honoris," 208–11.

39 Grund, "Templum Honoris," 212–17.

40 Christina Grimminger, "Die Aula Mariana des Jesuitenkollegs in Eichstätt. Bau und Ausstattung," in Appuhn-Radtke, *Die Schutzengelkirche*, 309–25, here 310.

41 Duhr, *Geschichte*, vol. 2.1, 438–39; Smith, *Sensuous Worship*, 143–55; Hildebrand Troll, "Die Neuburger Jesuitenchronik," *Archivalische Zeitschrift* 73 (1977): 51–57.

42 Surprisingly, few members were allowed to stay in Ingolstadt even though the town had a large Jesuit community and school.
43 For what follows, see Troll, "Jesuitenchronik," 55–57. The text source is the *Historia Collegii Societatis Jesu Neoburgi ad Istrum* (or the Neuburg Jesuit Chronicle), which records the local history from 1613 until 1732 in 599 folios. Troll notes (fifty-five) portions of the text were first published in a German translation, without citing its source, in 1820 and republished again in the *Neuburger Kollektaneenblatt* 57 (1893): 9–14.
44 Smith, *Sensuous Worship*, 150–54, figs. 138–40.
45 Jeffrey Chipps Smith, "Rubens, Bishop Veit Adam von Gepeckh, and the Freising *High Altar* (1623–25)," in *The Age of Rubens: Diplomacy, Dynastic Politics, and the Visual Arts in Early Seventeenth-Century Europe*, eds. Luc Duerloo and Malcolm Smuts (Turnhout: Brepols, 2016), 261–74, 302, here 270–71.
46 Duhr, Geschichte, vol. 2.1, 241.
47 Duhr, *Geschichte*, vol. 2.1, 420–21.
48 Paas, *Broadsheet*, vol. 6, no. P-1762 and see P-1763. The date in the print is given incorrectly as the sixteenth.
49 See Duhr, *Geschichte*, vol. 2.1, 423–28.
50 Duhr, *Geschichte*, vol. 2.1, 428 with illustration. Six Jesuits, three Franciscans, four Capuchins, two canons, two Cistercians from Fürstenfeldbruck, and four Augustinian Hermits made up the religious prisoners.
51 Smith, *Sensuous Worship*, 57–101.
52 Jeffrey Chipps Smith, "The Jesuit Church of St. Michael's in Munich: The Story of an Angel with a Mission" in *Infinite Boundaries: Order, Disorder, and Reorder in Early Modern German Culture*, ed. Max Reinhart (Kirksville, MO: Truman State University Press, 1998), 147–69; and Smith, *Sensuous Worship*, 68–75.
53 Duhr, *Geschichte*, vol. 2.1, 420.
54 Lorenz Seelig, "*Dieweil wir dann nach dergleichen Heiltumb und edlen Clainod sonder Begirde tragen. Der von Herzog Wilhelm V. begründete Reliquienschatz der Jesuitenkirche St. Michael in München*," in *Rom in Bayern. Kunst und Spiritualität der ersten Jesuiten*, ed. Reinhold Baumstark, exh. cat., Bayerisches Nationalmuseum, Munich (Munich: Hirmer, 1997), 199–262; Smith, *Sensuous Worship*, 72–73.
55 Seelig, "Heiltumb und edlen Clainod," 242–43; Baumstark, *Rom in Bayern*, 406–8, no. 108.
56 Seelig, "Heiltumb und edlen Clainod," 202.
57 Jeffrey Chipps Smith, "The Jesuit Artistic Diaspora in Germany after 1773," in *Jesuit Survival and Restoration. A Global History, 1773–1900*, eds. Robert A. Maryks and Jonathan Wright (Leiden: Brill, 2015), 129–47.
58 Joseph Braun, *Die Kirchenbauten des deutschen Jesuiten*, 2 vols. (Freiburg: Herder, 1908–10), vol. 2, 95–105; Duhr, *Geschichte*, vol. 2.1, 429–30; Rupert Jakob Reiter, "Die ehemalige Jesuitenkirche St. Ignatius zu Landshut" (PhD diss., Ludwig-Maximilians-Universität, Munich, 1976); Horst Nising, '*...In Keiner Weise Prächtig.' Die Jesuitenkollegien der süddeutschen Provinz des Ordens und ihre städtebauliche Lage im 16.–18. Jahrhundert* (Petersberg: Michael Imhof, 2004), 186–93; Smith, *Sensuous Worship*, 158–63.
59 Duhr, *Geschichte*, vol. 2.1, 429.
60 The statues depict Saints Augustine, Bruno, Norbert, Francis of Assisi, Benedict, Bernard of Clairvaux, Dominic, and Francis of Paola.
61 Smith, *Sensuous Worship*, 165–87.
62 Smith, *Sensuous Worship*, 6.

Further Reading

Duhr, Bernhard. *Geschichte der Jesuiten in den Ländern deutscher Zunge*, 4 vols. Freiburg and Regensburg: Herder, 1907.

Harms, Wolfgang, ed. *Deutsche illustrierte Flugblätter des 16. und 17. Jahrhundrts.* Vol. 1, *Die Sammlungen des Herzog August Bibliothek in Wolfenbüttel.* Munich: Kraus, 1980.

Mortimer, Geoff. *Eyewitness Accounts of the Thirty Years War 1618–48.* Baskingstoke: Palgrave, 2002.

Niemetz, Michael. *Antijesuitische Bildpublizistik in der Frühen Neuzeit.* Regensburg: Schnell & Steiner, 2008.

Paas, John Michael. *The German Political Broadsheet 1600–1700*, 9 vols. Wiesbaden: Otto Harrassowitz, 1985–2007.

Smith, Jeffrey Chipps. *Sensuous Worship: Jesuits and the Art of the Early Catholic Reformation in Germany.* Princeton: Princeton University Press, 2004.

5 The Seventeenth-Century Culture of War

Three Commanders and Their Legacy in the Arts

Kristoffer Neville

A convention of twentieth-century cultural history is that the United States be-came a full player on the international stage through its intervention in World War II. This took a number of different forms. Many practicing artists were displaced by the conflict, and through one means or another made their way to the United States. Piet Mondrian's *Broadway Boogie Woogie* (1942–43) was a version of an ongoing series of paintings that he had developed in the 1910s in the Netherlands and Paris. However, it was painted in New York in direct re-sponse to his experience of his new home. Ludwig Mies van der Rohe began his pioneering work in architecture in Berlin, but developed it to its conclusion in Chicago, where he worked after 1938 as director of the architecture program at the Illinois Institute of Technology. Along with his teaching duties, he designed a series of landmark works of modernist architecture.[1]

Mondrian and Mies were hardly alone. Alongside these and other practicing artists, a diverse group of writers including Heinrich and Thomas Mann, Bertolt Brecht, Theodor Adorno, and Hannah Arendt came to the United States; some returned to Europe after the war, and others settled permanently in their new home. Many prominent academics, dismissed from university posts or seeing little future in Nazi Germany, also sailed west across the Atlantic. In art history, Richard Krautheimer, Rudolf Wittkower (via England), Kurt Weitzmann, and others became foundational figures in their fields and trained a generation of scholars at the Institute of Fine Arts (IFA) at New York University, Columbia, and Princeton, and were joined by colleagues elsewhere. Erwin Panofsky, who took a post at the Institute for Advanced Study in 1935, and who taught oc-casionally at Princeton and the IFA, reshaped the discipline as a whole without directing PhD theses in the United States.[2] Walter Cook, the director of the IFA, joked that "Hitler is my best friend; he shakes the tree and I pick up the apples."

This flow of academic talent into the United States had a pendant. Many of the more prominent American art historians of the next generation first encountered European art and culture as soldiers stationed in Italy, Germany, France, and elsewhere. The young James Ackerman, a future PhD student of Krautheimer and subsequently a professor at Berkeley and Harvard, encountered northern

DOI: 10.4324/9781003157700-7

Italy at the end of the war, and, as he described it, liberated Mantua. David Coffin, later a professor at Princeton, was also in Europe in the course of the war.[3] Both returned to Italy a few years later as doctoral students and drove around Rome in Ackerman's Jeep to look at Renaissance architecture.[4]

If the arrival of the United States as an international cultural power in the aftermath of World War II—and in large part *because of* World War II—is largely acknowledged, the Thirty Years War (1618–48) is more often seen as an unmitigated cultural disaster. A bleak passage by Joachim von Sandrart, a German painter and writer who lived through the war, has long shaped our view of the effects of the war.

> Queen Germania saw her palaces and churches, decorated with magnificent pictures, go up in flames time and again; while her eyes were so blinded by smoke and tears that she no longer had the power or will to attend to Art, from which it appeared to us that she wanted to take refuge in one long, eternal night's sleep. So Art was forgotten, and its practitioners were overcome by poverty and contempt, to such an extent…that they were forced to take up arms or the beggar's staff instead of the brush, while the gently born could not bring themselves to apprentice their children to such despicable people.[5]

Jacques Callot's *Grandes Misères de la Guerre* [Great Miseries of War], a suite of etchings published in 1633, provided graphic imagery to complement Sandrart's description, and Hans Jakob Christoffel von Grimmelshausen's bleak passages in *Simplicissimus* (1668) painted a dark literary picture of the period. Around 1800, Friedrich von Schiller took up the war in both a history and a trio of plays on the death of Albrecht von Wallenstein, one of the main protagonists of the conflict, who is discussed below.

Certainly the Thirty Years War was devastating on many levels, as the article in this volume by Jeffrey Chipps Smith shows.[6] In regard to the conflict's cultural impact, however, Sandrart himself provides contradictory evidence. While the passage above describes impoverished artists and talented young people pursuing other professions, the biographies that make up a substantial part of his book largely describe success stories, as gifted young painters and sculptors find commissions, court positions, and ennoblement for their efforts. The war is hardly mentioned in these passages.

Even as some princes and others from the patron class lost resources and standing in the war, and those employed by them struggled commensurately, other patrons and opportunities emerged. This essay argues that, within the destruction of war, there was a form of cultural bloom that we have tended to overlook. The Thirty Years War and subsequent conflicts gave rise to a number of ambitious military figures who grew in wealth and stature through war, so that they had the means and justification to commission significant works of art. In some cases, war itself provided a kind of cultural education that guided the ways in which they pursued the cultural aspects of their ambitions.

The notion that a career soldier could become an important cultural figure was not new in the seventeenth century. The *condottieri* of the fifteenth century have long played a fundamental role in the narrative of the Italian Renaissance. Among them, Erasmo da Narni (called Gattamelata; 1370–1443) and Federigo da Montefeltro (1422–82) have become figureheads of the phenomenon, in large part because they were associated with major cultural projects that have become part of the canon of Italian Renaissance art. Federigo was the illegitimate son of the lord of Urbino. He seized his father's title after the assassination of his half-brother, Oddantonio da Montefeltro, and in 1474 he was made a duke.[7] Although Federigo continued his mercenary activities even as a territorial ruler, the vibrant court centered in the ducal palace in Urbino was arguably more closely bound to his role as a prince than as a warrior. Indeed, Federigo embodies several characteristics frequently found in *condottieri*, who were often illegitimate or younger sons of prominent noble families, or who seized political power. Although in each case they were unlikely to become a ruler or to be unconditionally accepted within noble society, neither were they entirely foreign to it, or to its trappings. Bartolomeo Colleoni (1395–1475) built a burial chapel for himself and left his fortune to the Venetian state with the stipulation that an equestrian statue should be raised in his honor, as had been done half a century earlier for Gattamelata, who was also a mercenary in Venetian service.[8] Both have become well known in art history, largely because the two bronze equestrian monuments, erected by Donatello and Andrea del Verrocchio, were the first two such monuments realized since antiquity (Figure 5.1). They became models for many subsequent works, most often commemorating rulers. Federigo is remembered in no small part for his employment of Piero della Francesca and other artists.

The *condottiere* tradition is often thought to have run its course by the sixteenth century, although we will see that a similar phenomenon was at play in the seventeenth and early eighteenth centuries.[9] In the sixteenth century, the conventions were rather different. Although Charles V and Philip II employed Fernando Álvarez de Toledo y Pimentel, the duke of Alba (1507–82), as commander of their armies, he seems to have been placed in charge of the troops because of his high station inherited from his family, complemented by his tactical abilities, rather than attaining his social rank through military prowess. Mercenary commanders and military entrepreneurs were present in the German lands, but they were bound to the feudal structure of the Holy Roman Empire in a way that their more famous earlier Italian colleagues were not, and thus not true free agents. To some extent, the question of whether these men were *condottieri* is a matter of definition, but the consequences of the political differences were significant. Sebastian Vogelsberger, a German military entrepreneur in the service of the French king, was executed in 1548 at the order of Charles V, and Albrecht von Wallenstein was assassinated in 1634 with the consent of Emperor

Figure 5.1 Andrea del Verrocchio, Bartolomeo Colleoni (cast by Alessandro Leopardi),
   1480–88. Photo: Didier Descouens/Wikipedia Commons/CC BY-SA 4.0.

Ferdinand II.[10] Certainly, relatively few men with little more than military talent
rose to positions of great power in Central Europe in the sixteenth century. In
contrast, those seventeenth-century commanders presented here attained their
success and positions primarily on the basis of their extraordinary tactical skills,
often complemented by a talent for organization and self-promotion.

Albrecht von Wallenstein

Albrecht von Wallenstein (also Waldstein or Valdštejn; 1583–1634) was born
in the village of Heřmanice (Hermanitz) in north-eastern Bohemia, near the
Silesian border.[11] Both of his parents came from minor noble families, but neither was wealthy; his father inherited the estate of a childless uncle, but otherwise would have owned no land. Nonetheless, because of the early deaths
of his parents and older siblings, Wallenstein inherited the title of *Freiherr* at
twelve years of age.

Wallenstein married two women who were substantially richer. In 1609 he married the widowed Lucretia Nikeß (or Nekeš) von Landek, and inherited a substantial estate in Moravia when she died five years later. In 1623, he married Isabella von Harrach, the daughter of a senior Austrian nobleman with close ties to the imperial house; because of her family's importance, Emperor Ferdinand II was present at the wedding.[12] These two alliances with substantial financial and social advantages suggest a personal ambition that is otherwise somewhat difficult to trace in the young Wallenstein.

In Central Europe in the 1620s and 1630s, any further ambitions on Wallenstein's part had to take into consideration the widespread and ongoing conflict sparked in Prague in 1618. This tension erupted with the Battle of White Mountain outside the city in 1620, in which the rebellious Protestant Bohemian nobles were routed by imperial forces. However, even before the opening of the Thirty Years War, Wallenstein had identified his route to prosperity. In 1617, he announced that he was setting out from Moravia with 200 cavalrymen, recruited and outfitted at his own initiative and expense, to aid Ferdinand, who was then still king of Bohemia, in his war against Venice.[13] Opportunities for such entrepreneurship increased enormously over the decades of hostilities. Wallenstein raised and financed armies to fight on behalf of the emperor. In addition, he became a financier to the imperial house, providing Ferdinand both with his own resources and money borrowed from other sources. The money owed to him was secured by the titles to estates. The acquisition of land was a primary goal for Wallenstein, and following the seizure of property from Bohemian Protestants, he was able to acquire vast tracts very cheaply. By the mid-1620s, he was the largest landowner in Bohemia and one of its richest men.[14] This entrepreneurial spirit combined with exceptional military skill; he was named commander of all imperial troops, along with Johann Tserclaes, Count of Tilly. Wallenstein's financial and military strength gave him tremendous leverage with the emperor. In 1623, Ferdinand granted an imperial patent joining many of Wallenstein's smaller estates to form a larger entity of Frýdlant (Friedland) in Bohemia, and eight months later raised it to a principality, making Wallenstein a prince of the Empire.[15] Five years later, Wallenstein was granted the title of duke of Mecklenburg, one of the great hereditary duchies of the Empire, after the reigning dukes were stripped of their title as punishment for allying themselves with Christian IV of Denmark against the emperor.[16]

Wallenstein's power grew so great that Ferdinand eventually became wary of him. The general was forced into retirement in the late 1620s, only to be recalled in desperation following the Swedish entry into the war and the victories of King Gustavus Adolphus. This reintroduced the challenge of Wallenstein's strength and ambition, however, and in 1634, with the emperor's consent, he was assassinated.

In a relatively short period before his death, Wallenstein repositioned himself as something of a prince, with all of the cultural trappings associated with the role. He became a major patron in Prague, and to some degree in

Bohemia.[17] He remade the town of Jičín (Jitschin) as a capital for Frýdlant, planning an enormous residence for himself, a church, and other representative buildings.[18] Although unrealized, he planned to build a new palace as duke of Mecklenburg as well. After considering Schwerin and Wismar, he settled on Güstrow, a longstanding ducal residence city, for the new palace. However, work was interrupted by the Swedish occupation of the territory, and the project was abandoned at his death.[19]

In Prague, Wallenstein built a huge palace at the base of the hill on which the imperial palace stands.[20] The imperial complex looms above Wallenstein's residence, and the juxtaposition is especially clear from the gardens, which offer the most attractive view of the residence.[21] Giovanni Pieroni, a fortifications engineer, came to Vienna from Florence in 1622 and subsequently worked on numerous projects throughout the Habsburg lands.[22] In 1628, he is documented designing parts of Wallenstein's palace, including the salon and the large loggia facing the garden. Although private, and not visible from the street, this was the true show façade of the palace, for Wallenstein seems to have privileged the view from the gardens, with the imperial castle framing his residence (Figure 5.2). The gardens were ornamented with large bronze sculptures by Adriaen de Vries (1556–1626), who had earlier been the imperial sculptor employed by Emperor Rudolf II (until his death in 1612), and subsequently for a group of princes, including King Christian IV of Denmark. The group of statues De Vries made for Wallenstein was hardly less impressive than the works he had produced for the

Figure 5.2 Giovanni Pieroni and others, Wallenstein Palace, Prague, 1623–29; with statues by Adriaen de Vries, 1620s, copies. Photo: Author.

emperor and was more than the equal of anything he made for Christian or any other imperial prince.[23]

The De Vries statues were relatively private, accessible only to those who passed the palace walls and entered Wallenstein's gardens. (Private palaces were however more accessible than is often understood.[24]) The commander's imperial image was propagated more broadly in an engraving presenting him in battle on horseback, produced in 1627 (Figure 5.3). This was a clear variant of an

ALBERTVS D.G.DVX FRIDLANDIÆ SAC²CÆS⁂MA⁂CONSILIARI,
BELLIC,CAMERARI, SVPREM,COLONELL ,PRAGENSIS ET
EIVSDEM MILITIÆ GENERALIS

Figure 5.3 Probably Sebastian Furck, Albrecht von Wallenstein on Horseback, 1627. Photo: National Archives, Prague. Národní archiv (National Archives/NA), Valdštejniana–rytiny (Valdštejniana–engravings/VL-Rytiny), Inv.-Nr. 13.

engraving of Emperor Rudolf II produced around 1603 by Aegidius Sadeler after a design by De Vries (Figure 5.4). The image would soon be revised to represent Emperor Ferdinand III on horseback, perpetuating a common trope of ruler imagery into which Wallenstein was interpolated, either at his own insistence or by an entrepreneurial publisher.[25]

Figure 5.4 Aegidius Sadeler after Adriaen de Vries, Emperor Rudolf II on Horseback, ca 1603. Photo: Harris Brisbane Dick Fund, 1953, www.metmuseum.org.

Certainly, in the course of his relatively short career, Wallenstein provided a center of patronage in Prague not just for De Vries, but for a large number of painters and others who worked for him in his palace or elsewhere. In the absence of the imperial court, which had returned to Vienna after the death of Rudolf II in 1612, he was the most outstanding patron in the city and did much to sustain its viability as a cultural center through the first phase of the Thirty Years War.

Carl Gustaf Wrangel

Carl Gustaf Wrangel (1613–76) was a career soldier, as his father had been, and rose to the post of field marshal and commander of all Swedish troops in Germany in 1646, at thirty-two years of age.[26] As with other Swedish commanders, success brought titles and social position, but always within the context of the military. Thus, for instance, he was named general-governor of Swedish Pomerania in 1648, a position somewhere between a military post and a semi-princely role that he tried to approximate.[27]

In the later 1640s and early 1650s, Wrangel employed two ambitious young painters sequentially. The first, Matthäus Merian the Younger (1621–87), was a very capable professional painter.[28] The son of a printmaker and publisher of the same name, he studied for six years in Frankfurt and Amsterdam with Joachim von Sandrart, of whom he spoke with deep respect and admiration in his autobiography.[29] He then worked in England and traveled in Italy before returning north. He met Wrangel in Germany in 1647 and was soon attached to his large retinue. The general also asked him to come to the peace banquets in Nuremberg in 1649–50 to paint a number of portraits of the celebrants, especially of Wrangel and other Swedish generals. There Merian was reunited with his former teacher, Sandrart.

In 1650, after the Nuremberg festivities, Merian accompanied Wrangel to the port in Wismar, and was prepared to board the ship to Stockholm with the general when he received news of his father's death. He felt obliged to remain in Germany and take over the family publishing house, and specifically the running chronicle of contemporary Europe, *Theatrum Europaeum*.[30] The fifth part of this work, published in 1647, had been dedicated to Wrangel, whom Merian noted had provided many "beautiful sketches and drawings" that were used for the production of the chronicle. In addition to these practical contributions, Wrangel had also recognized the dedication with a gift of one hundred ducats.[31] Now, given Merian's family obligations, the general reluctantly allowed him to remain, and Merian noted with pride that he received a gold chain for his loyalty. The two men remained in close contact, and Merian returned to Wrangel's "court" in Wolgast in 1651–52 and 1661–62 to paint portraits for the general.[32]

Following Merian's departure, Wrangel employed another young German painter, David Klöcker, who would be ennobled with the name Ehrenstrahl in

1674.[33] Working from information supplied by Ehrenstrahl himself, Sandrart wrote that Ehrenstrahl had been in the Swedish service as a scribe in the later 1640s, but that he had encountered Wrangel later, in the Low Countries.[34] There he had recognized immediately the advantages of joining the general's retinue, and in 1654 had received support for an extended period of study in Italy, France, and England. Wrangel's involvement in this award is uncertain. Documents suggest that the support came from Maria Eleonora of Brandenburg, the dowager queen of Sweden, but the nature of the overlapping relationships between the painter, the queen, and the general is unclear.[35] Certainly, Wrangel was more ambitious than Maria Eleonora in the imagery that he had the young painter produce. In 1652, Ehrenstrahl painted a large portrait of Wrangel on horseback, rearing before a troop train (Figure 5.5). Much as the engraving of Wallenstein discussed above has direct roots in imperial portraiture, the equestrian portrait of Wrangel relates closely to two royal portraits from the Scandinavian courts. The

Figure 5.5   David Klöcker Ehrenstrahl, Carl Gustaf Wrangel on Horseback, 1652. Photo: Skokloster Palace.

first is a portrait of the Danish king, Christian IV, made by Karel van Mander III in 1642–43, with the sitter brandishing a sword in a similar position on a pacing horse with a battlefield behind.[36] The other is a large portrait of Carl Gustaf of Pfalz-Zweibrücken (a general and the future King Carl X Gustaf of Sweden) on a rearing horse, painted in Nuremberg by Sandrart in 1650, soon after Carl Gustaf's confirmation as crown prince and successor to Queen Christina. Sandrart presented the picture to "[His] Royal Highness," and later described the picture itself as a "crown-worthy rider," making clear the royal nature of the image.[37]

Upon his return to Sweden in 1661, Ehrenstrahl almost immediately received a post as a painter to the king. As the king was then five years old, Ehrenstrahl was appointed by the new dowager queen and head of the regency, Hedwig Eleonora of Schleswig-Holstein-Gottorf.[38] Although she appears to have made the decision autonomously, we may wonder if Ehrenstrahl's work for Wrangel, who served on the regency, figured into her decision.

It is difficult to say whether Merian was truly disappointed to return to Frankfurt in 1650. Certainly, the Merian press was prosperous. However, he may well have been aware of the opportunity foreclosed by his return to Germany. Ehrenstrahl attained a position that might, in other circumstances, have gone to Merian. The likelihood of a future court position may have been apparent to Merian already upon his departure from Wrangel's retinue, for many of those artists and craftsmen who attained court positions in Sweden before the 1680s were first employed by military commanders, who brought them to the kingdom. The movement of the troops brought these commanders into contact with skilled craftsmen of all kinds, many of whom were prepared to move in search of better opportunities. Quite a number of these turned up in Sweden.[39] Carl [X] Gustaf wrote from Prague that he had met many outstanding Protestant craftsmen who were eager to follow him to Sweden.[40]

Both Carl Gustaf and Wrangel were in Prague in 1648 because the Swedish troops under Hans Christopher von Königsmarck had breached the city walls. They had then stormed the *Malá Strana*, the small side of the city where the residence and Wallenstein's palace were found, and looted many art treasures from both. The De Vries statues from Wallenstein's gardens and the imperial palace were packed and shipped to Sweden, along with hundreds of paintings and sculptures by outstanding masters. Many of these went into the royal collections. Queen Christina took some with her to Rome when she abdicated in 1654, and gave away others.[41] Wrangel arrived on November 24, well after the looting was completed, but some objects from the imperial collection may have come into his possession.[42] (Looting after the ratification of the peace in October 1648 was illegal in seventeenth-century law.) Ultimately, however, his direct experience of Wallenstein's outsized stature and cultural presence in Prague may have been more formative for Wrangel than any looted goods.[43]

Christina routinely gave land to favorites and as a reward for service to the crown. In 1648, while he was in Prague, she gave a large plot in Stockholm to

Wrangel, presumably with the expectation that he would build an imposing palace on the site (Figure 5.6). With the help of a local architect, he began immediately to plan the project. It went through numerous iterations and partial rebuilds under the leadership of various architects and engineers, with Wrangel himself always a decisive voice. In particular, the general liked the noble residences he had seen in Warsaw in the course of the Swedish campaigns in Poland in 1655–56, and wanted to include elements of the tradition he encountered there.[44]

In 1646, two years before work began on the Stockholm residence, Wrangel began assembling materials to build a new house at his ancestral estate at Skokloster, north of Stockholm (Figure 5.7) and began construction after the peace.[45] The conception and design of this building, too, may be informed directly by Wrangel's experiences as a soldier. Wrangel was in Aschaffenburg, near Frankfurt, twice, and possibly three times, in the course of the Thirty Years War, and saw the palace of the elector of Mainz there. Each building has a large square courtyard with towers at the corners. More tellingly, a written description of the region from the 1660s explicitly compares Skokloster to Aschaffenburg.[46] This was in an early draft for *Suecia antiqua et hodierna* [Ancient and Modern Sweden], a large-scale topography of Sweden that was published after 1715 without a text, following numerous failed efforts to produce one.[47] Like most other elites in the kingdom, Wrangel was eager to see his own estates included in

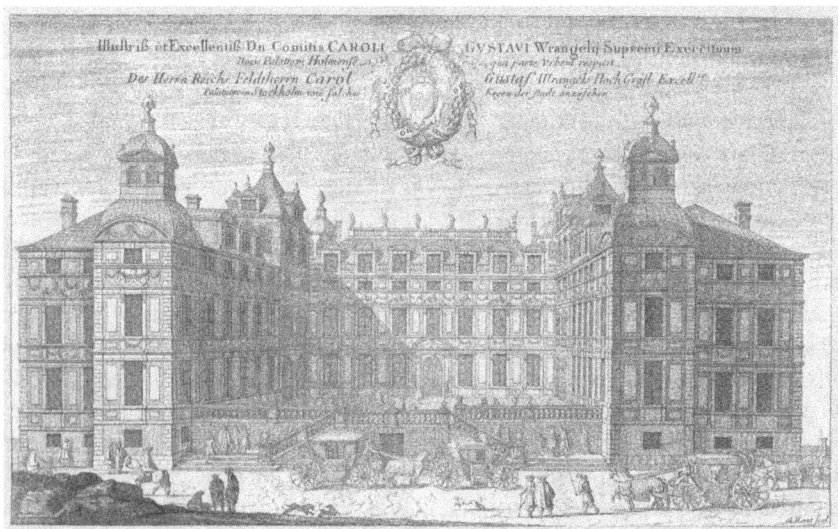

Figure 5.6 Nicodemus Tessin the Elder and Others, Wrangel Palace, Stockholm, from 1648. Print from *Suecia antiqua et hodierna* (after 1715). Photo: The Royal Library, Stockholm.

Figure 5.7 Caspar Vogel, Nicodemus Tessin the Elder, and others, Skokloster, near Stockholm, begun 1653. Photo: Author.

the project, and followed its development closely. The comparison of Skokloster to Aschaffenburg may well have come from him.

Wrangel appears several times in Sandrart's well-known encyclopedic study of art and art history, *Teutsche Academie* [German Academy] (1675–79). In one passage, on his troops taking Landshut, in Bavaria, in 1648, Sandrart writes that "he immediately visited these two altarpieces, sat down before them, looked long at them, and highly praised both the artist and the art."[48] The painter praised here rather immodestly is Sandrart himself. The note that the altarpieces were in the Jesuit church excludes any possibility that the commander for a militantly Lutheran state was there for pious reasons. Rather, Wrangel is presented as an arbiter of taste and artistic quality. If we accept this at face value, it implies a high level of cultural or artistic authority on the part of someone from an exclusively military background. Aside from a relatively short journey to complete his schooling, that cultural education took place entirely in the field, and was in effect a completion of his educational travels that were cut short by the war.

Sandrart's description of Wrangel's visit to the Jesuit church in Landshut is the only source describing that episode, and it may be a fabrication. If it is an invention on Sandrart's part, however, designed as a vehicle to promote his own work, it is a telling one. It supposes a readership that was prepared to accept a soldier as an arbiter of quality in the arts, filling a role that otherwise might have been occupied by a ruler or a revered painter. It is supported by the importance given to the general in Sandrart's description of the life of Merian, which mirrors Wrangel's presentation in Merian's own autobiographical sketch. By extension, it also suggests that Carl [X] Gustaf and others were seen, or could have been seen, as substantial cultural arbiters as well. Certainly, this view is supported by the membership of Wrangel, Carl Gustaf, and virtually all of the other Swedish high commanders in the *Fruchtbringende Gesellschaft,* the most prestigious literary society in the German lands.

Prince Eugene of Savoy

Wallenstein and Wrangel were hardly alone in their twin military and cultural ambitions. Nor did the circumstances and conventions that enabled their rise soon dissipate. At the end of the seventeenth century, Prince Eugene of Savoy (1663–1736) likewise rose to great status through military and organizational brilliance in service to the Habsburgs, first in the ongoing wars against the Ottoman Turks, and then in the War of the Spanish Succession (1701–14).[49] The parallels to the earlier commanders were close enough that Emperor Charles VI worried that Eugene would become another Wallenstein: too talented, ambitious, and powerful to control. This concern was ultimately unjustified, for Eugene's ambitions remained aligned with those of his employer.[50]

Just as Wallenstein came from the Bohemian borderlands, and Wrangel was born in Sweden to a Baltic-German family, Eugene moved easily from one region to another. His mother, Olimpia Mancini, was an Italian who came to Paris in the retinue of her uncle, Cardinal Jules Mazarin (Giulio Mazzarino), Louis XIV's powerful chief minister. His father, Eugène Maurice, was a younger son of a prominent soldier (himself an illegitimate younger son of a prince) and received much of his military advancement through his marriage to Olimpia. Eugene embraced this multifaceted background, signing his name tri-lingually: Eugenio von Savoy. Like many who pursued military careers, he was a younger son. His family expected him to join the church, but at the age of nineteen, he announced his intention to join the military. However, Louis XIV refused his petition to command a French company, apparently because Olimpia had become implicated in a scandal and fallen into disfavor. Eugene thus offered his services to the Habsburgs and soon distinguished himself, much to Louis's regret when French and Imperial armies met in the War of the Spanish Succession.

Like Wallenstein and Wrangel, Prince Eugene turned military capital into cultural capital. His collections were exceptional. His paintings were widely admired both for their number and quality. After his death, they went to his cousin, Carlo Emmanuele III, king of Sardinia and duke of Savoy, and now form part of the Galleria Sabauda in Turin. His library was no less extraordinary. It was acquired by Emperor Charles VI and forms a large part of the historical collections of the Austrian National Library.[51] The prints and drawings were also bought by the emperor, and form a core of the Albertina Museum in Vienna.[52]

Because the collections have been absorbed into other institutions, Eugene's cultural legacy is now most visible in architecture. His winter palace in Vienna and its pendant, the Upper and Lower Belvedere (or summer palace), then just outside the city walls, are landmarks in the city. Farther afield, he rebuilt an earlier castle near Bratislava, Schloss Hof, as an elegant country estate, and constructed Ráckeve Palace on an island in the Danube near Budapest.

The winter palace in Vienna, which Eugene used as an office for his many duties, was built by Johann Bernhard Fischer von Erlach, the imperial architect (Figure 5.8). However, the interior decoration was handled by Johann Lukas von Hildebrandt. Of the two architects, Fischer has been more highly esteemed by most critics, and this switch has often been regarded with some curiosity. (Hildebrandt was however not yet in Vienna when the impatient commander began work on the residence.) Although details are few, Eugene's preference for Hildebrandt may be explained at least in part by circumstance, for the architect worked as an engineer under the commander early in his career, during campaigns in Piedmont in 1695–96. Moreover, Eugene had strongly supported Hildebrandt's father, Christoph, for the post of captain in the imperial army.[53] Eugene's promotion of Hildebrandt helped set in motion a major architectural career, not just in Vienna, where he became Fischer von Erlach's main competitor,

Figure 5.8 Johann Bernhard Fischer von Erlach, Palace of Prince Eugene, Vienna, begun 1695. Print from Fischer von Erlach, *Entwurff einer historischen Architectur* (1721). Photo: Getty Research Institute/Image Archive.

but elsewhere in the German lands as well.[54] Certainly, Eugene preferred him for his subsequent projects. The most notable of these is the Belvedere complex in Vienna.[55] From 1714–16, following the conclusion of the War of the Spanish Succession, Eugene had Hildebrandt build a long, low garden palace—the Lower Belvedere—that is generally comparable to the *Lustgartengebäude* (garden pleasure houses) built by other noble families.[56] Very soon thereafter, in 1717, he began work on the much larger Upper Belvedere, which sits on a hill above the Lower Belvedere, with formal gardens providing an axial link joining the two (Figure 5.9).

Although the War of the Spanish Succession resulted in a political compromise, Eugene's status and wealth grew substantially through his participation in it, and this was manifested in his cultural imprint. As a complex, the Upper and Lower Belvedere compare favorably to the imperial residences, and the Habsburgs and the Savoys fought mercilessly to acquire them.[57] Ultimately, they were purchased (along with Schloss Hof) by Empress Maria Theresia from Eugene's niece and heir, Maria Anna Victoria of Savoy. Initially, Maria Theresia planned to unite them with other nearby properties and to form a new court and government center. However, in 1776 she decreed that the Belvedere should house the imperial painting collections, a function it retained until the construction of the present Kunsthistorisches Museum in the late nineteenth century.[58]

By all accounts, Eugene's engagement with the arts and culture was deep and sincere, rather than a perceived obligation for a successful and ambitious figure. Even from the field, he maintained close contact with Hildebrandt regarding his architectural projects, rather than designating an assistant to oversee the work.[59] Likewise, when war brought him to the Netherlands, he took time from

Figure 5.9 Johann Lukas von Hildebrandt, Upper Belvedere, Vienna, 1717–23. Photo: Author.

his crushing duties to go to Amsterdam, a major center in the art trade, where he personally chose a number of paintings for his growing collection.[60]

The prince's personal investment in high culture may perhaps be seen most closely in the development of his library and the joy he seems to have taken in the company of literati. The playwright and poet Jean-Baptiste Rousseau (1671–1741) came to Vienna in 1712 after being exiled from France. Through Eugene's mediation, he was named imperial court historiographer, with a very substantial annual income of 2,800 gulden. Eugene found this sum insufficient, however, and supplemented it both monetarily and with gifts.[61] To some degree, Rousseau was then something of a courtier to Eugene. He has also been described as Eugene's court poet, a designation supported by a verse that describes the prince as a *philosophe guerrier* (philosopher-warrior).[62] Aside from such panegyric, however, each man seems to have admired the other genuinely. Rousseau wrote privately of his astonishment that Eugene, who bore tremendous political and military responsibilities, appeared to have read, or at least perused, virtually everything in his immense library, even before the books went to the binder. Eugene, for his part, seems to have been personally stung when Rousseau, for reasons of his own, supported a faction that aimed to remove him as governor-general of the Habsburg Netherlands.[63]

From 1712, Eugene was a friend and supporter of the polymath Gottfried Wilhelm Leibniz (1646–1716).[64] He promoted Leibniz's effort to establish an academy of the arts and sciences in Vienna. The effort failed, but the friendship and mutual admiration of the two men endured. Leibniz, who was, among many other roles, an important bibliographer who oversaw the exceptional library in Wolfenbüttel, was very impressed by Eugene's personal collection of books.[65] Eugene supported Leibniz even when the imperial house did not, and while their discussions must have ranged widely, given the broad interests of both men, we may imagine that the library stood at the center of their relationship.

Conclusion

Wallenstein, Wrangel, and Prince Eugene were very different figures and approached their careers in very different ways, but all three found ways to turn war—a virtually continuous state in their generations—into a means for personal advancement. For them, personal advancement implied substantial cultural patronage at a level that could offset some of the cultural destruction of war.

These three men represent a particular kind of ambition that, in the seventeenth and eighteenth centuries, was often channeled through military activity, but manifested in part in cultural investment. In some cases, this may have been perceived as a kind of crass self-promotion. In others, it was considered appropriate and underwritten at the highest levels. In the course of the War of the Spanish Succession, the Battle of Blenheim in 1704 was a major victory of the allied British and Austrians, and a check on French ambition. Eugene's co-commander in the battle, John Churchill (1650–1722), had risen through military skill from a gentry background to become the earl and then duke of Marlborough. The victory was so important that Queen Anne had Sir John Vanbrugh build Blenheim Palace near Oxford for him beginning in 1705, during the heart of the conflict.[66] It was the only secular, non-royal residence in Britain granted the title of "palace." In its own way, it was a British parallel to the Vienna Belvedere, and it points to a broader and more variable group of figures who represent the kind of positive cultural effects of militarized society described in this essay.

Notes

1 Terence Riley and Barry Bergdoll, eds., *Mies in Berlin* (New York: Abrams, 2001); Phyllis Lambert, ed., *Mies in America* (New York: Abrams, 2001).
2 See especially Panofsky's own comments on his move to the United States, in Erwin Panofsky, "Epilogue–Three Decades of Art History in the United States: Impressions of a Transplanted European," in *Meaning in the Visual Arts: Papers in and on Art History* (Garden City, NY: Doubleday, 1955), 368–95. See also Michael Ann Holly, *Panofsky and the Foundations of Art History* (Ithaca, NY: Cornell University Press, 1984). For Weitzmann's memoirs, see Kurt Weitzmann, *Sailing with Byzantium from Europe to America: The Memoirs of an Art Historian* (Munich: Editio Maris, 1994).
3 In addition to various interviews, see James S. Ackerman, *Origins, Invention, Revision* (New Haven: Yale University Press, 2016), esp. 42–55.
4 David Coffin, *Pirro Ligorio. The Renaissance Artist, Architect, and Antiquarian* (University Park: Pennsylvania State University Press, 2004), xiii.
5 Joachim von Sandrart, *Teutsche Academie der Bau-, Bild-, und Mahlerey-Künste* (Nuremberg: Johann-Philipp Miltenberg, 1675–79). Quoted from Dieter Graf, *German Baroque Drawings* (London: Heim Gallery, 1975), introduction.
6 For a nuanced examination of the destruction of the war, see Thomas DaCosta Kaufmann, "War and Peace, Art and Destruction, Myth and Reality: Considerations on the Thirty Years' War in Relation to Art in (Central) Europe," in *1648. War and Peace in Europe*, vol. 2, ed. Klaus Bussmann and Heinz Schilling (Münster: Westfälisches Landesmuseum für Kunst und Kulturgeschichte, 1998), 163–72. On the war and the arts more generally, see this volume and Jacques Thuillier and Klaus Bussmann, eds., *1648. Paix de Westphalie: L'art entre la guerre et la paix / Westfälischer Friede: Die Kunst zwischen Krieg und Frieden* (Paris: Musée du Louvre, 1999); and Claudia Brink et al., eds., *Bellum & Artes: Central Europe in the Thirty Years' War* (Dresden: Sandstein, 2021).
7 Walter Tommasoli, *La vita di Federigo da Montefeltro (1422–1482)* (Urbino: Argalìa, 1978).
8 Jeanette Kohl, *Fama und Virtus: Bartolomeo Colleonis Grabkapelle* (Berlin: Akademie Verlag, 2004).
9 See however Stig Förster, Christian Jansen, and Günther Kronenbitter, eds., *Rückkehr der Condottieri? Krieg und Militär zwischen staatlichem Monopol und Privatisierung: von der Antike bis zur Gegenwart* (Paderborn: Schöningh, 2010), which traces a more or less continuous line of *condottieri*-like operators from antiquity to the present.
10 Reinhard Baumann, "Die deutschen Condottieri. Kriegsunternehmertum zwischen eigenständigem Handeln und 'staatlicher' Bindung im 16. Jahrhundert," in *Rückkehr der Condottieri? Krieg und Militär zwischen staatlichem Monopol und*

Privatisierung: von der Antike bis zur Gegenwart, eds. Stig Förster, Christian Jansen, and Günther Kronenbitter (Paderborn: Schöningh, 2010), 111–25, here 119. For Wallenstein, see endnote 11.

11 See Golo Mann, *Wallenstein, his Life Narrated by Golo Mann*, trans. Charles Kessler (New York: Holt, Rinehart and Winston, 1976); Geoff Mortimer, *Wallenstein. The Enigma of the Thirty Years War* (New York: Palgrave Macmillan, 2010); Robert Rebitsch, *Wallenstein: Biografie eines Machtmenschen* (Vienna: Böhlau, 2010).

12 Mann, *Wallenstein*, 187–89; Mortimer, *Wallenstein*, 68.

13 Mortimer, *Wallenstein*, 18. For the entrepreneurial aspect of recruiting and developing armed forces, see David Parrott, *The Business of War: Military Enterprise and Military Revolution in Early Modern Europe* (Cambridge: Cambridge University Press, 2012).

14 Mortimer, *Wallenstein*, 35–51.

15 Ibid., 69.

16 Ibid., 106–7.

17 See generally Eliška Fučíková and Ladislav Čepička, eds., *Waldstein–Albrecht von Waldstein: Inter arma silent musae?* (Prague: Academia, 2007).

18 See the essays by Rudolf Anděl, Jaromír Gottlieb, Barbora Klipcová, and Petr Uličný in Fučíková, *Waldstein*, 201–44.

19 Petr Fidler, "Waldstein als Bauherr, Mäzen und 'Hausvater'," in *Wallenstein: Mensch–Mythos–Memoria*, eds. Birgit Emich et al. (Berlin: Duncker & Humblot, 2018), 281.

20 Mojmír Horyna, *The Waldstein Palace in Prague* (Prague: Gema Art, 2002).

21 This is pointed out in Thomas DaCosta Kaufmann, *Court, Cloister, and City. The Art and Culture of Central Europe 1450–1800* (Chicago: University of Chicago Press, 1995), 249–55; Eliška Fučíková, "Inspiration durch die Burg–Albrecht von Waldstein und das Waldsteinpalais," in Fučíková, *Waldstein*, 70–78.

22 Jarmila Krčálová, "Giovanni Pieroni–Architekt?" *Umění* 36, no. 6 (1988): 511–42; Guido Carrai, "Nuovi documenti su Giovanni Pieroni e un'ipotesi per Palazzo Wallenstein" *Umění* 52, no. 6 (2004): 537–42; Guido Carrai, "Architektur und Diplomatie: Giovanni Pieroni, Berichterstatter der Medici bei General von Waldstein," in Fučíková, *Waldstein*, 312–19.

23 Lars Olof Larsson, *Adrian de Vries* (Vienna: Verlag Anton Schroll, 1967), 91–98.

24 Michaela Völkel, *Schloßbesichtigungen in der Frühen Neuzeit. Ein Beitrag zur Frage nach der Öffentlichkeit höfischer Repräsentation* (Munich: Deutscher Kunstverlag, 2007).

25 The format was used by other military commanders, though perhaps rarely derived so directly from an imperial portrait. See Zdeněk Munzar, Tomáš Kykal, and Michal Hokynek, "Das Abbild von Albrecht Eusebius von Waldstein auf den Stichen des 17. Jahrhunderts und alte Drucke aus der Waldstein-Zeit, nicht nur aus der Bibliothek des Militärhistorischen Instituts," in Fučíková, *Waldstein*, 354–63.

26 Arne Losman, *Carl Gustaf Wrangel och Europa. Studier i kulturförbindelser kring en 1600-talsmagnat* (Stockholm: Almqvist & Wiksell, 1980).

27 Ivo Asmus, "Carl Gustav Wrangel–Schwedischer Generalgouverneur und 'pommerscher Fürst'," in *Unter der schwedischen Krone. Pommern nach dem Westfälischen Frieden*, ed. Ivo Asmus (Greifswald: Publikationen der Stiftung Pommersches Landesmuseum, 1998), 53–75.

28 Daniela Nieden, *Matthäus Merian der Jüngere (1621–1687)* (Göttingen: Cuvillier, 2002).

29 Merian the Younger's autobiographical essay is reproduced in Rudolf Wackernagel, "Selbstbiographie des jungern Matthäus Merian," *Basler Jahrbuch* (1895): 227–44. It provides the basis for the following discussion.

30   Lucas Heinrich Wüthrich, *Das druckgraphische Werk von Matthaeus Merian d. Ae. Vol. 3, Die grossen Buchpublikationen I* (Hamburg: Hoffmann und Campe Verlag, 1993), 113–272.

31   Wüthrich, *Das druckgraphische Werk*, vol. 3, 130, 190–95. Although a causal relationship is not spelled out, this is evident from a letter from Matthäus Merian the Elder to Wrangel, in which he thanks the latter for the gift. Riksarkivet, Stockholm, Skoklostersamling E 8420, June 11, 1647.

32   Asmus, "Carl Gustav Wrangel," 70.

33   For Ehrenstrahl, see Axel Sjöblom, *David Klöcker Ehrenstrahl* (Malmö: Allhems, 1947); Allan Ellenius, *Karolinska bildidéer* (Uppsala: Almqvist & Wiksell, 1966); Bengt Dahlbäck, ed., *David Klöcker Ehrenstrahl* (Stockholm: Nationalmuseum, 1976); Kjell Wangensteen, *Hyperborean Baroque: David Klöcker Ehrenstrahl (1628–98) and the First Swedish School of Art* (PhD diss., Princeton University, 2019).

34   Sandrart, *Teutsche Academie*, vol. 2, 334. Sandrart's biography of Ehrenstrahl may have been composed by Ehrenstrahl's friend (and Uppsala University professor), Blasio Ludovico Teppati. See Doris Gerstl, "Joachim von Sandrarts *Teutsche Academie der Edlen Bau-, Bild- und Mahlerey-Künste – zur Genese*," in *Künste und Natur in Diskursen der Frühen Neuzeit*, vol. 2, ed. Hartmut Laufhütte (Wiesbaden: Harrassowitz, 2000), 883–98.

35   Wangensteen, *Hyperborean Baroque*, 152–65.

36   Povl Eller, *Kongelige portrætmalere i Danmark 1630–82. En undersøgelse af kilderne til Karel van Manders og Abraham Wuchters' virksomhed* (Copenhagen: Selskabet til Udgivelse af danske Mindesmærker, 1971), 155–56. Both Van Mander's and Ehrenstrahl's portraits were reproduced in print and copied elsewhere.

37   Christian Klemm, *Joachim von Sandrart. Kunst-Werke u. Lebens-Lauf* (Berlin: Deutscher Verlag für Kunstwissenschaft, 1986), 177–78, 183–91.

38   Kjell Wangensteen, "Hedwig Eleonora as Patron of David Klöcker Ehrenstrahl" in *Queen Hedwig Eleonora and the Arts. Court Culture in Seventeenth-Century Northern Europe*, ed. Kristoffer Neville and Lisa Skogh (London: Routledge, 2017), 78–87.

39   Kristoffer Neville, *The Art and Culture of Scandinavian Central Europe, 1550–1720* (University Park: Pennsylvania State University Press, 2019), 95–112.

40   Gerhard Eimer, *Carl Gustaf Wrangel som byggherre i Pommern och Sverige. Ett bidrag till stormaktstidens konsthistoria* (Stockholm: Almqvist & Wiksell, 1961), 69.

41   Görel Cavalli-Björkman, "La collection de la reine Christine à Stockholm" in Thuillier, *1648: Paix de Westphalie*, 295–317.

42   Losman, *Carl Gustaf Wrangel och Europa*, 31–32; Wangensteen, *Hyperborean Baroque*, 131.

43   For context, see Parrot, *The Business of War*, 241–59.

44   Osvald Sirén, *Gamla Stockholmshus af Nicodemus Tessin d.ä. och några samtida byggnader*, vol. 1 (Stockholm: Norstedt, 1912), 43–114; Gerhard Eimer, *Carl Gustaf Wrangel som byggherre*, 121–38; Kristoffer Neville, *Nicodemus Tessin the Elder. Architecture in Sweden in the Age of Greatness* (Turnhout: Brepols, 2009), 104–8.

45   Erik Andrén, *Skokloster. Ett slottsbygge under stormaktstiden* (Stockholm: Nordisk rotogravyr, 1948); Gerhard Eimer, *Carl Gustaf Wrangel som byggherre*, 138–50; Carin Bergström, ed., *Skoklosters slott under 350 år* (Karlstad: Votum, 2017).

46   Hans-Bernd Spies, *Schloß Johannisburg in Aschaffenburg und Schloß Skokloster am Mälarsee in Schweden* (Aschaffenburg: Geschichts-und Kunstverein Aschaffenburg e.V., 1986).

47 The production and publication history of *Suecia antiqua* is exceptionally complex. See Samuel E. Bring, "Sveciaverket och dess text," *Lychnos* 2 (1937): 1–67; Börje Magnusson and Jonas Nordin, *Drömmen om stormakten. Erik Dahlberghs Sverige* (Stockholm: Medströms bokförlag, 2015); Börje Magnusson, "Tryckningen av Suecia Antiqua," *Biblis* 73 (2016): 19–49.

48 Joachim von Sandrart, "Lebenslauf und Kunst-Werke des woledlen and gestrengen Herrn Joachims von Sandrart," in *Teutsche Academie*, vol. 1, 18. "…hat Er sofort diese zwey AltarBlätter besuchet/ sich davor niedergesetzet/ sie lang beschauet/ und sowol den Künstler/ als die Kunst/ sehr gerühmet." For more on this church and other examples of Swedish commanders–notably King Gustavus Adolphus (Gustaf II Adolf)–appreciating the arts they encountered, see the essay by Jeffrey Chipps Smith in this volume.

49 The literature on Eugene is large. See especially Alfred Arneth, *Prinz Eugen von Savoyen*, 3 vols (Vienna: Typogr.-literar.-artist. Anstalt, 1858); Franz Herre, *Prinz Eugen: Europas heimlicher Herrscher* (Stuttgart: Deutsche Verlags-Anstalt, 1997).

50 Herre, *Prinz Eugen*, 238–39.

51 Gabriele Mauthe, "The Bibliotheca Eugeniana in its Contemporary European Context," in *Prince Eugene: General-Philosopher and Art Lover*, ed. Agnes Husslein-Arco and Marie-Louise von Plessen (Munich: Hirmer, 2010), 190–97.

52 Christian Benedik, "Prince Eugene's Collection of Engravings in the Albertina," in Husslein-Arco, *Prince Eugene,* 155–59.

53 Bruno Grimschitz, *Johann Lukas von Hildebrandt* (Vienna: Heroldt, 1959), 7–8, 14.

54 Grimschitz, *Hildebrandt*; Peter Heinrich Jahn, *Johann Lucas von Hildebrandt (1668–1745): Sakralarchitektur für Kaiserhaus und Adel. Planungsgeschichtliche und projektanalytische Studien zur Peters-und Piaristenkirche in Wien sowie dem Loreto-heiligtum in Rumburg* (Petersberg: Imhof, 2011).

55 Ulrike Seeger, *Stadtpalais und Belvedere des Prinzen Eugen. Entstehung, Gestalt, Funktion und Bedeutung* (Vienna: Böhlau, 2004); Peter Stephan, *Das Obere Belvedere in Wien: Architektonisches Konzept und Ikonographie: Das Schloss des Prinzen Eugen als Abbild seines Selbstverständnisses* (Vienna: Böhlau, 2010).

56 Hellmut Lorenz, "Das 'Lustgartengebäude' Fischers von Erlach–Variationen eines architektonischen Themas," *Wiener Jahrbuch für Kunstgeschichte* 32 (1979): 58–76.

57 Karl Schütz, "Prince Eugene's Cultural Legacy in Vienna," in Husslein-Arco, *Prince Eugene*, 268.

58 Michael Yonan, "Kunsthistorisches Museum/Belvedere, Vienna: Dynasticism and the Function of Art," in *The First Modern Museums of Art. The Birth of an Institution in 18th- and 19th-Century Europe*, ed. Carole Paul (Los Angeles: The J. Paul Getty Museum, 2012), 167–89.

59 Herre, *Prinz Eugen,* 265.

60 Georg Lechner, "Pieter van den Berge–Prince Eugene Visiting the Art Dealer Zomer in Amsterdam," in Husslein-Arco, *Prince Eugene*, 84.

61 Peter Faber, *Prinz Eugen von Savoyen: Feldherr und Staatsmann Europas* (Gilching in Fünfseenland: Druffel & Vowinckel-Verlag, 2011), 175–76.

62 Herre, *Prinz Eugen*, 294, 300–5.

63 Faber, *Prinz Eugen,* 176.

64 Herre, *Prinz Eugen*, 295–300.

65 Hans G. Schulte-Albert, "Gottfried Wilhelm Leibniz and Library Classification," *The Journal of Library History* 6, no. 2 (1971): 133–52.

66 Vaughan Hart, *Sir John Vanbrugh: Storyteller in Stone* (New Haven: Yale University Press, 2008), 127–46.

Further Reading

Brink, Claudia, Susanne Jaeger, and Marius Winzeler, eds. *Bellum & Artes: Central Europe in the Thirty Years' War*. Dresden: Sandstein, 2021.

Bussmann, Klaus and Heinz Schilling, eds. *1648. War and Peace in Europe*. 3 vols. Münster: Westfälisches Landesmuseum für Kunst und Kulturgeschichte, 1998.

Fučíková, Eliška and Ladislav Čepička, eds. *Waldstein–Albrecht von Waldstein: Inter arma silent musae?* Prague: Academia, 2007.

Husslein-Arco, Agnes and Marie-Louise von Plessen, eds. *Prince Eugene: General-Philosopher and Art Lover*. Munich: Hirmer, 2010.

Losman, Arne. *Carl Gustaf Wrangel och Europa. Studier i kulturförbindelser kring en 1600-talsmagnat*. Stockholm: Almqvist & Wiksell, 1980.

Thuillier, Jacques and Klaus Bussmann, eds. *1648: Paix de Westphalie. L'art entre la guerre et la paix/Westfälischer Friede. Die Kunst zwischen Krieg und Frieden*. Paris: Musée du Louvre, 1999.

Part II
Ideas and Ideologies of War

6 The Wars of Louis XIV and the Language of Europe

Daniel Riches

The wars of Louis XIV (r. 1643–1715) placed his opponents under intense military, diplomatic, fiscal, and intellectual pressure.[1] While mobilizing the enormous power-political resources needed to combat the Sun King accounted for the lion's share of the task confronting allied statesmen spread across Europe, the development of an effective rhetoric of resistance to Louis formed another crucial piece in the struggle against France and itself posed two substantial challenges. The first was the imperative to articulate a sufficiently ecumenical language to give common cause to the complicated alliances of remarkably diverse polities, ranging from Britain, to the Dutch Republic, to the Holy Roman Emperor, and to many of the Empire's territories that needed to stand united against Louis but that were themselves divided by fundamentally different individual and regional interests, confessional alignments, national cultures, governmental forms, and specific grievances against France. The second was the necessity of simultaneously speaking in a language sufficiently powerful and evocative to domestic audiences to motivate and sustain the staggering outlays of blood and treasure, as well as patience with the interruption of trade and other painful sacrifices, called forth by the wars. These two goals could stand at cross-purposes—an appeal to defend Protestantism, for instance, might animate a portion of a domestic population but could risk alienating a Catholic ally, whereas discourses on the sanctity of international law might appeal to allied statesmen but resonate only tepidly with the man or woman on the street. On the levels of both foreign and domestic policy, the wars of Louis XIV were ordeals of the highest order for his opponents. A united front against France needed to be constructed and serviced, and domestic support for the war effort needed to be produced and sustained. Political rhetoric played a central role in each of these processes.

This chapter focuses on an unintended consequence of the rhetorical war against Louis. Its central argument is that the multifarious pressures unleashed by Louis's wars induced late seventeenth- and early eighteenth-century Europeans to articulate language that contributed in unintentional yet important ways to emergent understandings of what Europe itself was and how its various parts should relate to one another. Viewed primarily through the prism of the

DOI: 10.4324/9781003157700-9

extensive pamphlet literature that Louis's wars gave rise to, the chapter brings into dialogue two distinct historiographies that have not often spoken to one another—one dealing with changing conceptualizations of Europe, the other centered on resistance to Louis—in order to explore the ways in which anti-Louis polemicists contributed to a new "language of Europe" that lived on long after Louis's long reign itself was finally over, and through this to gain a fresh appreciation of the impact of the wars of Louis XIV on important developments in European self-understanding.[2]

The new image of Europe that Louis's opponents helped construct during their decades-long struggle against the French king was built from individual elements that were anything but original. Indeed, the pamphleteers' toolkit was filled with implements that (for the most part) would have struck their craftsmen as comfortingly tried and true rather than daring or innovative, and wringing original, abstract thought from their aggregation would have been the furthest thing from the authors' minds when there were all-too-real wars to be waged and won. Appeals to tropes such as resistance to tyranny and universal monarchy, preservation of a European balance of power, and fidelity to an overarching notion of Western Christendom could (and did) appear in European print culture long before the Sun King had risen to prominence. The innovation lay instead in the frequency and intensity of their combination that produced subtly transformed shades of meaning. As Steve Pincus has written, the struggle against Louis called forth a "sophisticated and well-informed" public discussion that utilized building blocks that were old but with "uses and meanings" that were "decidedly novel."[3] The creative force of this discussion fed upon the intensity of the moment, and its novelty was made no less important by the fact that it was unintentional.

The contours of this language of Europe themselves become visible only on a broad geographic scale that eludes the nationally focused approach of most scholarship on anti-Louis polemic.[4] Whereas Pincus argues that pressure applied by Louis "goaded" forth nationalism in the polities opposing him, I would add that the challenge of resisting Louis also drew out a new understanding of Europe itself within which notions of particular nations and their interests were nested.[5] Connecting anti-Louis polemic to foundational changes in thinking about Europe requires looking beyond the confines of distinct national cultures (even while related pressures may have helped constitute those cultures) and reveals how serious individuals seeking to address pragmatically a specific, concrete contemporary problem could collectively, if unwittingly, contribute to the advance of more general, abstract ideas.

Discussing Europe during the Wars of Louis XIV

Though differing substantially from Pincus in important respects, Tony Claydon shares Pincus's emphasis that early modern Englishmen viewed their struggles with Louis XIV within a fundamentally European framework.[6] Indeed, people not just in England but all across Europe were fully aware that their own particular conflicts with France took place within a much broader European context. But what exactly did Europe mean to these early modern Europeans?

Existing scholarship suggests that thinking about Europe underwent a substantial transformation in the early modern period when the long process of defining Europe primarily as a moral and cultural rather than geographic entity reached its culmination.[7] Indeed, the terms "Europe" and "European" appear to have been used more frequently in the seventeenth and eighteenth centuries than in any other historical period.[8] Several scholars note a fulcrum of change centered in the decades surrounding 1700—a period corresponding with the second half of Louis's reign that witnessed the largest of his wars.[9] The concept of Europe became a nearly ubiquitous presence in European peace treaties (notably excluding those involving France) starting in the 1690s.[10]

An explicitly public discussion in Europe's burgeoning pamphlet and periodical literatures drove these developments. Historians have long noted the late seventeenth and early eighteenth centuries as a high point in the history of European political pamphlet production. The foreign policy discussions that dominated these pamphlets focused largely on the figure of Louis.[11] Jean Schillinger has cited roughly 300 anti-Louis pamphlets published in the Holy Roman Empire alone.[12] A similar explosion took place in Europe's political periodical press, which like its pamphlet counterpart was obsessed not only with Louis, but also with the concept of Europe. Many of the continent's new periodicals carried the terms "Europe" or "European" in their titles.[13] Both pamphlets and periodicals drew from a common stock of ideas that, despite the efforts of censors, flowed more or less freely across Europe and encouraged a broad public conversation that was in many respects about Europe. Scholars like James van Horn Melton follow the lead of Benedict Anderson in highlighting how the growing sphere of public print culture "transcended locally embedded identities and thereby made it possible for people to imagine themselves as members of a larger political community," an integrative process that contributed to the rise of nationalism.[14] I would suggest that public discussion in the era of Louis XIV could and did also lead to an expanded and more clearly articulated sense of a European political community, a parallel integrative process that ran alongside the rise of national awareness with neither negating the other.

Early modern governments were acutely aware that something important was taking place in connection to these public conversations—hence their intense efforts to participate in and influence them. This was true regardless of regime type, with absolutist monarchies as involved as more constitutionally-limited principalities and even republics in efforts to guide public discussion.[15] Louis's France in fact was one of the most sophisticated and insistent state influencers, and went to tremendous lengths to craft the king's public image for both domestic and foreign audiences.[16]

The efforts of French propagandists, artists, and diplomats to portray Louis as the solar center around which the rest of Europe orbited were met with biting satire by his opponents, who delighted in inverting French imagery (the eclipse motif making frequent appearances) and in sarcastic mockery of the gaping distance between glittering French pretensions and the sober and often ugly reality of an aging monarch suffering from embarrassing ailments and presiding over decreasingly successful armies as his long reign ground forward and his own physical body deteriorated.[17] Beneath the invective that found bursts of catharsis in a cutting turn of phrase or a moment of *Schadenfreude* ran a deeply serious discussion of how Louis's transgressions violated an evolving sense of what it meant to be European.

This sense of European identity was constructed in relation to warfare in general and to Louis's wars in particular. Peter Burke has noted the centrality of military pressure, and especially conditions of invasion, to articulations of the term "Europe" going back to antiquity.[18] Burke's structuralist reading of identity congealing in opposition to a (military) other resonates with theories of interactive identity formation that argue that the definition of Europe continues to this day to be constituted through cross-border interactions with those lying outside Europe's embrace.[19] What happens, then, when "the other" lies within?[20] The remainder of this chapter will argue that those fighting against Louis came to describe him as the great internal other—the center of European politics, diplomacy, culture, and courtly life whose transgression of communal norms made him no European at all, and thus the perfect foil against which the definition of Europe could be constructed. The undeniable fact that Louis was both an imminent (perhaps existential) threat to Europe's polities and also most definitely *not* an external threat meant that he could neither be easily shrugged off nor blithely explained away as an exotic outsider.[21] He rather presented a serious and immediate concrete problem that required substantial mental work to reckon with. Processing Louis though the language of heinously violated norms cast those norms in a new light. Doing so within the context of major ongoing wars with profound real-world stakes helped reify the norms as something more meaningful than mere abstraction, but rather as constitutional elements of an altered understanding of Europe.

What Did Europe (Come To) Mean During the Wars of Louis XIV?

The European press played a central role in producing "the remarkably uniform ideology of opposition to Louis XIV that underlay the anti-French coalitions of the later wars."[22] Over the course of a full half-century, authors from across Europe produced a polyglot corpus of texts of remarkably varying scale, tone, and genre in resistance to the French king. The diversity of these writings in Dutch, English, French, German, and Latin did not prevent them from sharing an underlying message built from a body of unifying themes. A first and perhaps most funda-mental point in common dealt with labeling Louis a tyrant. Though attacking one's enemy as tyrannical is hardly a surprising tactic in wartime polemic, the way in which the authors framed their attacks provides a first window into the connec-tions between criticisms of Louis and larger understandings of Europe.

In his sprawling anti-French volume *t'Verwerd Europa* [Weather-Beaten Eu-rope] (1675), the Dutch lawyer and diplomat Pieter Valkenier argued that some, most notably "the peoples of *Europe*," were inclined by their very nature "to love golden liberty," whereas others, namely the peoples of Asia and the African Moors, were "servile by nature." The one exception to the intrinsic European hatred of tyranny and servility were the French, who "like donkeys willingly pulled the weight of an all-powerful ruler" and reveled in their own servility to a mode of power that Valkenier went out of his way to code as oriental and which he argued "had no place" in freedom-loving Europe.[23] A chorus of voices from all quarters shared Valkenier's condemnation of Louis's tyrannical power. From England, Gilbert Burnet wrote that "[w]hen I have named France, I have said all that is necessary to give you a complete idea of the blackest tyranny over men's consciences, persons and estates, that can possibly be imagined."[24] Scores of German pamphlets echoed Valkenier's equation of Louis's tyranny to oriental despotism.[25]

Some were more sympathetic to the plight of the French people than Valk-enier himself had been, placing them front and center on the list of Louis's victims. The author of one German tract railed against "the inhuman cruelty this monarch has used his entire life long both within and outside his king-dom against both his subjects and other peoples he has subjected, whom he has burned, murdered and devastated without distinction," adding that arms featured in French royal iconography "displayed the ruthlessness with which he sup-presses his people and the tyranny he has inflicted on so many innocent subjects whom he has trampled underfoot."[26] Another German pamphlet (whose subtitle stated that it was written "to give good warning to an almost shackled *Europe*") agreed that "one can say with truth that France makes no distinction between its subjects and its neighbors… it plunders and robs the one as well as the other whenever the opportunity presents and whenever it pleases."[27] A satirical pam-phlet from 1707 that described with mocking wit funeral ceremonies for Louis's failed designs at universal dominion went out of its way to depict the poverty and suffering of the French countryside, showing how the French people had paid dearly for their monarch's boundless ambition.[28] Indeed, authors from across Europe were united in calling for the liberation of the French people. As a German pamphlet of 1690 declared, "an all-too-powerful king is a tyrant to his own subjects. Not only [the French] townspeople and peasants, but indeed all of the sovereign Parlements of the kingdom, all princes, great lords, and noblemen, and especially the clergy, sigh for their deliverance."[29] The French, though bitter wartime enemies, were nevertheless Europeans whose subjection to a regime of

Louis's nature was unnatural and left them in need of rescue. This image of the breadth of French society sighing for freedom from Louis's tyranny in fact resonated perfectly with the language of French resistance pamphlets, including the famous *Les Soupirs de la France esclave, qui aspire après la liberté* [The Sighs of Enslaved France, Longing for Liberty] published in fifteen installments from August 1689 to September or October 1690.[30] Historians of French thought have in fact long noted that through domestic opposition to Louis's arbitrary rule, the terms "despot" and "despotism" emerged into continuous public use, denoting not just a flawed regime but rather a type of regime that should not exist.[31]

Both Louis's foreign and domestic opponents therefore described his rule as unnatural and in important ways un-European.[32] Pamphleteers linked these ideas closely to resistance to universal monarchy, by far the most ubiquitous theme in anti-Louis publications.[33] Authors from across Europe stressed that designs on universal dominion had now shifted from Habsburg Spain to Louis's France, contributing to what some scholars have described as a new French "black legend."[34] Regardless of their land of origin, virtually every author to write against Louis since the late 1660s decried his insatiable appetite for conquest.[35] Louis's drive to conquer was an obsession without limit, and would be satisfied only when he had placed himself "on the throne of Europe."[36] Author after author stressed the audacious scope of Louis's designs that threatened all of Europe. The German lawyer and diplomat Philipp Wilhelm von Hörnigk (1640–1714), for instance, examined the seemingly limited policy of French territorial aggrandizement along the Rhine known as *réunions* that argued that territories that had once been under French power remained the rightful property of the French crown in perpetuity. Hörnigk argued that if *any* part of the French justifications for this policy would be taken to their logical conclusions, they would lead inevitably to France gobbling up not only the entire Holy Roman Empire, but also all of Italy, Switzerland, the Dutch Republic, Denmark, and Poland on a march toward universal monarchy.[37]

Authors sounding the alarm against French designs for universal monarchy shared a particularly vitriolic reaction to Antoine Aubéry's *Des iustes pretentions du Roy sur l'Empire* [The Just Pretentions of the King to the Empire] (1667). Aubéry, a French historian and *avocat* at the parlement of Paris, wrote in fawning terms of a French royal power that he claimed stood in direct line of descent from Charlemagne, thus entitling the French king to Charlemagne's full dominions as the "universal king" of "all of Europe" and the "absolute monarch of the world."[38] An anonymous Latin summary of the book's contents that sought to expose Aubéry's claims to the European powers as a hideous monstrosity quickly followed.[39] The continental tidal wave of resistance to Aubéry began in earnest, however, with the publication of the Habsburg diplomat Franz Paul von Lisola's (1613–74) *Bouclier d'estat et de justice, contre le dessein manifestement découvert de la Monarchie Universelle, sous le vain pretexte des pretentions de la Reyne de France* [The Buckler of State and Justice Against the

Manifestly-Discovered Design for Universal Monarchy, Under the Vain Pretext of the Pretentions of the Queen of France] (1667).[40] Lisola became the single most influential anti-Louis polemicist, and the *Bouclier* was his most influential text. It resonated deeply with Louis's enemies across the Empire, the Dutch Republic, and Britain and helped form a continental-wide image of the king "as a rapacious aggressor bent on universal monarchy."[41] Lisola sparked a discussion of resistance to French designs for universal monarchy that stretched across Europe and contained a great deal of thinking about Europe.

For Lisola and other authors, resistance to universal monarchy was equivalent to resistance to collective enslavement. A committee of the British House of Commons writing in the immediate aftermath of the coronation of William and Mary claimed that "all Europe in general" was endangered by Louis's plan for universal monarchy "which threatens all Christendom with no less than absolute slavery."[42] The Dublin Protestant clergyman William King agreed that Louis's "[d]esign was universal, and aimed at the destruction and enslaving all the Kingdoms and States of *Europe*."[43] A German pamphlet of 1689 made clear that peace on the terms sought by Louis's France would in fact be "servitude."[44]

Just as subjection to domestic tyranny was portrayed as an unnatural state for Europeans, the slavery threatened by universal monarchy illuminated a core principle of collective European identity through its violation. Innumerable anti-Louis texts argued that freedom-loving Europe as a whole shared an organic aversion to dominance by a single ruler. Authors incessantly invoked the need to defend the "Liberties of Europe"—a concept with a wide and varying embrace but that in any constellation encompassed resistance to universal monarchy.[45] Louis's opponents from all across Europe had long since made recourse to a language of collective liberty a central element in their polemical resistance to the Sun King. In the Dutch Republic, William III had been using the phrase since the early 1670s to characterize the struggle against Louis, and well into the eighteenth century the Republic's government stressed that they did not wage war for honor, glory, or conquest, but rather for the "liberty of Europe."[46] Out of England, Archbishop of Canterbury John Tillotson declared that the war against France was being fought "in the public cause of the rights and liberties of almost all Europe" for "the vindication of the common liberties of mankind, against tyranny and oppression."[47] Sir Richard Cocks added that Louis must be resisted "to preserve the liberties and properties of Europe against the monstrous, avaricious and barbarous tyranny of the French King."[48] German authors mixed ceaseless references to the particular German liberties endangered by Louis with allusions to the common cause they shared with all of Europe to defend their collective liberties.

One way for Europe to defend its imperiled liberties was to present a united front against France. The *Wahre Abbildung* [True Representation] presented Louis as a raving lion seeking to devour the other European powers, and called for "all of Europe to finally stir and pull together in complete union to take up arms together in order to check such a powerful enemy."[49] Another German pamphlet spoke confidently that Louis could be readily defeated if "the princes of *Europe*" were to cast aside the bonds of fear that have held them, for "if you resist the Devil, he will flee from you, but if you fear him, he will bring you under his yoke."[50] Other pamphlets brimmed with confidence of the nearly limitless potential that Europe, "the Queen of the World," could hold if it were to "concentrate its power."[51] The Habsburg secretary and librarian Eberhard Wassenberg stated plainly that Louis's behavior "had left him few friends in all of Europe," and that he would succumb to the effects.[52]

Woven into the anti-Louis polemic, therefore, was a suggestion of collective European responsibility to ensure mutual security through cooperation against shared threats. Alongside this ran a growing commitment to balance-of-power thinking as another avenue to preserve the liberties of Europe. Scholars have long pointed to the wars of Louis XIV as a turning point in the long history of Western balance-of-power discourse.[53] Several single out Lisola as the cornerstone figure in Europe-wide debates over balance during Louis's wars.[54] Some pamphlet authors intimated that French designs for universal monarchy could only have arisen in the first place through a kind of systemic failure where the individual European states were too disjointed and dysfunctional to join together to preserve balance.[55]

Implicit to arguments that European powers shared interlocking security obligations and a collective imperative consciously to maintain balance was an image of Europe itself as a system of multiple sovereignties.[56] One of the strongest elements of the language of Europe to come to the fore during Louis's wars was in fact the intrinsic value of pluralism as not only healthy *for* Europe but as definitional *of* Europe. The equation of Europe with political pluralism stretched back at least as far as Machiavelli and had become even stronger by the time of Leibniz and Montesquieu, whose critique of Louis XIV in the *Lettres persanes* (*Persian Letters*) was grounded in a defense of pluralism as prototypically European.[57] The conclusion of the final of Louis's wars in the Peace of Utrecht contained a full embrace of the idea of Europe as a whole made out of autonomous parts held together by balance.[58] As Pincus has written, the struggle against Louis "pitted the entire community of European nations" versus the gravest threat to the pluralism that was Europe in "a universal defense of particularism."[59]

For Louis's European contemporaries, the defense of pluralism was as much a spiritual as it was a political, military, and diplomatic enterprise. As one author declared:

> The lust for conquest and self-interest of the king of France is a river whose course neither blood relation, nor alliances, nor peace treaties, nor armistices, nor mutual oaths can stop. To go further: not even the boundary stones themselves [can stop him,] which God in His wise prudence has set on the borders of every monarchy as if to declare *Non plus ultra*.[60]

The borders that embodied European pluralism were thus a gift of Divine Providence, and their violation by a would-be universal monarch a crime against God. For some anti-Louis authors, the sacred mandate to preserve European pluralism appears to have included the necessity to defend religious pluralism as well, at the very least in the form of robust inter-confessional cooperation for a shared spiritual goal. Claydon has written of the widespread belief during the wars of Louis XIV in the inherently anti-Christian nature of universal monarchy, and the moral and spiritual moorings upon which the balance of power thinking consequently rested.[61] An avalanche of publications condemned Louis for seeking to sow dissension among his European neighbors—the work of the Antichrist, of whom Louis was the firstborn son, wrote Valkenier—along religious lines.[62]

Louis's opponents responded to his un-Christian and un-European project to divide and conquer through religious tension by breathing new life into the very old concept of Christendom as the marker of European spiritual unity. Christendom and Europe had long been closely linked if never fully synonymous terms.[63] Heinz Duchhardt has argued that the wars of Louis XIV played a key role in recentering images of Western solidarity from the language of Christendom to that of Europe, since the former could not serve as the focal point for resistance to the Sun King.[64] A careful reading of contemporary publications, however, lends support to Claydon's argument that Christendom remained a remarkably potent and indeed foundational element in European thought on the struggle against French universalism in ways that were both spiritual and cultural.[65] Throughout the period of Louis's wars, European authors grafted their new language of Europe onto an underlying understanding of Christian unity that was explicitly ethical and non-confessional, building a vision of Europe as a spiritually united yet fundamentally cultural entity anchored in common values and standards of conduct that transcended denominational fault-lines. Scores of pamphlets presented Louis as the enemy of all Christians, a merciless shedder of Christian blood who cynically cloaked his evil designs behind an ostentatious veneer of piety and whose absolutist self-representation was nothing short of blasphemous idolatry.[66] He was an atheist; an apostate and idolater; the scourge of God; the disciple of Lucifer; the Antichrist himself.[67] He was, as William III's declaration of war declared, "the Common Enemy of the Christian World."[68] Wars against him were wars "for the common good of Christendom."[69] Due to Innocent XI's well-known conflicts with Louis, authors from across the religious spectrum—Protestant as well as Catholic—could write without irony of the papacy, an institution so profoundly divisive over the previous two centuries, as a rallying point for a common European resistance to the dark forces threatening to rend Christendom asunder.[70] A Protestant sermon delivered to the British House of Commons openly cited the pope's statement identifying Louis as the "common enemy of West-Europe" with great approval.[71]

The collective struggle against Louis therefore affirmed a definition of Europe as a spiritual and cultural as well as political entity built upon the preservation of pluralism and shared norms that transcended confessional divides. Louis's tyrannical behavior and un-Christian striving for universal monarchy placed him outside the political and moral/spiritual community of Europe, and he in turn served as the perfect "other" against whom his opponents could articulate a clearer sense of what Europe meant. This use of Louis was cemented by a final recurrent motif in wartime polemic: references to the king as the internal Turk, the great existential threat to European Christendom from within. Discussion of the Ottomans saturates wartime anti-Louis polemic. Pamphleteers ruthlessly and relentlessly pointed to Louis's alliance with the Turks as conclusive proof of his complete depravity.[72] Some portrayed the Ottomans as Louis's guard dogs, ready to be unleashed upon Europe at his behest.[73] Others were especially savage in their critique of Louis's inaction, or, even worse, aggressive opportunism, during the 1683 siege of Vienna, when the rest of Christian Europe had gloriously rallied against the great arch-enemy of the West.[74] As one pamphlet noted, once the siege was lifted "the universal joy of *Europe*" knew no bounds as celebratory fires were lit and all voices were raised in a collective "*Te Deum Laudamus* that rang out across the air" of Christendom while "no one other than France sat silent," marking Louis's regime as both un-Christian and un-European.[75]

Even more damning were those pamphlets that drew direct parallels between Louis and the Turks as not just allies, but rather as regimes that were similar by nature. Many attacked Louis for practicing Turkish-style politics according to Turkish maxims.[76] Gottfried Lange wrote that Louis had in fact invited the Turks to serve as godparents of his envisioned French universal monarchy.[77] A German pamphlet of 1690 declared that "one can rightly say that as Sultan Mehmet was a persecutor of Christendom in the Orient… so has France earned the title of a persecutor of Christendom in the Occident."[78] Another focused on Louis's forces plundering and pillaging "in Turkish style," and argued that "while the king has taken the title of 'Most Christian'" he had actually taken the name "of his good friend and ally the Great Sultan," with each producing Christian martyrs and defiling Christian virgins and holy places.[79] The *Wahre Abbildung* wrote of Louis working with the Ottomans as they led Christians off into slavery and transformed churches into mosques.[80] Some authors went so far as to detect elements of Islam in his policies, and to argue that Louis himself must be Muslim.[81] Even burlesque attacks on Louis's sexual immorality drew parallels to the sultan's harem.[82] A Dutch pamphlet of 1688 identified Gog and Magog from the Book of Revelations with the French *fleur-de-lis* and the Turkish crescent moon respectively.[83] Louis, an oriental despot at home and aspiring universalist abroad, was the ultimate internal other, the antithesis of what it meant to be European.[84]

Conclusion: The Wars of Louis XIV and the Language of Europe

Taken collectively, those speaking and writing against Louis XIV throughout the decades covered by his wars presented an image of Europe as a naturally freedom-loving and pluralistic system of polities bound together in a political and moral/spiritual community that could still be described as Christendom and whose health was to be preserved through mutual security obligations and the conscious balancing of power. As stated above, not one of these elements was original to the time period of Louis's reign. The language of Europe put forth by his opponents was instead cobbled together from raw materials long at hand under conditions of extremity and immediate necessity. The crisis of Louis's wars accelerated lines of thinking already under development, raised the profile of certain ideas and brought them more clearly to the fore, and above all synthesized disparate elements together into an increasingly coherent whole. While no single piece of this wartime language of Europe represents a radical departure from older streams of Western thought, the overall discussion of what Europe meant had become decidedly different by the end of Louis's reign than what it had been at its beginning.

Europe had come to mean something subtly yet importantly different throughout the course of the wars of Louis XIV. Europeans from across the continent engaged in substantial intellectual work to construct a language of Europe that articulated what Louis's exceptionally diverse array of opponents held in common and what they were collectively defending in resisting him. Recognizing these things helps explain why Europeans living through the last decades of the seventeenth century and the first decades of the eighteenth century felt the compulsion that scholars have long noted to discuss Europe with such unprecedented frequency. The imperative to preserve wartime solidarity provided the necessity, and Louis himself the perfect vehicle—though as Europe's negative impression. Louis's opponents formed their image of what they had in common and what they were fighting for against the anti-image of whom they were fighting against. It was a logical, if unintentional, step from describing what Europe was not, to realizing what Europe might actually be.

Notes

1   The conflicts often referred to as the wars of Louis XIV include the War of Devolution (1667–68), the Franco-Dutch War (1672–78), the Nine Years' War (1688–1697), and the War of the Spanish Succession (1701–14). In each conflict, and especially the latter three, France faced a coalition of opponents.

2   One excellent recent study that does deal directly with anti-Louis rhetoric in the explicit context of conceptualizations of Europe is Nils Grüne and Stefan Ehrenpreis, "Liberty and Participation: Governance Ideals in the Self-Fashioning of Sixteenth- to Early-Eighteenth Century Europe," in *Contesting Europe: Comparative Perspectives on Early Modern Discourses on Europe, 1400–1800,* eds. Nicolas Detering, Clementina Marsico, and Isabella Walser-Bürgler (Leiden: Brill, 2020), 275–316. Discussions of Louis, however, form only a portion of Grüne and Ehrenpreis's study.

3   Steve Pincus, *1688: the first modern revolution* (New Haven: Yale University Press, 2009), 307.

4   The literature on the rhetoric of resistance to Louis is large and contains a number of excellent works, the majority of which focus primarily on a single national context. For a sampling of works centered on the Holy Roman Empire, see Hubert Gillot, *Le Règne de Louis XIV et l'Opinion Publique en Allemagne* (Paris: Édouard Champion, 1914); Johannes Haller, *Die Deutsche Publizistik in den Jahren 1668–1674: ein Beitrag zur Geschichte der Raubkriege Ludwigs XIV.* (Heidelberg: Carl Winter's Universitätsbuchhandlung, 1892); Alexandre Yali Haran, "Le dénigrement de la France en Allemagne à la fin du XVIIᵉ siècle, à travers les ouvrages d'expression française," *Histoire, Économie et Société* 15, no. 2 (April-June 1996): 203–19; Michael Rohrschneider, "'Holland kan die Tyranney Franckreichs nicht gnung beschreiben...' Die französisch-niederländischen Beziehungen 1672–84 im Spiegel antifranzösischer deutscher Flugschriften," *Zeitschrift für Geschichtswissenschaft* 56, nr. 2 (2008), 101–22; Jean Schillinger, *Les pamphlétaires allemands et la France de Louis XIV* (Bern, Switzerland: Peter Lang, 1999); Rohrschneider, "La Monarchie universelle française dans les pamphlets allemands entre 1672 et 1715," in *Littérature de contestation: Pamphlets et polémiques du règne de Louis XIV aux Lumières*, eds. Jean-Jacques Tatin-Gourier and Pierre Bonnet (Paris: Éditions Le Manuscrit, 2011), 89–116; and Martin Wrede, *Das Reich und seine Feinde. Politische Feindbilder in der Reichspatriotischen Publizistik zwischen Westfälischen Frieden und Siebenjährigem Krieg* (Mainz: Philipp von Zabern, 2004), esp. 330–483. For examples of works that focus on the Dutch Republic, see Donald Haks, *Vaderland en vrede 1672–1713. Publiciteit over de Nederlandse Republiek in oorlog* (Hilversum: Verloren, 2013); Haks, "Publieke opinie, buitenlandse politiek en het einde van de Spaanse Successieoorlog," *Tijdschrift voor Geschiedenis* 127, no. 4 (2014): 673–94; and P.J.W. van Malssen, *Louis XIV d'après les pamphlets répandus en Hollande* (Amsterdam: H.J. Paris, 1937). Important works on English history that include substantial discussion of anti-Louis polemic include Tony Claydon, *Europe and the Making of England, 1660–1760* (Cambridge, UK: Cambridge University Press, 2007), esp. 152–210; and Pincus, *1688*, esp. 305–65. For works on criticism of Louis within France itself, see Friedrich Kleyser, *Der Flugschriftenkampf gegen Ludwig XIV. zur Zeit des pfälzischen Krieges* (Berlin: Dr. Emil Ebering, 1935); Lionel Rothkrug, *Opposition to Louis XIV: the political and social origins of the French Enlightenment* (Princeton: Princeton University Press, 1965); Damien Tricoire, "Attacking the Monarchy's Sacrality in Late Seventeenth-Century France: the underground literature against Louis XIV, Jansenism and the Dauphin's court faction," *French History* 31, nr. 2 (2017), 152–73; and several essays in Tatin-Gourier and Bonnet. Among the comparatively small number of works that deal with a broader geographic scale are Pierre Bonnet, "De la critique à la satire: trente années d'opposition pamphlétaire à Louis XIV," *Bulletin de la Société de l'Histoire du Protestantisme Français* 157 (Janvier-Février 2011), 27–64; and Hendrik Ziegler, *Der Sonnenkönig und seine Feinde. Die Bildpropaganda Ludwigs XIV. in der Kritik* (Petersberg: Michael Imhof, 2010). The linguistic limitations of the author prevented engagement with Spanish-language anti-Louis pamphlets and secondary literature in this chapter.

5   Pincus, *1688,* 348.

6   A central theme in Claydon's *Europe and the Making of England* as well as Pincus's *1688.*

7   Federico Chabod, *Der Europagedanke; von Alexander dem Großen bis Zar Alexander I.*, trans. Stefan Burger (Stuttgart: W. Kohlhammer, 1963), 134. Other important

works on conceptions of Europe include August Buck, ed., *Der Europa-Gedanke* (Tübingen: Max Niemeyer, 1992); and Heinz Gollwitzer, *Europabild und Europagedanke. Beiträge zur deutschen Geistesgeschichte des 18. und 19. Jahrhunderts*, 2nd ed. (Munich: C.H. Beck'sche Verlagsbuchhandlung, 1964).

8 Klaus Malettke, "Konzeption kollektiver Sicherheit in Europa bei Sully und Richelieu," in Buck, *Europa-Gedanke*, 83–106, here 83.

9 The period around 1700 is the explicit focus of Peter Burke, "Did Europe Exist Before 1700?," *History of European Ideas* 1, no. 1 (1980): 21–29; and Fritz Wagner, "Europa um 1700–Idee und Wirklichkeit," *Francia. Forschungen zur westeuropäischen Geschichte* 2 (September 1974): 295–308. Wagner (295) claimed that "around 1700 there was no more frequently used word, no more widespread term in political, social, literary, and journalistic parlance, than Europe." See also Nicolas Detering, Clementina Marisco, and Isabella Walser-Bürgler, "Contesting Europe: Comparative Perspectives on Early Modern Discourses on Europe, 1400–1800–an Introduction," in Detering, *Contesting Europe*, 1–10.

10 Heinz Duchhardt, *Gleichgewicht der Kräfte, Convenance, Europäisches Konzert. Friedenskongresse und Friedensschlüsse vom Zeitalter Ludwigs XIV. bis zum Wiener Kongreß* (Darmstadt: Wissenschaftliche Buchgesellschaft, 1976), 28–29; Duchhardt, "'Europa' als Begründungs- und Legitimationsformel in völkerrechtlichen Verträgen der Frühen Neuzeit," in *Faszinierende Frühneuzeit. Reich, Frieden, Kultur und Kommunikation 1500–1800. Festschrift für Johannes Burkhardt zum 65. Geburtstag*, eds. Wolfgang E.J. Weber and Regina Dauser (Berlin: Akademie, 2008), 51–60; Duchhardt, "Europabewußtsein und politisches Europa–Entwicklungen und Ansätze im frühen 18. Jahrhundert am Beispiel des Deutschen Reiches," in Buck, *Europa-Gedanke*, 120–31, here 130. Duchhardt has noticed more than other scholars the deep connection between the wars of Louis XIV and thinking about Europe, though this connection has never stood at the absolute center of his studies, and he has drawn more heavily from official state documents than from pamphlets and periodicals.

11 Markus Baumanns, *Das publizistische Werk des kaiserlichen Diplomaten Franz Paul Freiherr von Lisola (1613–1674): ein Beitrag zum Verhältnis von Absolutischem Staat, Öffentlichkeit und Mächtepolitik in der frühen Neuzeit* (Berlin: Duncker & Humblot, 1994), 78.

12 See Schillinger, *Les pamphlétaires allemands*. Number cited in Andreas Brandtner, Franckreichs Geist (1689)*: Argumentatives Handeln in der Frühaufklärung* (Frankfurt am Main: Peter Lang, 2013), 12.

13 Heinz Duchhardt notes that the wave of periodicals with the term "Europe" in their titles "swelled to a hurricane" in the three decades centered around 1700, including titles such as the *Mercure historique et politique contenant l'Etat présent de l'Europe* (founded 1686); *Europäische Staats-Cantzley* (founded 1697); and *Die europäische Fama* (founded 1702). Duchhardt, "Europabewußtsein," 120.

14 James van Horn Melton, *The Rise of the Public in Enlightenment Europe* (Cambridge, UK: Cambridge University Press, 2001), 274.

15 On the need for absolutist regimes to justify themselves publicly, see Andreas Gestrich, "Politik im Alltag: zur Funktion politischer Information im deutschen Absolutismus des frühen 18. Jahrhunderts," *Aufklärung* 5, nr. 2 (1991): 9–27; and Gestrich, *Absolutismus und Öffentlichkeit: politische Kommunikation in Deutschland zu Beginn des 18. Jahrhunderts* (Göttingen, Germany: Vandenhoeck & Ruprecht, 1994).

16 Seminal works include Peter Burke, *The Fabrication of Louis XIV* (New Haven: Yale University Press, 1992); and Joseph Klaits, *Printed Propaganda Under Louis XIV: absolute monarchy and public opinion* (Princeton: Princeton University Press, 1976).

17 For examples of eclipse imagery, see Ziegler, *Sonnenkönig*, 34–42. On mockery of Louis's physical body, including maladies like his well-known anal fistula, see Pierre Bonnet, "Préface. Figures et configurations de la littérature politique de contestation du règne personnel de Louis XIV au premier XVIIIe siècle," in Tatin-Gourier, *Littérature de contestation*, 13–71, esp. 43–49.

18 Burke, "Did Europe Exist Before 1700?," 22–26.

19 Lars-Erik Cederman, "Political Boundaries and Identity Trade-Offs," in *Constructing Europe's Identity: the external dimension*, ed. Lars-Erik Cederman (Boulder and London: Lynne Rienner, 2001), 1–32.

20 See Grüne and Ehrenpreis, "Liberty and Participation," 298–307.

21 It is worth keeping in mind that many of the same European courts that hated, rejected, and "othered" Louis continued to wish desperately to emulate aspects of him. His magnetic force on other European regimes was simultaneously attractive and repulsive.

22 Klaits, *Printed Propaganda*, 20–22.

23 Petrus Valkenier, *t'Verwerd Europa, ofte Politijke en Historische Beschryvinge Der waare Fundamenten en Oorsaken van de Oorlogen en Revolutien in Europa, voornamentlijk in en omtrent de Nederlanden zedert den jaare 1664. gecauseert door de gepretendeerde Universele Monarchie der Franschen* (Amsterdam: Hendrik en Dirk Boom, 1675), 15–16. Italics in original. See also 62–63.

24 Gilbert Burnet, *A Sermon Preached before the House of Peers*, 5 November 1689 (London: Richard Chiswell, 1689), 29. Cited in Pincus, *1688*, 336.

25 See Schillinger, *Les pamphlétaires allemands*; and Wrede, *Das Reich.*

26 *Wahre Abbildung Des/ Durch die* Europaeischen *Potentaten/ unter Ludwig den XIV. Bekriegten Frankreichs* (Cölln: Pierre Marteau, 1690), 8.

27 Johannes Liberius, *Der bißhero künstlich Bedeckte Aber anitzo klärlich Entdeckte Und mit lebendiger Farbe vorgestelte Geist Von Franckreich/ Nebenst denen* Maximen *und Grundregeln wodurch König Ludewig der XIV. Die* Monarchie *und allgemeine Herrschafft über gantz* Europam *endlich zu Erlangen verhoffet. Dem fast angefesselten* Europa *zu guter Warnung ans Licht gestellet und zum Druck befordert* (Freystadt, 1689), C3r-v.

28 Gottfried Lange, *Kurtzer Entwurff und Beschreibung Was bey* solenner *Beerdigung Der fünfften Monarchie In dem* Castro-Doloris, *In der* Procession, *In der* Parentation &c. *zu Pariß merckwürdig vorgegangen* (Cologne: Peter Marteau, 1707), A6v.

29 *Wahre Abbildung*, 24.

30 The sizeable literature on the *Soupirs* often focuses on attribution of authorship. See, for example, Antony McKenna, "*Les Soupirs de la France esclave, qui aspire après la liberté*: la question de l'attribution," in Tatin-Gourier, *Littérature de contestation*, 229–68; Tricoire, "Attacking the Monarchy's Sacrality"; Guy Howard Dodge, *The Political Theory of the Huguenots of the Dispersion: with special reference to the thought and influence of Pierre Jurieu* (New York: Columbia University Press, 1947), 140–56; Kleyser, *Flugschriftenkampf*, 28–44; and Gotthold Riemann, *Der Verfasser der "Soupirs de la France Esclave qui aspire après la liberté" (1689–1690): ein Beitrag zur Geschichte der politischen Ideen in der Zeit Ludwigs XIV.* (Berlin: Dr. Emil Ebering, 1938).

31 See Rothkrug, *Opposition*; Phyllis K. Leffler, "French Historians and the Challenge to Louis XIV's Absolutism," *French Historical Studies* 14, no. 1 (Spring 1985): 1–22; R. Koebner, "Despot and Despotism: Vicissitudes of a Political Term," *Journal of the Warburg and Courtauld Institutes* 14, nos. 3 and 4 (1951): 275–302, here 293–302.

32 See Grüne and Ehrenpreis, "Liberty and Participation," 300–7, for a similiar argument with different emphases.
33 A fact noted by many scholars. See in particular Schillinger, "La Monarchie universelle française."
34 Everard Wassenberg, *Maroboduus In Serenissimo & Potentissimo Ludovico XIV. Galliarum Rege, redivivus, principus Europae demonstratus. Et, si esse perseveret, suo Arminio destinatus* (1672), 8; Liberius, *Der bißhero künstlich Bedeckte*, A4r; Valkenier, *t'Verwerd Europa*, 75; Charles-Édouard Levillain, *Le Procès de Louis XIV. Une guerre psychologique. François-Paul de Lisola, citoyen du monde, ennemi de la France* (Paris: Tallandier, 2015), 133–34; Wrede, *Das Reich*, 375; Pincus, *1688*, 313; Claydon, *Europe*, 189. On the French "black legend," see Haran, "Le dénigrement," 204; Schillinger, "La Monarchie universelle française," 94. On the re-purposing of Dutch artistic depictions of Spanish atrocities in anti-French works, see Haks, *Vaderland*, 38ff. On earlier French anti-Spanish polemic providing the tools that were later adapted and used by German publicists against Louis, see Schillinger, *Les pamphlétaires allemands*, 307.
35 German texts brim with references to Louis's "Regiersucht" or "Ländersucht"; Latin texts to his "libido dominandi." Valkenier's Dutch text speaks of his "Heers-sucht."
36 Liberius, *Der bißhero künstlich Bedeckte*, D4r.
37 [Philipp Wilhelm von Hörnigk], *Francopolitæ Wahrer Bericht von dem alten Königreich Lothringen/ Und klarer Beweiß/ Daß die Frantzösische von denen Carolinischen Fränckischen Königen anmaßlich hergeführte Sprüch auff die Uber=Rheinische Reichs=Länder allerdings nichtig und untüchtig seyn. Ferner auch Wann ihnen einige Krafft zugelegt werden solte/ sie alsdann zugleich das gesamte disseitige Teutschland nebenst Italien/ Schweitz und denen Vereinigten Niederlanden nach sich ziehen müsten* (1682).
38 Antoine Aubery, *Des ivstes pretentions dv Roy svr l'Empire* (Paris: Antoine Bertier, 1667), 159.
39 *Chimæra Gallicana Continens Axiomata Politica Imperij Gallicani deducta ex tractatu Des justes pretentions du Roy sur l'Empire. Par le* Sieur Aubery *Advocat au Parlament, & aux Conseils du Roy. Monstrum horrendum informe ingens cui lumen ademptum. Regibvs, Principibvs et Potestatibvs* (Paris: Antoine Bertier, 1667). An identical text with different title, title page, and pagination was published as *Axiomata Politica Gallicana, Ex. Dn. Aubery, Advocati Parlamenti Parisiensis & Consiliarij Regij Tractatu, quem de Justis Prætentionibvs Regis super Imperium, et de Præeminentia Regis Svper Imperatorem* [1667?], with the title page stating that the text "tùm inprimis Germanicæ Nationi ad considerandum proposita."
40 [Lisola], *Bouclier d'estat et de justice, contre le dessein manifestement découvert de la Monarchie Universelle, sous le vain pretexte des pretentions de la Reyne de France* (1667).
41 Quotation from Klaits, *Printed Propaganda*, 23. On Lisola, see Baumanns, *Das publizistische Werk*; and Levillain, *Le Procès de Louis XIV*. The *Bouclier* went through at least eight printings in the original French from 1667–1701 and was translated into six languages, bringing the total number of editions to at least twenty. Baumanns, *Das publizistische Werk*, 165. Claydon has written that the English translation of the text "was an instant success, and it burned an image of French universal monarchy onto the public mind." Claydon, *Europe*, 156. Donald Haks similarly traces the Dutch language of resistance to Louis's drive to universal monarchy to a line of argumentation stretching back to Lisola's *Bouclier*. Haks, *Vaderland*, 66.
42 Quoted in Pincus, *1688*, 340.

43   William King, *Europe's Deliverance from France and Slavery: a sermon preached at St. Patrick's Church, Dublin, On the 16th of November, 1690* (London: Tim. Goodvin, 1691), 9. Italics in original.

44   *Entdeckung Der listigen Kunst-Stücke Womit Die Franzosen die Catholische und Protestirende Stände an einander zu hetzen gedencken/ auff daß Sie durch ihre Trennung endlich allein herrschen/ und in gantz Europa die Meisterschafft und Oberhand behalten mögen* [1689], A2v.

45   Duchhardt has noted a shift in emphasis from "tranquilitas" to "libertas" in the language of European peace treaties that occurred by the end of the era of the wars of Louis XIV. Duchhardt, "Europa' als Begründungs- und Legitimationsformel," 56–57.

46   Generale Petitie of 1706, quoted in Haks, *Vaderland*, 70.

47   John Tillotson, *A Sermon Preached before the King and Queen at Whitehall, 27 October 1692* (London: Brabazon Aylmer, 1692), 25, 32. Cited in Pincus, *1688*, 343.

48   Cited in Pincus, *1688*, 354.

49   *Wahre Abbildung*, 4.

50   Liberius, *Der bißhero künstlich Bedeckte*, C4r, K2v.

51   *Curiöse Staats=Vorstellungen/ Uber Den gegenwärtigen Zustand in Europa/ dessen Reiche/ Republiqven/ und Häupter/ deren ietziges Staats=*Interesse, Dessein und *Absehen* (Cölln: Peter Marteau, 1701), 1.

52   Wassenberg, 24.

53   Duchhardt, *Gleichgewicht der Kräfte*, 68–69; Duchhardt, "Europabewußtsein," 122–23. Claydon argues that "from the mid-1690s the 'balance of power' became the key discourse in discussion of England's foreign relations." Claydon, *Europe*, 194.

54   See, for example, Wrede, *Das Reich*, 386–87; Duchhardt, "Europabewußtsein," 122.

55   See, for example, Lange, *Kurtzer Entwurff*, B5r-B6r.

56   Pincus argues convincingly that envisioning the wars of Louis XIV as the defense of a Europe of multiple (he would say national) sovereignties was central to English Whig thought.

57   Chabod, *Der Europagedanke*, 67, 71ff.; Gollwitzer, *Europabild*, 44–45.

58   Wagner, "Europa um 1700," 296–97.

59   Pincus, *1688*, 339.

60   Liberius, *Der bißhero künstlich Bedeckte*, H3r-v.

61   Claydon, *Europe*, 199–200.

62   Valkenier, *t'Verwerd Europa*, 66. For a sampling of pamphlets that condemned Louis's stoking of religious tensions, see *Entdeckung Der listigen Kunst-Stücke*, passim; Liberius, *Der bißhero künstlich Bedeckte*, C4r, I3r; Valkenier, *t'Verwerd Europa*, 169, 196–97; *Wahre Abbildung*, 34.

63   Gollwitzer, *Europabild*, 25ff.

64   Duchhardt, "Europabewußtsein," 121–22.

65   The enduring power of Christendom as a concept is one of the central arguments in Claydon's *Europe and the Making of England*.

66   To give only singular examples of these very widespread tropes: on the shedding of Christian blood, see *Der Christlich=Teutschen Wahrheit gehabte Audienz, Bey dem Aller=Christl. König Ludwig dem XIV. zu Versailles am Tage des H. Apostels Thomæ, den 21. Decembr. 1689. Worinnen Im Nahmen deß Christl. Europæ, der König seines bißherigen un=Christlichen Verfahrens erinnert/ und deßwegen Rechenschafft von ihm begehrt wird* (Freyburg, 1690), B2r-v; on the cynical use of religion as a cover for worldly designs, see *Entdeckung Der listigen Kunst-Stücke*, C1r; on Louis's self-representation as idolatry, see *Wahre Abbildung*, 3–4, 31.

67   Schillinger, *Les pamphlétaires allemands*, 190ff.; Wrede, *Das Reich*, 480–82; Haks, *Vaderland*, 128; Haran, "Le dénigrement," 219.
68   *Their majesties declaration against the French king, 7 May, 1689*. Cited in Claydon, *Europe*, 187.
69   A phrase used multiple times in William's instructions to his diplomats. Cited in Pincus, *1688*, 342.
70   *Christlich=Teutschen Wahrheit*, B3vff.; *Wahre Abbildung*, 12–16, 34–35; Liberius, *Der bißhero künstlich Bedeckte*, C4r; *Entdeckung Der listigen Kunst-Stücke*, A4r.
71   Thomas Tenison, *A Sermon against Self-Love* (London: Richard Chiswell, 1689). Cited in Pincus, *1688*, 342. Pamphleteers also went to great lengths to depict Louis's atrocities against Catholic populations and the Catholic clergy. See, for example, *Der Vermeinte/ Und von Franckreich erdichtete/ Religions=Krieg. Worinnen enthalten Was eigentlich ein Religions-Krieg seye? Wenn dergleichen in Europa geführet worden? Und ob gegenwärtiger Krieg davor zu halten?* (Bonn, 1689), 69–70; *Christlich=Teutschen Wahrheit*, C1r-v; *Wahre Abbildung*, 34–35.
72   Schillinger, *Les pamphlétaires allemands*, 491–552; Wrede, *Das Reich*, 474–83. Some pamphleteers pointed to an even longer history of French cooperation with the Ottomans dating back to François I. See, for example, Valkenier, *t'Verwerd Europa*, 44.
73   Schillinger, *Les pamphlétaires allemands*, 510–12. Louis's support for the Hungarian rebels fighting against Habsburg rule was also frequently depicted as benefitting the Turks. See, for example, *Franckreich schäme dich! Das ist: Heimlich und unverhoffte Entdeckung derer Französischen/ fast an allen* Europæischen *Höfen geschmiedeten/ und sehr übel gelungenen Rathschlägen lesens=würdig der* Curieusen *Welt vor Augen gestellt* (1685), E3v–E4r.
74   *Wahre Abbildung*, 5–6, 10–11; Liberius, *Der bißhero künstlich Bedeckte*, E1r–E2r; Haran, "Le dénigrement," 216–219; Claydon, *Europe*, 177–79.
75   Liberius, *Der bißhero künstlich Bedeckte*, E2r.
76   Schillinger, *Les pamphlétaires allemands*, 466–75; Rohrschneider, "Holland," 116, 118.
77   Lange, *Kurtzer Entwurff*, B7r.
78   *Christlich=Teutschen Wahrheit*, E3r.
79   Liberius, *Der bißhero künstlich Bedeckte*, E1v.
80   *Wahre Abbildung*, 14–15. The same text (19–20) wrote of Louis's peace with the Algerian corsairs, "welche sich allein vom Christen = Blut nähren," in similar terms.
81   Claydon, *Europe*, 181.
82   Schillinger, *Les pamphlétaires allemands*, 140.
83   Audax Philalethes, *De Lydsaamheid en het Gelove der Heiligen. Onder so vele sware en Bittere Vervolgingen, Die Gods Kerke nun alomme moet uitstaan, en wel meest in Vrankryk* (Amsterdam, 1688). Cited in Wrede, *Das Reich*, 478. Imperial wartime imagery often represented Louis and the Turks together as the sun and the moon respectively. See Ziegler, *Sonnenkönig*, 65.
84   For additional examples, see Grüne and Ehrenpreis, "Liberty and Participation," 302–6.

Further Reading

Claydon, Tony. *Europe and the Making of England, 1660–1760*. Cambridge, UK: Cambridge University Press, 2007.
Detering, Nicolas, Clementina Marsico, and Isabella Walser-Bürgler, eds. *Contesting Europe: Comparative Perspectives on Early Modern Discourses on Europe,*

1400–1800. Intersections: Interdisciplinary Studies on Early Modern Culture, vol. 67. Leiden: Brill, 2020.

Pincus, Steven C.A. *1688: the First Modern Revolution*. New Haven: Yale University Press, 2009.

Schillinger, Jean. *Les pamphlétaires allemands et la France de Louis XIV*. Série II– Gallo-germanica, vol. 27. Bern, Switzerland: Peter Lang, 1999.

Wrede, Martin. *Das Reich und seine Feinde. Politische Feindbilder in der Reichspatriotischen Publizistik zwischen Westfälischen Frieden und Siebenjährigem Krieg*. Veröffentlichungen des Instituts für Europäische Geschichte Mainz, vol. 196. Mainz: Philipp von Zabern, 2004.

7 Calvinism and the Thirty Years War

Abraham Scultetus and the Palatinate

Howard Louthan

The connection between religion and violence is one of the most fraught scholarly questions today. From the Crusades of the eleventh and twelfth centuries to our contemporary moment of terrorism and the Islamic State, scholars from across the disciplinary spectrum have invested significant energy seeking to understand the complicated relationship between the two phenomena.[1] Specialists of early modern Europe have quite naturally focused on the Reformation and the impact of these religious changes on the society and culture of the day. Historians such as Natalie Davis helped pioneer the field by borrowing from the toolkits of anthropologists and sociologists to analyze moments of violence between rival confessional groups. Over the years there has been growing interest in Reformed communities in particular. This was a tradition that in some respects could trace its origins to the battlefield with the story of a heroic Zwingli dying as a soldier. Scholars have often contrasted what they perceived as the more quiescent nature of Lutheranism, the reluctance of its princes to defy a Catholic sovereign openly, with a Calvinism that was restless, restive, and at times even revolutionary. Christine Kooi has recently observed, "By the second half of the sixteenth century Calvinism and war were, it appeared, yoked together. Wherever Calvinism planted its metaphorical flag in early modern Europe, disruption and disorder seemed to follow."[2]

While it may be tempting to search for a causal link between Calvinism and violence, the situation is actually far more complicated. Kooi continues by reminding us that Calvinists were at times both model citizens submitting to the laws of their societies and violent rebels challenging the regimes under which they lived. Context is all-important, and we should be cautious before making any broad pronouncements on a topic as thorny and complex as religion and violence. Still, it is hard to deny that Calvinism contributed in some way to the conflicts of the period, especially when considering the greatest of these catastrophes, the Thirty Years War. By accepting the throne of Bohemia, the Calvinist elector of the Palatinate, Frederick V, helped trigger a series of cascading events that led to the rapid expansion of hostilities and its tragic consequences. More generally, political Calvinism immediately before the war was cresting

DOI: 10.4324/9781003157700-10

in Central Europe. In 1620, two of the four secular electors of the Empire were Calvinists. Five dukes in Silesia, one in Anhalt, the landgrave of Hesse-Cassel, and seventeen imperial counts alongside several cities also belonged to the Reformed camp.[3]

But was Calvinism, especially in its political manifestation, a coherent and cohesive movement? Was it more than simply the sum of its parts? Can we draw broader conclusions simply from the actions of individuals such as Christian von Anhalt who shaped the aggressive policies of the Protestant Union or Georg Erasmus von Tschernembl who rallied the Upper Austrian estates against the Habsburgs? These are open questions worth careful consideration, but for many in the late sixteenth century, both friend and foe alike, Calvinism did have an ideological center, a coordinating hub, and a potential military platform from which it could support its co-religionists across the continent. Its home was the Palatinate governed by the most senior of the Empire's secular electors. Wittelsbach Elector Frederick III (1559–76) introduced the Reformed faith to his territory in 1562, and except for a short eight-year period (1576–84), the territory remained Calvinist up to the Thirty Years War. Heidelberg also boasted one of the great universities of Central Europe, an institution that eventually supplanted Geneva as the premier Calvinist academy on the continent. Its international faculty featured the Reformed's leading theologians—Pierre Boquin and Daniel Tossanus from France, Immanuel Tremellius and Girolamo Zanchi from Italy, and Zacharias Ursinus and David Pareus from Silesia. The student body soon mirrored the faculty. Traditionally, a third of its students came from outside the Empire. The French and Dutch dominated initially, but in its later years, more and more came from the growing Reformed communities of Eastern Europe.[4]

Critically spanning both banks of the Rhine, the Palatinate occupied a position of singular importance both geographically and politically. Historically, it had strong ties with both France and the Low Countries. These connections took on a special meaning confessionally in the second half of the sixteenth century, as the Calvinist electorate offered military support to Reformed communities in both regions. They sent troops to assist the Huguenots in 1568, 1576, and 1587 while the elector dispatched an army to aid William of Orange in 1578.[5] Outside events contributed to Heidelberg's sense of urgency and the need to confront the Catholic threat. With Spain and France publicly agreeing to execute the decrees of Trent, and with Alba's brutal campaign in the Low Countries and growing hostility to the Huguenots culminating in the St. Bartholomew nightmare, there seemed real reason to fear that these developments were but the first stage of a broader plan to destroy Protestantism altogether. The status of the Reformed within the Empire also seemed in jeopardy since Calvinism was not covered by the guarantees of the Augsburg Peace. The Palatinate faced hostility not only from Catholics but also from many Lutheran princes who openly questioned the community's legality.[6] There was also a growing personal rivalry between the Wittelsbach and the Habsburg families. Frederick III's predecessor, the Lutheran Otto Henry, had suffered for his faith even before he became elector. His support of the Schmalkaldic League cost him dearly as imperial troops occupied his lands after the war. His politics became increasingly radical as he assumed a leading role at the imperial diet of 1556/57 where he demanded complete religious liberty for the Empire's Protestants. His successors amplified these tendencies, and refugees displaced by religious conflict radicalized the situation yet further. In the early 1560s, Frederick III welcomed Dutch and French exiles at nearby Frankenthal. All these factors contributed to the aggressive anti-Catholic policies that were developing in Heidelberg.

This situation reached its climax during the reign of Elector Frederick V (1610–23). Frederick worked assiduously to cultivate his image as the leader of a united Protestant front ready to challenge the Catholic Habsburgs. In one famous illustrated broadsheet celebrating his 1613 marriage to Elizabeth Stuart, daughter of English king James I, four lions representing the great Protestant powers of England, the Low Countries, Bohemia, and the Palatinate prance in front of the couple while in the background Martin Luther, John Calvin, and Jan Hus place their hands together on an open Bible in a sign of friendly union.[7] But what was distinctly Calvinist about Frederick's understanding of himself and his place in the Empire? How did such notions contribute to his policies, and in what ways, if any, did they influence his decision to defy the Habsburgs and accept the Bohemian throne thus setting in motion events that ultimately led to the greatest conflict in Europe before World War I? To probe this question more deeply, we will turn to one of the most influential personalities in the Palatinate in the decade immediately prior to the outbreak of the Thirty Years War.

Abraham Scultetus: A Silesian Calvinist at the Palatine Court

Abraham Scultetus (1566–1625) was a court preacher, university professor, and adviser to Elector Frederick. His family came from what is today Lower Silesia. His father was the *Schultheiss* or mayor of Grünberg (Zielona Góra), hence the Latinized name Scultetus. Raised a Lutheran, the precocious Scultetus began his studies at the *Gymnasium* in Breslau (Wrocław) before moving on to Wittenberg where crypto-Calvinist currents pushed him beyond Luther. He eventually finished his studies in Heidelberg where he completed his theological transition to the Reformed church. Scultetus has not figured prominently in the literature on the Thirty Years War. Surprisingly, there is no mention of him in C.V. Wedgwood's grand narrative of the war nor in Claus-Peter Clasen's perceptive overview of the Palatinate. Where he does appear, it is usually in a cameo role thundering from the pulpit in Heidelberg or inspiring an iconoclastic mob in Prague to rid the cathedral of its "idols." Despite this lack of attention today, however, Scultetus was one of the best-known personalities of the war's first stage and more generally one of the most influential figures in the Reformed world of the early seventeenth century.[8]

Scultetus served two electors in Heidelberg, Frederick IV and V, as a court preacher. His influence reached its height during the reign of Frederick V as he oversaw a Calvinist reform of many of the Palatinate's institutions. From a review of schools to parish visitations, the vigilant Scultetus devoted significant energy to ensure that proper standards of doctrine and worship were maintained across the Palatinate. His attention, though, was not limited to his home region. In the decade immediately prior to the Thirty Years War, Scultetus was one of the Reformed community's most prominent leaders advising the church at critical moments of decision. From April to September 1610, for example, he was the army chaplain of Christian von Anhalt during the succession crisis of Jülich and Cleves. Seeking to exploit a power vacuum, an aggressive Calvinist faction championed by Anhalt and Scultetus endeavored to expand their power and eliminate Spanish influence in these two strategic Lower Rhenish duchies. A few years later he was in England with a delegation to seal a broader Protestant alliance with James I through the marriage of his daughter Elizabeth with Frederick. In 1614, he headed east to Brandenburg for a six-month stay at the court of Elector Johann Sigismund. The previous year Johann Sigismund had converted to Calvinism, a decision that sent political tremors across Central Europe, for his defection from Lutheranism shifted the confessional balance of the seven imperial electors. Scultetus came as an adviser to complete the Calvinist Reformation of the elector's territory. He was also present at the Synod of Dort, that theological conference which in fundamental ways defined the doctrinal contours of the Reformed church. He arrived in November 1618 representing the Palatinate at the synod and quickly emerged as one of the most influential of the foreign delegates. Across an active career of four decades, he spent significant time working and strengthening the Calvinist cause in at least a dozen different locations.

The scope of his intellectual activities was equally broad. Apart from his public roles as preacher and diplomat, Scultetus wrote an influential church history to celebrate the centenary of Luther's break with Rome. As a theologian, he produced a critical assessment of the church fathers. He lectured as a professor in Heidelberg and carried on an active correspondence with Isaac Casaubon and other learned humanists. Some of his poetry has even survived, including an apocalyptic denunciation of Rome in the published work of Ireland's archbishop and primate James Ussher. If his rise to prominence was sudden, his fall occurred just as swiftly. He figured conspicuously in the satirical literature produced in the wake of the Winter King's spectacular disaster at Prague. In broadsheets he appeared in a variety of guises: the devil seducing Frederick with the riches of Bohemia, the snake in the primeval garden tempting the elector and his wife, or simply a Calvinist buffoon listening to the regrets of a now-disgraced king.[9] For the Catholics, he was an object of derision. For the Calvinists, he became one of embarrassment or a symbol of hubris. Whatever the case, when the long war finally reached its end, the Reformed community was heading in a very different direction than the one charted by Scultetus at its beginning.

Though Scultetus has become a footnote in the study of the Thirty Years War, he occupies a critical intersection between politics and Calvinist theology and offers important insights into the growth and development of a militant ideology that reached its zenith in this period. There is a real need, then, to bring together the various aspects of his career for a fuller understanding of this dynamic that contributed to Elector Frederick's resolution to accept the Bohemian throne, but before turning to this topic, a brief consideration of his background is necessary. The connections that developed in the Palatinate between England, the Low Countries, and France have tended to overshadow an equally important association that helped shape Scultetus's confessional outlook and remained an enduring influence throughout his active career. He was a Silesian, a region that has played an underappreciated role in the growth of a Reformed identity during this period. Two of the university's most prominent professors, Pareus and Ursinus, were Silesians as was the Old Testament scholar in Basel, Amandus Polanus, whose German Bible was the first expressly Calvinist translation of Scripture.[10]

Although we need to be careful in assessing the importance of geography, Silesia is an intriguing region to consider and served as a critical link between East and West. Primarily German speaking, this patchwork territory of more than twenty duchies belonged to the Bohemian crownlands. With its rich natural resources and intersecting trade routes, Silesia and its bustling towns thrived economically. Its schools were also superb. Those of Breslau, Goldberg (Złotoryja), Troppau (Opava), and Hirschberg (Jelenia Góra) produced a remarkable group of humanist scholars.[11] Attending primary and secondary schools in Freistadt (Fryštát), Breslau, and Görlitz (Gorlice), Scultetus was a product of this cultured world, which also molded and informed his politics. Towards the end of his life, he wrote an autobiography in which he wistfully recalls the Habsburg emperor at the time of his birth, Maximilian II (1564–76). As Scultetus relates, Silesia prospered during Maximilian's beneficent rule. More importantly, he recounts what he saw as a lost opportunity when the Habsburgs themselves nearly converted to Protestantism. Encouraged by his court preacher, Johann Sebastian Pfauser, Maximilian came so very close to abandoning the Roman faith of his family.[12] While Scultetus may have exaggerated the likelihood of Maximilian's conversion, this passage does reflect his clear hopes of Protestantism's ultimate triumph across the region.

More generally, Scultetus's Silesian background contributed to his broader perspective of the Protestant world. In the *Autobiography* he also discussed the implementation of the *Formula of Concord* (1577), the doctrinal statement intended to bring uniformity to Luther's church. He clearly saw this settlement as a narrow and parochial expression of a Lutheran belief that excluded many. Instead of unifying and expanding a potential Protestant alliance that could confront the Catholic church directly, this doctrinal statement stripped Luther's message of its vigor, leaving its churches isolated, tepid, and politically neutered.[13] Though Scultetus certainly colored his account with his own theological convictions, these formative days in Silesia influenced his understanding of the Christian community. A number of his early teachers were sympathetic to Calvinism, and while studying in Wittenberg, he encountered a group of crypto-Calvinists. Concurrently, Scultetus had a growing interest in Slavic Christianity. Silesia was ethnically mixed, and over time he developed other contacts. He lived briefly in Dessau with Melanchthon's son-in-law, Caspar Peucer. It may have been there that he had his first interactions with Protestant groups to the east such as the Bohemian Brethren. Peucer, whose mother was Sorbian, had played an influential role in Wittenberg facilitating contact with Slavic churches.[14] Throughout his career, Scultetus regularly looked east where he saw isolated Protestant communities. Fearing their exposure and vulnerability, he advocated consistently for a wide confessional union. He also had a growing audience in the East well before his relocation to Prague as publishers eagerly translated his work into Polish, Czech, and Hungarian.[15]

Abraham Scultetus: Theologian and Historian

Scultetus's Silesian background shaped in profound ways his view and understanding of the world around him. It may have also contributed to his work as a theologian and the practical needs he perceived of Protestant communities surrounded by hostile Catholics. He was a prolific writer and recognized authority in the Reformed community publishing more than sixty treatises, tracts, and sermon collections in his lifetime. As a theologian, he enjoyed his greatest success as a patristics scholar. Historical theologians interested in the reception of the church fathers during the Reformation have devoted passing attention to Scultetus. Though the court preacher was an able if not brilliant scholar, his accomplishment was truly significant. His *Medulla Theologia Patrum Syntagma*, a four-volume collection that appeared between 1598 and 1613, was the first broad overview of patristic theology produced by a Protestant reformer.[16]

Scultetus compiled the *Syntagma* during an important moment of scholarly change. Attitudes towards the church fathers were shifting during his lifetime. In earlier decades of the sixteenth century, Reformers, both Catholic and Protestant, frequently used patristic texts in a more haphazard and polemical fashion to attack the views of their confessional rivals. Such practices were now giving way to a more disciplined and critical approach to these writings. Among the Protestants, Isaac Casaubon, Thomas James, and James Ussher helped promote more rigorous scholarly practices in reading and interpreting the Fathers. Scultetus, too, reflected these new methodological developments. He was critical of the New Testament Apocrypha, noting how books such as the *Gospel of Nicodemus* and the *Protogospel of James* did not match the stylistic conventions of the apostles. Likewise, in his treatment of Basil of Caesarea, he radically cut the number of texts directly attributable to the Cappadocian bishop and theologian through a careful reading of both Protestant and Catholic commentators.[17] It would, however, be a mistake to credit Scultetus's interpretation merely to his appreciation of these new text-critical tools, for his confessional convictions remained keen and his commitment to the Calvinist cause constant.

Here it may be helpful to look more closely at the *Syntagma* and examine Scultetus's treatment of a single individual. Irenaeus (c. 130–c. 202 CE), bishop of Lyon, is best known for his writings against the gnostics. For Scultetus, Irenaeus was particularly significant for his understanding of church tradition as a source of doctrine. Irenaeus was a follower of Polycarp who in turn claimed to have heard the preaching of John the Evangelist. Hence, Irenaeus was an important source and early compiler of apostolic teaching and tradition. Scultetus turned to him on the question of clerical celibacy. There his reading was highly selective as he focused on those passages which did not support the practice while ignoring other sections that actually praised chastity and abstinence. Scultetus could also be direct and confrontational when he encountered views he saw antithetical to Reformed teaching. He devoted a substantial section of the *Syntagma* attacking those "contradictory doctrines, dangerous opinions, and troublesome phrases" in Irenaeus. In their defense of church teaching, Catholic apologists such as the Italian cardinal Cesare Baronio had argued that the church fathers spoke with a single voice. Scultetus, in contrast, attacked and ultimately destroyed the perceived unity of this tradition by highlighting its inconsistencies and irregularities, arguably his greatest scholarly accomplishment.[18]

Scultetus's use of Irenaeus also illustrates how he examined church figures of the past to attack confessional rivals of the present. His sharp and emphatic denunciation of Mary was actually a response to his Catholic opponent, the Jesuit Robert Bellarmine.[19] Irenaeus had claimed that the obedience of Mary had compensated for the sin of Eve, an argument that Bellarmine and other Catholic theologians used to elevate Mary as a heavenly intercessor but one that Scultetus naturally sought to counter. Scultetus's understanding and reaction to Irenaeus reflects as well a transition that was occurring within the Reformed community. Though Calvin had not studied Irenaeus extensively, his understanding of the bishop was certainly more subtle and nuanced than that of Scultetus. Calvin's views were actually closer to Erasmus who was more open and positive to church tradition as a source of authority. Scultetus, on the other hand, knew Irenaeus better but removed any sense of ambiguity. Scultetus's more simplistic treatment of Irenaeus was a hallmark of the *Syntagma* as a whole and an expression of a Reformed scholasticism that was beginning to emerge in the seventeenth century.[20]

Scultetus thus stands out in several important respects from other contemporary scholars of the early church. The *Syntagma* was, in large part, derivative. Scultetus compiled material from others in this synthetic handbook of the early church. Unlike Isaac Casaubon or Thomas James, the first librarian of the

Bodleian, Scultetus was first and foremost a churchman. Scholarship was ultimately a means to an end. The *Syntagma* was also selective. It was not intended to offer an exhaustive overview of the church fathers. There were, in fact, curious omissions. For the Greeks, he considered neither Gregory of Nazianzus nor John Chrysostom. Even more glaring was his omission of both Augustine and Jerome. His aim was above all practical, not encyclopedic. He sought to retrieve the church fathers from the errors of the Catholics and demonstrate that the Reformed church represented no doctrinal deviation akin to the errors of the Arians and gnostics. As such, the *Syntagma* condensed and distilled patristic theology for pragmatic ends. In his *Autobiography,* Scultetus asserted that this scholarship was a necessary parallel to the day-to-day work he pursued on the church council of the Palatinate ensuring that the churches and schools of the region maintained proper doctrine and good order.[21] His efforts in extracting the critical teachings of the early church provided a model or even blueprint for completing a full and thorough reform within the Palatinate.

Scultetus's understanding of the more recent past mirrored his approach to the earlier patristic period. Apart from the *Syntagma* he was also slowly compiling a broad history of the Reformation's first century as a way to commemorate the 1517 anniversary. He took advantage of his travels around the continent to collect material for this endeavor stopping at local archives and libraries along his route. Most of this material remained in manuscript form. It accompanied him to Prague, and much of it was lost when after the battle of White Mountain, the approaching imperial army forced his precipitous flight with Frederick. With the fragments that remain, we can, however, recreate the rough outlines of the project.[22] Two points are particularly salient in our investigation of a more militant Calvinist ideology that was developing in Heidelberg.

The first is the decidedly Reformed character of his ecclesiastical history. Scultetus approached his work with none of the nuance and cautious assessment we normally associate with the study of the past. The reform movement that he chronicled was heading in a very definite direction best represented by the practice and doctrine of the Palatine church. We see this most clearly with the issues he highlighted, what he considered the most important themes of the Reformation. Somewhat surprisingly, he placed minimal emphasis on a new understanding of justification or the elevation of Scripture as the sole rule of faith. Indeed, he never mentioned Luther's German translation of the Bible. Instead, he focused squarely on the Eucharist. From Scultetus's perspective, the belief that Christ's body was physically present in the bread constituted the great error on which the papacy was established.[23] The story of reform, then, was one that centered around this original sin of the papists. To him, the Catholic understanding of this sacrament was a form of idolatry that true believers everywhere were called to challenge.

Scultetus began his history by briefly describing the various currents of reform that circulated prior to Luther. In France, the Waldensians bore patient witness to the truth. In England, John Wycliffe and the Lollards struggled against the idolatry of the papists. In Bohemia, Jan Hus and his followers faced great persecution for the gospel while in the Low Countries, Wessel Gansfort "long before Luther was born" condemned the errors of the Roman church.[24] When Luther finally did appear in 1517, he bravely continued their work confronting the false teaching of indulgences. But the Luther of Scultetus's history, though important, was just one of many in a long line of prophets to call the church to repentance and restore true doctrine and practice, and as Scultetus reminded his readers, the prophets of the Old Testament opposed pagan religion selectively. Some destroyed golden calves while leaving sacred groves of foreign gods untouched. Others tore down those high places but tolerated household idols. Luther, too, had his blind spots such as the Eucharist, which even Wycliffe understood better than he. Scultetus did not even spare one of his great favorites, Philip Melanchthon, for though Luther's lieutenant was more receptive and sympathetic to Reformed theology, he was "not yet completely cleansed from papal filth" as evident at Augsburg in 1530, when the timid Melanchthon was all too ready to compromise with the Catholics.[25]

The second distinguishing feature of this history is its scope. Though his narration followed the traditional model of a chronicle, a series of short entries corresponding to a specific year, geographically Scultetus spanned the breadth of the continent. He recounted developments ranging from Italy to the Baltic, from Ireland to Poland. Not surprisingly, Silesia featured prominently in this overview, a region to which he consistently returned. He gave the greatest attention to those areas where Calvinist ideas flourished. His history traced the contours of a broader Reformed alliance he imagined and worked so hard to foster. Scultetus was the consummate networker. As we have seen, his work took him to centers of Reformed activity across the continent, and he used those opportunities and personal connections to gather material for his project. For Scultetus there was but a thin line between scholarship and advocacy. His history was a means to connect scattered Calvinist communities together by creating a common narrative and origin story they could reflect on and celebrate collectively. He used the various opportunities of his travels to promote this aspect of his project. When he accompanied Elector Frederick to England for the 1613 marriage with Elizabeth, he used his time to cultivate a number of potential allies including the Biblical scholar and translator Henry Savile and Thomas James. He also met George Abbot, the mild Calvinist archbishop of Canterbury. It is no coincidence that Scultetus dedicated the second part of his history to Archbishop Abbot, the titular leader of the English church.

Abraham Scultetus: Chaplain and Preacher

Despite his accomplishments as both a patristics scholar and a historian, Scultetus was best known for his work in the pulpit. Approximately half of his publications were homiletic in nature: individual sermons, sermon collections, or homiletic handbooks. As we have already observed, he calibrated his work as both a historian and to some degree as a theologian to a popular audience. In an effort to reach a broader public, he enlisted his son-in-law Reinhard Wolff to prepare a German translation of his Latin history. In the *Autobiography* he proudly pointed to one of his most successful publications, the *Kirchenpostill*, which ran through seven German editions alone in the seventeenth century. He recounted how he and his enterprising publisher marketed the sermon collection, noting with obvious satisfaction that it had been "translated in a variety of languages and still today is read in the German lands and beyond for the edification and encouragement of pious hearts."[26]

Some of the most dramatic and celebrated moments of his career came in the pulpit. There was in 1613 the famous "Thanksgiving sermon" he preached to celebrate the arrival of Frederick and Elizabeth in Heidelberg when Scultetus exhorted the newlyweds to complete the building of a new Jerusalem that would one day vanquish "popish Babilon."[27] Four years later during the Reformation centenary, he delivered the New Year's sermon that recounted the historical lessons of the last century. He preached again on November 2 for the actual anniversary. Before a packed audience at the Church of the Holy Spirit, Heidelberg's central house of worship, Scultetus compared the work of the current Reformation to that of King Josiah restoring true worship at the temple in Jerusalem. In 1618, he assumed center stage during a critical moment at Dort. At one of the early sessions of the synod, he appealed for unity among the council's feuding delegates. Under this pretense, however, Scultetus was actually rallying support against the Remonstrants, the Arminian faction of the Dutch Reformed Church.[28] The following year he was in Prague where he gave perhaps the most famous sermon of his career as he called upon the godly to remove all traces of papal devotion from the city's grand cathedral. His words precipitated one of the great iconoclastic incidents of the Reformation period when Calvinists destroyed much of the sacred art of St Vitus Cathedral. Even after the collapse of Frederick's regime, Scultetus's heated rhetoric continued unabated. In 1621, he published a sermon calling all true believers to continue to proclaim the gospel and hold firm against all slanders, libels, and persecution.[29]

To understand Scultetus aright and the more militant ideology he advocated, we must appreciate this performative aspect of his career. His greatest strengths were rhetorical, and he used these skills to singular effect in the pulpit and beyond. In the *Autobiography* we catch intriguing glimpses of Scultetus the preacher. He recounted how traveling through Polish territory as a teenager, he visited a Jesuit church in Poznań where he carefully observed the style and technique of the cleric. He listened to Jesuit sermons on a number of occasions to become more effective himself. He spoke with pride when in April 1613 he learned that a distinguished panel of Catholic theologians including his old nemesis Robert Bellarmine publicly condemned and proscribed one of his sermon collections.[30] Additionally, he boasted about his opportunity to accompany Frederick to Frankfurt for the 1619 imperial election. For two months he preached "before eager crowds of French, Dutch, and German listeners along with the citizens of Frankfurt."[31] Scultetus's confessional rivals saw him as a dangerous opponent in this regard even characterizing his rhetorical skills as diabolic. In one famous broadsheet of the period an artist depicted Scultetus soaring high above Prague with an entranced Frederick in tow. As Satan accosted Christ in the wilderness with promises of wealth and influence, Scultetus, too, enticed the prince by showing him the riches of Bohemia and with honeyed words persuaded him to grasp for power. In another illustration, he appeared as the serpent of Eden whispering seductively in the ears of Frederick and Elizabeth while extending a pair of royal crowns to the couple.[32]

To better understand Scultetus's effectiveness as a preacher and his broader vision of reform, we will look more closely at his two well-known sermons commemorating the centenary of the Reformation. Though written texts obviously cannot capture the aural and visual dimension of Scultetus's oratory, they do provide some hints that help explain his public appeal. In the pulpit, he was simple and direct. He avoided technical discussions of complicated doctrines. Instead, he told stories often in dramatic fashion and with clever turns of phrase that communicated clear and unambiguous lessons. The New Year's Sermon offers one of the best examples in this regard. In it, Scultetus developed a series of parallels between the growth and development of primitive Christianity with that of the Reformation church. As John the Baptist announced the coming of Christ, so God prepared the way of reform through the books of Luther and the writings of other godly men.[33] Through these types of parallels, Scultetus carefully crafted a compelling narrative that culminated in the ultimate victory of his church. Victory, though, only came through suffering. Scultetus's treatment of these intertwining themes were the key motif of the New Year's Sermon and pointed to an important apocalyptic dimension of the chaplain's thought.

Faithful Christian communities inevitably encountered suffering. Such trials, though, were necessary, for they facilitated the growth of the church. For the early church, the fall of Jerusalem led to the dispersal of believers across the Mediterranean while recent persecutions in France, Italy, and the Low Countries contributed to a rapid evangelical advance. Scultetus, though, moved beyond general observations and personalized the story of persecution and suffering as he brought it home to the Palatinate and Frederick's family. He directed attention to his patron's predecessors, Otto Henry and Frederick III. For the sake of the gospel, they confronted the emperor and his Catholic allies who threatened both with loss of possession and prestige.[34] But if suffering was a central component of the faith, so was victory for those who persevered until the end. Masterfully, Scultetus exploited the tension between these twin dynamics of suffering and triumph as he told his story to great dramatic effect. Every defeat leads to an even greater victory. When the church is at its lowest state, God intervenes wondrously to raise it above its enemies. And in the end, as history draws to a close, a grim cosmic reckoning awaits those who persecuted the faithful. Scultetus concluded his homily with a grisly role call recounting the frightful deaths of notorious opponents of the gospel.[35]

If the New Year's Sermon provided a grand triumphal narrative that could justify the Palatinate's aggressive policies in support of international Calvinism, the sermon Scultetus delivered ten months later to celebrate the actual Reformation anniversary offered a more specific model of Reformed rule. In it, he evoked the Old Testament king of Judah, Josiah. Perhaps more than any other Old Testament figure, Josiah was of special political significance. Calvinist leaders from Ireland to Hungary regularly held him up as an archetype of the ideal prince.[36] Building his sermon around II Kings 23:1–4, Scultetus highlighted Josiah's actions proclaiming God's Word, renewing the covenant, and abolishing idolatry from his kingdom. Scultetus must have seen a tractable Frederick as the perfect pupil, and indeed the prince seemed to take those lessons to heart. Just a few years later, he followed his chaplain's advice by supporting the campaign in Prague to cleanse its churches from idols. There was an additional aspect of Josiah's reign that must have seemed especially compelling from Scultetus's perspective. The Hebrew king was not merely content with the reform of his own domains but had ambitions to bring godly change to lands beyond his borders. As it turned out, Frederick, too, was willing to take the risk of war to follow Josiah's precedent.

There is one final dimension of the 1617 sermons that ties together many of the themes we have already examined and provides some concluding insights into the situation of the Palatinate on the eve of the Thirty Years War. Scultetus crowded both these homilies with references to prominent Evangelical leaders. In the New Year's Sermon, he identified approximately one hundred teachers from Scotland to Italy "who have revealed the abomination of the papacy." In the preface of the Anniversary Sermon, he listed more than eighty "evangelical kings, princes, and nobles" who stood loyal to the gospel.[37] The prosopographical component of these two sermons was typical of his writing as a whole. He crammed both the *Autobiography* and his history with references to scores of princes and religious leaders whom he saw working within a wide but well-defined tradition of reform that was ineluctably heading to a complete and final restoration of the church. Such work matched the real networks he had built over his career as a churchman and diplomat. Few if any within the Reformed community had as many actual contacts or those imaginative powers that could envision a united Calvinist alliance stretching across the entire continent.

Looking back on Scultetus, what broader conclusions, if any, can we make concerning the relationship between Calvinism and conflict? The first lessons are primarily negative. While some have pointed to ideological distinctives of the Reformed tradition, the hatred of idolatry, the cause of the elect, the active presence of Satan, there is good reason to remain skeptical that these beliefs predisposed it to violence. Intolerance was endemic, there was a range of theological diversity within the Reformed camp, and many of the convictions that could lead to violence were equally shared among Catholic, Lutheran, and Calvinist. Context is critical, but even here when we narrow our focus to Heidelberg in the decade before the Thirty Years War, we should be careful. Like his Catholic critics, it may be tempting to cast Scultetus as a villain quietly whispering in the ears of his princely patrons to seize power whatever the cost. Such an assessment is overly simplistic. Catholic leaders were also plotting, and there were a host of other factors at play in this complicated crisis.

At the same time, it is hard to deny that Scultetus contributed in some way to the looming cataclysm that would quickly engulf Central Europe. It is difficult not to read his sermons, theology, and history without sensing his absolute confidence that divine providence favored the Reformed. His reading of the Christian past could be generous and expansive but moving only in one direction. He actually praised a Spanish grand inquisitor, Cardinal Cisneros, as a forerunner of the Reformation. He confidently asserted that Habsburg Emperors Charles V, Ferdinand I, and Maximilian II had died in the true faith, and with conviction, he maintained that Archbishop Stephan Gardiner, England's implacable opponent of Protestantism, confessed on his deathbed that he had betrayed the gospel.[38] Scultetus, at the very least, was an exemplar of Calvinist triumphalism, an individual whose unwavering belief in the final victory of his church helps explain the counsel he gave Frederick. What other Protestant princes deemed risky and rash was for him simple and sure. Scultetus, of course, was not alone. There were others close to Frederick who advocated a similar course. Christian von Anhalt, Ludwig Camerarius, and Elizabeth Stuart, undoubtedly exercised significant influence. And so in the end, they reached a decision. With his chaplain at his side, the young and impressionable elector left the castle and gardens of Heidelberg and *with wandering steps and slow* made his way to Prague and the approaching cataclysm.

Notes

1 The literature here is immense. For two prominent examples, see Philippe Buc, *Holy War, Martyrdom, and Terror. Christianity, Violence and the West* (Philadelphia: University of Pennsylvania Press, 2015); Mark Juergensmeyer, *Terror in the Mind of God: The Global Rise of Religious Violence*, 4th ed. (Berkeley: University of California Press, 2017).
2 Christine Kooi, "Calvinism and War," in *Cultures of Calvinism in Early Modern Europe*, eds. Crawford Gribben and Graeme Murdock (Oxford: Oxford University Press, 2019), 171.
3 Henry Cohn, "The Territorial Princes in Germany's Second Reformation, 1559–1622," in *International Calvinism*, ed. Menna Prestwich (Oxford: Oxford University Press, 1985), 135.

4   Notker Hammerstein, "Vom 'Dritten Genf' zur Jesuiten-Universität: Heidelberg in der frühen Neuzeit," in *Die Geschichte der Universität Heidelberg* (Heidelberg: Heidelberger Verlagsanstalt, 1986), 34–44; Eike Wolgast, "Geistiges Profil und politische Ziele des Heidelberger Späthumanismus," in *Späthumanismus und reformierte Konfession*, ed. Christoph Strohm, et al. (Tübingen: Mohr Siebeck, 2006), 1–25.

5   For an insightful overview of the region, see Claus-Peter Clasen, *The Palatinate in European History 1559–1660* (Oxford: Blackwell, 1963). Here p. 5.

6   For this complicated dynamic between Lutheran and Reformed communities, see Howard Hotson, "Irenicism in the Confessional Age: The Holy Roman Empire, 1563–1648," in *Conciliation and Confession. The Struggle for Unity in the Age of Reform*, eds. Howard Louthan and Randall Zachman (South Bend, IN: University of Notre Dame Press, 2004), 228–85.

7   On this image carefully cultivated around Frederick at home and abroad, see Jaroslav Miller, "Between Nationalism and European Pan-Protestantism. Palatine Propaganda in Jacobean England and the Holy Roman Empire," in *The Palatine Wedding of 1613: Protestant Alliance and Court Festival*, eds. Sarah Smart and Mara Wade (Wiesbaden: Harrasowitz, 2013), 61–82; also useful for context is Andrew Thomas, "The Culture of the Palatine Court in Heidelberg at the Dawn of the Seventeenth Century," in *Churfürstlicher Hochzeitlicher Heimführungs Triumph*, eds. Nichola Hayton, Hanns Hubach, Marco Neumaier (Ubstadt-Weiher: Regionalkultur, 2020), 327–50.

8   The work of G.A. Benrath remains standard. For an overview, see his "Abraham Scultetus (1566–1624)," in *Pfälzer Lebensbilder*, vol. 2, ed. Kurt Baumann (Speyer: Verlag der Pfälzischen Gesellschaft zur Förderung der Wissenschaften, 1970), 97–116.

9   For a sampling of the broadsheets, see G.A. Benrath, ed., *Die Selbstbiographie des Heidelberger Theologen und Hofpredigers Abraham Scultetus (1566–1624)* (Karlsruhe: Verlag Evangelischer Presseverband, 1966), 130–38. Rudolf Wolkan, ed., *Deutsche Lieder auf den Winterkönig* (Prague: J.G. Calve, 1898).

10 Robert Letham, "Amandus Polanus: A Neglected Theologian?," *Sixteenth Century Studies Journal* 21 (1990): 463–76; Susann El Kohli, "Schlesisch-pfälzische Beziehungen am Beispiel des Heidelberger Theologen Abraham Scultetus (24.8.1566–24.10.1624)," in *Früchte vom Baum des Wissens: Eine Festschrift der wissenschaftischen Mitarbeiter (100 Jahre Heidelberg Akademie der Wissenschaften)*, eds. Ditte Bandini and Ulrich Kronauer (Heidelberg: Universitätsverlag, 2009), 223–27.

11 Manfred Fleischer, "Institution of Humanism in Protestant Silesia," *Archiv für Reformationsgeschichte* 66 (1975): 256–74.

12 *Selbstbiographie*, 12.

13 *Selbstbiographie*, 13–15.

14 Ernst Benz, *Wittenberg und Byzanz. Zur Begegnung und Auseinandersetzung der Reformation und der östlich-orthodoxen Kirchen* (Marburg: Elwert-Gräfe und Unzer Verlag, 1949), 129–40. More recently see Martin Roebel, *Humanistische Medizin und Kryptocalvinismus: Leben und medizinisches Werk des Wittenberger Medizinprofessors Caspar Peucer (1525–1602)* (Herbolzheim: Centaurus Verlag, 2015).

15 Benrath provides a useful bibliography of Scultetus's work in the *Selbstbiographie*, 131–43. For translations of his homiletic material, see the following entries: 16d (Hungarian), 19a (Hungarian), 19f (Polish), 36e (Hungarian), 47d (Czech).

16 Irena Backus, *Historical Method and Confessional Identity in the Era of the Reformation (1378–1615)* (Leiden: Brill, 2003), 218–27; Irena Backus, "The Fathers and Calvinist Orthodoxy: Patristic Scholarship. The Bible and the Fathers according to Abraham Scultetus (1566–1624) and André Rivet (1571/73–1651). The case of Basil

of Caesarea," in *The Reception of the Church Fathers in the West. From the Carolingians to the Maurists*, vol. 2, ed. Irena Backus (Leiden: Brill, 2001), 839–65; G.A. Benrath, *Reformierte Kirchengeschichtsschreibung an der Universität Heidelberg im 16. und 17. Jahrhundert* (Speyer: Zechner, 1963), 21–27.

17 Backus, *Historical Method*, 226; Backus, "The Fathers and Calvinist Orthodoxy," 238–39.
18 Wilhelm Schmidt-Biggemann, *Apokalypse und Philosophie* (Göttingen: Vandenhoeck and Ruprecht, 2007), 92–94.
19 The second volume of the series published in 1605 directly attacked Bellarmine in its title.
20 Irena Backus, "Irenaeus, Calvin, and Calvinist Orthodoxy: The Patristic Manual of Abraham Scultetus (1598)," *Reformation and Renaissance Review* 1 (1999): 49–53.
21 *Selbstbiographie*, 39–40; Also see his *Delitiae evangelicae Pragenses* (Hanau, 1620) which he dedicated to the mild Calvinist bishop of Bath and noted preacher, Arthur Lake. Here once more, he stressed the practical benefits of the church fathers.
22 Fundamental here is Benrath, *Reformierte Kirchengeschichtsschreibung*, 27–37.
23 Of the 800 pages of this project that survive, 130 consider this topic. Benrath, *Reformierte Kirchengeschichtsschreibung*, 35.
24 Abraham Scultetus, *Historischer Bericht, wie die Kirchenreformation in Teutschlandt vor hundert jahren angangen* (Heidelberg: Johann Lancellot 1618), 9–10.
25 Scultetus, *Historischer Bericht*, 11–12; Benrath, *Reformierte Kirchengeschichtsschreibung*, 35.
26 *Selbstbiographie*, 54.
27 Abraham Scultetus, *A Sermon Preached before the two high borne and illustrious Princes Fredericke the 5. Prince Elector Palatine, Duke of Bavaria, &c. And the Princesse Lady Elizabeth* (London: John Beale, 1613), 20.
28 Abraham Scultetus's "Sermon on Peace," in *Acta et Documenta Synodi Nationalis Dordrechtanae*, vol. 2, eds. Donald Sinnema, Christian Moser, and Herman Selderhuis (Göttingen: Vandenhoeck and Ruprecht, 2018), 474–86.
29 Abraham Scultetus, *Kurtzer, aber Schriftmässiger Bericht von den Götzenbildern an die christlicher Gemein zu Prag* (Prague, 1619); Abraham Scultetus, *Daß man sich weder an dem Unmenschlichen Schmehen, Lästern, Verleumden der Rechtgläubigen noch an der großen Verfolgung...* (Heidelberg, 1621).
30 *Selbstbiographie*, 21–22; 68.
31 *Selbstbiographie*, 21–22, 77.
32 *Selbstbiographie*, 130–38.
33 Abraham Scultetus, *Newe Jahrs Predigt* (Heidelberg: Johann Lancellot, 1617), 6.
34 Scultetus, *Newe Jahrs Predigt*, 16–17; 22–23.
35 Scultetus, *Newe Jahrs Predigt*, 28–30.
36 Graeme Murdock, "The Importance of Being Josiah: An Image of Calvinist Identity," *Sixteenth Century Journal* 29 (1998): 1043–59.
37 Scultetus, *Newe Jahrs Predigt*, 12–14; Abraham Scultetus, *Evangelische Jubel-Jahrs Predigt* (Heidelberg: Johan Lancellot, 1617), iiiv–ivr.
38 Scultetus, *Newe Jahrs Predigt*, 6, 11, 28–29.

Further Reading

Benrath, G.A. "Abraham Scultetus (1566–1624)." In *Pfälzer Lebensbilder*, vol. 2, edited by Kurt Baumann, 97–116. Speyer: Verlag der Pfälzischen Gesellschaft zur Förderung der Wissenschaften, 1970.

Benrath, G.A., ed. *Die Selbstbiographie des Heidelberger Theologen und Hofpredigers Abraham Scultetus (1566–1624)*. Karlsruhe: Verlag Evangelischer Presseverband, 1966.

Clasen, Claus-Peter. *The Palatinate in European History 1559–1660*. Oxford: Blackwell, 1963.

Cohn, Henry. "The Territorial Princes in Germany's Second Reformation, 1559–1622." In *International Calvinism*, edited by Menna Prestwich, 135–66. Oxford: Oxford University Press, 1985.

Kooi, Christine. "Calvinism and War." In *Cultures of Calvinism in Early Modern Europe*, edited by Crawford Gribben and Graeme Murdock, 171–85. Oxford: Oxford University Press, 2019.

Murdock, Graeme. "The Importance of Being Josiah: An Image of Calvinist Identity." *Sixteenth Century Journal* 29 (1998): 1043–59.

8 Troubles Concerning Religion

Causes, Parties, and Armed Conflict in the French Wars of Religion[1]

Brian Sandberg

King Charles IX lamented in March 1563 that "our kingdom has been afflicted and worried by many troubles, seditions, and tumults between our subjects, provoked and aroused by the diversity of opinions concerning religion and the doubts of their consciences."[2] The king's edict of pacification formalized a negotiated religious settlement, the peace of Amboise, that aimed to end the fighting in France between the armed forces of Catholics and Huguenots (French Calvinists) that had engaged in town seizures, sieges, and clashes across the kingdom. This peace, which ended the first of the kingdom's wars of religion, had come after a royal field army composed of Catholics won a victory over the main Huguenot army at the battle of Dreux in December 1562, in which both sides had suffered heavy losses.[3] Even then, the royalist and Catholic siege of the Huguenot-controlled city of Orléans had dragged on for months before the assassination of the Catholic army commander finally led both sides to come to terms. After so much bloodshed and the shattering of so many communities, many French people could agree with their king in describing their experiences of religious conflict and civil war as "troubles." The troubles concerning religion could not be calmed, however, and religious warfare would continue for decades.

A strong scholarly consensus has emerged that the French Wars of Religion (1559–1629) were indeed primarily *about* religion. The rapid spread of the Calvinist Reformation into France in the 1550s attracted many converts, including a significant proportion of the French nobles, who acted as protectors for Reformed commoners in their regions. Sharp religious divisions and confessional politics then increasingly destabilized French political culture. The groundbreaking historical research of Natalie Zemon Davis, Denis Crouzet, Barbara Diefendorf, Mack Holt, Philip Benedict, and others demonstrates that religion infused crowd violence, urban politics, noble culture, polemical debates, and political culture during this long period of persistent civil warfare.[4] Recent research has also emphasized the particular dynamics of religious violence, confessional politics, and religious peacemaking during the French Wars of Religion.[5]

What is still missing from this historical literature is an examination of the intersections between the language and the practices of religious warfare. This

DOI: 10.4324/9781003157700-11

chapter draws on an array of manuscript and printed sources to consider not just the religious dimensions of the waging of civil warfare in late sixteenth- and early seventeenth-century France, but also the ways in which this religious conflict was described and represented by contemporaries. Royal edicts, judicial acts, administrative correspondence, military records, and municipal records allow us to consider the language that participants in the religious wars utilized. Local chronicles, histories, and pamphlets recount episodes of religious war, while articulating polemical positions supporting and attacking particular armed groups involved in the fighting.[6]

These diverse sources present interpretive challenges, forcing us to consider how religious warfare transformed the everyday lives of combatants and non-combatants. Studies of war, culture, and society have greatly expanded our understanding of the social history of military organizations and wartime societies.[7] However, these studies have often ignored the religious dimensions of war or relegated them to a general backdrop of culture. Religion has sometimes been seen as a cause of war or as a motivation for soldiers to enlist in military organizations, but not as an important dimension of strategic thought or military operations. Even sophisticated models of combat motivation and military effectiveness tend to limit their consideration of religious perspectives to religious causes, confessional propaganda, and sustaining motivations for soldiers.[8] Several recent studies of the European Wars of Religion offer suggestive approaches to interpreting religion and warfare in new ways.[9] However, more needs to be done to integrate methodologies of the history of war, culture, and society with those of the social history of religion in order to understand societies experiencing sustained religious conflict.[10]

I will argue in this chapter that the confessional nature of the troubles in late sixteenth- and early seventeenth-century France demands a robust analysis of the religious dimensions of military activity and civil warfare. Confessional divisions and animosities not only supplied the root causes of the French Wars of Religion, but religious identities and spiritual inspirations also motivated many nobles and commoners to join military units and engage in armed conflict. The Catholic and Huguenot nobles who organized religious and political parties, mobilized armed forces, and directed military campaigns represented key actors in the religious wars. Confessional justifications also shaped the war aims and strategies of the belligerents, as they attempted to establish control over urban centers, their populations, and their churches. Military operations then supported broader strategic aims to win converts, restore churches, and ultimately achieve religious union.[11] Negotiations among warring parties led to religious peaces that attempted to settle the troubles or at least to consolidate the confessional domination of regions. In this chapter, I will analyze contemporary descriptions of civil warfare and examine the language of troubles in order to demonstrate the ways in which these religious identities and confessional politics became embedded in the everyday practices of waging war and making peace. I will thereby show how the troubles concerning religion were closely connected with mobilization processes, war aims and strategies, wartime destruction, and peacemaking processes.

Language of Troubles

To consider the religious dimensions of sixteenth- and early seventeenth-century warfare, let us examine the contemporary language employed by participants in the conflicts. The precise phrases of "wars of religion" or "religious war" occasionally appear in documents of the period, but seem to have been relatively rare. A rich language of religious warfare nonetheless permeated the printed pamphlets, treatises, and histories from the period that we know as the French Wars of Religion. The spreading Reformation movement had attracted many converts in France by the 1550s, as roughly ten percent of French subjects and perhaps thirty percent of the noble elites embraced the Reformed faith.[12] Confessional tensions between this growing Calvinist minority and the Catholic majority produced crowd violence, iconoclastic attacks, religious riots, and massacres, especially after the death of King Henri II in a jousting accident in 1559. A series of religious wars between various armed parties of Catholics and Huguenots ensued, lasting with only brief pauses until 1629. Participants and observers embroiled in the conflicts employed the terms *troubles, mouvements, malheurs, séditions, émotions, guerres civiles,* and *rebellions* to describe religious warfare. Each of these terms could be used in diverse ways, but they often exhibited overlapping connotations and were employed alongside each other in descriptions of conflicts and grouped together in long lists of the disasters afflicting France.

French observers described the initial massacres and armed clashes between Catholics and Huguenots from 1559 to 1562 as "troubles" (*troubles*). The tragic death of King Henri II in a jousting accident in 1559 produced a breakdown in royal authority that exacerbated religious tensions, and French people described the resulting troubles as creating social disorder and chaos in urban centers and rural villages. Calvinists challenged the rituals of the Eucharist, refusing to participate in the Mass or Corpus Christi processions and sometimes attacking the Host itself, while Catholic theologians condemned Huguenots' disruption of Christian community and accused them of disturbing social order.[13] Local chroniclers and pamphleteers soon recounted these episodes of religious violence as the "first troubles" (*premiers troubles*), while ongoing religious conflicts could be described as the "present troubles" (*presents troubles*), and each new outbreak of violence was depicted as "new troubles" (*nouveaux troubles*). French chroniclers and historians also recounted each successive civil war as the "latest troubles" (*derniers troubles*) or "latest conflicts" (*derniers mouvements*), sometimes adding successive editions of their works as new religious conflicts broke out.

Provincial governors and administrators reported episodes of crowd violence, urban riots, and peasant revolts in their regions, describing them as
"riots" (*émeutes*), "agitations" (*mouvements*), "disturbances" (*émotions*),
"alarms" (*alarmes*), or "tumults" (*tumultes*). Royal officers and municipal leaders often described crowds as reacting to news and rumors with sudden anger
and spontaneous violence directed at confessional enemies and municipal leaders. They sometimes noted the size of crowds and the seriousness of the violence
by describing incidents as "small disturbances" or as a "large popular distur-
bance." Other accounts mentioned urban crowds beating residents, throwing
stones, and brandishing weapons to intimidate their opponents.[14] Such crowd
violence could also escalate into massacres or urban uprisings to overthrow municipal governments.[15] A 1622 pamphlet related the "troubles and movements
of this time," referring to the religious warfare then spreading across most of
southern France.[16] In this sense, a "movement" held simultaneous associations
with military operations and with earthquakes.

Royal edicts issued during religious conflicts labeled opposition movements
as seditions (*séditions*), civil wars (*guerres civiles*), revolts (*révoltes*), and rebellions (*rebellions*). Royal officers and administrators appropriated the language
of revolt to brand their regional enemies as unlawful "enemies of His Majesty"
and rebellious enemies were also accused of being criminals and disturbers of
the peace. Discussions of the state of the kingdom also employed medical meta-
phors to describe the problem of religious division and disunity within France,
as well as within broader Christendom. The diseased social body could be depicted as cancerous, leprous, or gangrenous, and troubles could be described as
a contagious disease that threatened to consume the entire body politic.[17]

Justifications for violence at different times contemplated the conversion, expulsion, prosecution, or elimination of heretics. Both Catholics and Calvinists
described unorthodox or incorrect religious practices as unnatural, dangerous,
seditious, or treasonous. Reformed minister Pierre Viret, for example, railed
against false religion in sermons and religious pamphlets that motivated Huguenot militants.[18] Many Catholic texts also depicted heresy and error as forms
of religious pollution, expressing intense fears that heresy might corrupt entire communities and the body politic.[19] Authors and artists gendered heresy
as feminine, representing the figure of the heretic as a multi-headed hydra or a
witch-like hag—embodiments of aberrant and dangerous forms of femininity.[20]
Allegories could emphasize abominable error, depicting heresy as an adulterous
wife or a prostitute, evoking Biblical images of the Whore of Babylon from the
Apocalypse.[21] Fears of heresy fueled powerful apocalyptic language and imagery in religious polemics and confessionalized political culture.[22]

The dramatic expansion of printing in France during the period of the religious wars created new forms of political culture and allowed for sustained
religious debates. Political and religious pamphlets disseminated the language
of troubles widely, constructing polemical discourses of hatred.[23] Warrior nobles

participated directly in these political debates as producers and consumers of print culture, even as they engaged in religious warfare. Some military commanders authored pamphlets and manifestos, for example, and many more nobles probably commissioned or sponsored publications to articulate their religious and political positions.[24] Noble courtiers, administrative officials, and military officers read and shared printed pamphlets—often sending them as enclosures along with their handwritten letters. A multimedia political culture emerged as news and information circulated through printed, manuscript, and oral communications about military operations, political activities, and current events. War news, religious polemics, and political debates were thus closely intertwined in this hybrid political culture.[25]

Through these multimedia sources, we see that combatants and civilians described religious divisions as the fundamental cause of the troubles and hostilities. Contemporaries often evoked religious differences, diversity of opinions, and innovations as prompting disorder and clashes in French communities. Clandestine Reformed worship services, public preaching, and iconoclasm provoked riots and massacres in many cities and suburbs in the early 1560s, leading to growing confessional animosities between Catholics and Calvinists that fueled mutual hatreds.[26] The Massacre of Wassy in 1562, in particular, prompted Huguenot nobles to mobilize to protect Reformed communities, leading to the First War of Religion. Chroniclers and historians would soon use the phrases "first disorders" (*premiers mouvemens*), "first troubles" (*premiers troubles*), or "the origins of the troubles" (*original des troubles*) to interpret religious differences as the underlying cause of the religious wars.[27]

In addition to seeing the mere existence of religious difference as sparking violence, however, many contemporaries specifically blamed the opposing confession for causing the troubles in France. At the onset of widespread religious warfare in 1562, for example, Louis I de Bourbon, prince de Condé, already a prominent Reformed protector, suddenly became a major Huguenot military leader and asserted "the innocence of the Reformed churches," contrasting them with "those who are the authors of these troubles," clearly referring to Catholic extremists.[28] Catholics, meanwhile, blamed the outbreak of religious warfare on royal leniency toward adherents of "the so-called Reformed religion" and their unwillingness to prosecute them as heretics. Thus, when the First War of Religion ended in the compromise peace of Amboise in 1563, which provided some limited protections for Reformed practice, many Catholics were displeased. Louis d'Orléans, a committed Catholic, later claimed that "since the year 1563, when [Charles IX] granted permission to preach the Calvinist religion in this kingdom, we have had nothing but troubles and civil wars."[29]

Widespread religious violence and civil conflict produced a proliferation of arms in French society, contributing to clashes in many communities as well as general political instability. Thus armed soldiers were deployed in Paris in June 1562 to provide security even for a religious procession that was organized to bring about the "pacification of the troubles."[30] Contemporaries described the successive religious conflicts and civil wars that engulfed all of France as a generalized disorder, and people often employed the language of troubles to express a desire for the restoration of law and order within the kingdom. One solution might be military reforms, which could be seen as remedies for the ills of religious conflict. Henri III, for example, described the nature of the religious conflicts in his realm: "Among all the evils that I recognize to be proceeding from the troubles which have long afflicted this realm, one of the principal ones and which has the most need of reforming, is the corruption of the discipline and order in all estates, and especially the gendarmerie."[31] The king and his advisors thus viewed military and administrative reform as a method of calming religious tensions and stemming civil strife in the kingdom.[32]

Nevertheless, mutual suspicions and distrust became ubiquitous during the religious wars, preventing all such attempts to calm the troubles.[33] A manuscript *mémoire,* for example, related that the mayor of a town in Bourgogne (Burgundy) was suspected of disloyalty for having traveled to Dijon to meet with Charles de Lorraine, duc de Mayenne, one of the principal Catholic League leaders, soon after hostilities had broken out once again.[34] Warrior nobles and other belligerents described the state of war through a language of troubles. A 1593 letter related that "the wars and troubles that have come about in France have constrained the seigneur de Saint-Antoine to withdraw to this comté de Bourgogne," where he took refuge in the village of Champlitte, near a benefice that his family apparently possessed.[35] The language of troubles was thoroughly embedded in participants' and observers' descriptions of the religious conflicts they were experiencing.

Armed Mobilization and Party Formation

French nobles and their clients swore oaths and undertook a series of rituals as they took up arms (*prises d'armes*) and formed militant religious and political parties (*partis*) at each outbreak of religious warfare.[36] The names and identities of such parties shifted over time, encompassing diverse armed associations of Catholics, Huguenots, and sometimes bi-confessional parties. Belligerent party members often made public declarations and sometimes published printed manifestos to enunciate their causes. The rhetoric of these manifestos, along with their oaths and declarations, defined their military opponents as also being religious adversaries. The derogatory terms used in such sources provided some of the best-known names for parties during the French Wars of Religion, many of which were eventually adopted even by the parties themselves: Huguenots, Malcontents, *politiques*, and Catholic Leaguers. Confessional militants asserted collective identities for their religious and political parties through complex rituals and practices. Indeed, Arlette Jouanna describes the nobles' "rituals of revolt" during the religious wars, but she sees nobles taking arms primarily as an expression of discontent toward the monarchy.[37] However, the self-identities of armed religious and political parties were often overtly religious and associated with particular confessional agendas.

From the outbreak of the religious wars, military mobilization processes were closely linked with the formation of religious and political parties that articulated confessional causes and aims. The language contained within military commissions, army ordinances, and logistical documents often associated military forces' pragmatic actions with religious aims and confessional politics. Royal commissions asserted religious aims while authorizing specific nobles to carry out armed mobilizations in designated regions and provinces. Yet, royal commissions only accounted for a small fraction of military forces raised during the religious wars. As armed mobilizations began, nobles raised cavalry companies rapidly through their clienteles, and simultaneously issued commissions empowering their clients to act as subordinate officers and recruit infantry companies and regiments. Nobles activated their friendships and clientele relationships as they gradually banded together in associations to form their own religious and political parties.[38]

The First War of Religion (April 1562 to March 1563) set a pattern for military mobilizations throughout the long religious wars. Huguenot nobles responded to the Massacre of Wassy by mobilizing rapidly in the Orléanais, Normandie, Guyenne, Languedoc, Dauphiné, and Provence. Armed Huguenot militants seized Tours, Blois, Orléans, Poitiers, Lyon, Rouen, Caen, and other cities across the kingdom. Louis I de Bourbon, prince de Condé, established a base at Orléans and assembled a field army to advance against Paris. The prince de Condé published a series of declarations and manifestos defining the Huguenot cause and articulating their war aims, claiming that he and his followers were fighting "against the calumnies and impostures of the enemies of God, the king, and him [the prince]."[39] Some of Condé's pamphlets included psalms, evoking the Reformed practice of psalm-singing which had already become popular.[40] Another pamphlet published daily prayers for the Reformed soldiers in the prince de Condé's field army to recite when marching and camping. The prayers were published in both French and German, a reflection of the growing support of Reformed soldiers from beyond the kingdom's borders.[41] When Catholic forces responded to the Reformed seizures of towns, Huguenots developed further justifications for their arming. The Huguenots who seized control of Lyon in 1563 claimed to be mounting a "just and saintly defense" of the city against besieging Catholic forces.[42] The main Huguenot field army under the prince de Condé marched toward Paris and then maneuvered against the main royal field army. Meanwhile, the prominent Reformed minister Theodore Beza aided in the organization of the defense of Orléans and developed theological arguments for the Huguenot cause in France.[43]

The Huguenot mobilization in the province of Normandie in the spring and summer of 1562 reveals some of the techniques that religious and political

parties employed when mobilizing in a regional context. The Huguenot seizure of Rouen in part represents a local reaction against Jacques Villebon
d'Estouteville, seigneur de Villebon, a client of François de Lorraine, duc de
Guise, but was simultaneously part of a much broader movement of Huguenot
military mobilization across the kingdom.[44] Huguenot manifestos signaled that
they were mobilizing in defense of Reformed churches and adherents. Huguenot
troops and crowds carried out iconoclasm to "cleanse" churches in Rouen and
beyond. For this reason, the royal commission's mention of disorders and seditions seems to have referred not only to the expulsion of Villebon, but also to the
wave of iconoclastic attacks by Huguenot militants across Normandie.

In response to the seizure of Rouen, Charles IX ordered the creation of a
royal field army in Normandie, after being "advised of the disorders, tumults,
and seditions that have been started in certain places in our province of Normandie, and principally in our city of Rouen, where those who claim to profess
the reformed religion have forced the seigneur de Villebon, our lieutenant general ... to leave the said city." In this order the king also accused the Huguenots
in Rouen of having "seized the keys of the gates, artillery, and munitions, taking
the château and the old palace."[45] Charles IX authorized Claude de Lorraine, duc
d'Aumale, to mobilize Catholic forces and re-establish order throughout Normandie.[46] The king's selection of the brother of the duc de Guise, the accused
instigator of the Massacre of Wassy, as army commander probably further an-
tagonized outraged Huguenots in Rouen. The duc d'Aumale's small field army
advanced to blockade Rouen in late May, beginning the first major siege of the
religious wars. Antoine de Bourbon, roi de Navarre, led additional royal forces
to join the ongoing siege. Royal artillery breached Rouen's walls in October and
Catholic soldiers sacked the city.[47]

The devastation of Rouen failed to accomplish Charles IX's stated aim of
"gently re-establishing everything in its previous state," but the language of
troubles in this and other military documents had already intensified the confessional nature of the religious wars.[48] François de Lorraine, duc de Guise,
organized the main royal field army near Paris, with the assistance of Anne
de Montmorency, *connétable* (constable) de France, and Jacques d'Albon,
maréchal (marshal) de Saint-André. This field army maneuvered against the
prince de Condé's Huguenot army and fought a major battle at Dreux (December
1562), where both Condé and Montmorency were captured and Saint-André
was killed.[49] Following this bloody battle, Huguenot forces withdrew to their
base at Orléans and the royal army under the duc de Guise pursued them and
besieged the city. Negotiations and the assassination of the duc de Guise in 1563
led to the end of the first religious war. However, the experience of waging religious warfare during the First War of Religion reinforced confessional party
affiliations and deepened animosities.

During subsequent religious wars, armed mobilizations followed the pattern already established during the initial conflicts, but with even stronger

confessional identities. When Huguenot militants mobilized and attempted to capture the royal family during the Surprise at Meaux in September 1567, their plot failed spectacularly, as the royal family was able to escape and return to Paris. This incident shocked Catholics and deepened suspicions and animosities among opposing confessional parties. As religious conflict re-erupted across the kingdom in the Second War of Religion (September 1567 to March 1568), a royal commission mobilized military units to counter "seditious" Huguenots and to "halt the troubles, seditions, rebellions, and enterprises."[50] Royalist rhetoric employed a hardened language of troubles against Huguenots, who were now treated more generally as seditious rebels. A Huguenot noble exclaimed that "to fight to maintain the honour of God, the peace of the realm, and extend the King's power over his enemies ... that is true virtue."[51] Meanwhile, the prince de Condé claimed that he and his Huguenots followers had "armed for our security against our enemies" because of their "determination to abolish the exercise of the Reformed Religion and exterminate and drive out of your kingdom all those who profess it."[52]

Throughout the religious wars, nobles and civic magistrates justified their military mobilizations as a "defense of religion" (*défense de la religion*), a "defense of the true religion" (*défense de la vraye religion*), or "a defense of Christianity" (*défense de la chrétienté*). Various armed parties could claim to be fighting for "God's cause" (*la cause de Dieu*). Calvinist theologian Theodore Beza gradually elaborated a theory of legitimate defense by the godly against tyrants.[53] This Huguenot resistance theory arguably drew directly from the language of troubles already being employed by Huguenot nobles in their justifications for taking up arms.[54] Some Reformed theologians and pastors promoted heroic martyrdom by Reformed faithful and claimed that God had no need of being defended with arms, but nonetheless justified armed mobilization in certain circumstances.[55]

Catholic nobles and their clienteles formed armed associations and mobilized cavalry companies and infantry regiments during each cycle of religious warfare, while urban militias in predominantly Catholic cities and towns sometimes became affiliated with broader armed parties. Such military units and militia companies carried flags, banners, and badges that identified both their confessional and political affiliations. These Catholic parties could claim to be furthering the king's service and simultaneously upholding God's honor. Groups of Catholic militants thus claimed to be "true Christians" (*vrayes chrétiennes*), "good Catholics" (*bon catholiques*), "good Frenchmen" (*bon français*), or members of a "Holy League" (*sainte ligue*). Henri de Lorraine, duc de Guise, and his militant Catholic allies provided powerful leadership for the movement of such Catholic Leagues that grew up in the 1570s and 1580s. The Catholic Leagues then became even more radicalized following the assassination of the duc de Guise and Louis II de Lorraine, cardinal de Guise, at Blois in December 1588. As the religious wars progressed, Catholic Leaguers formed armed parties claiming to uphold "God's honor" (*l'honneur de Dieu*) or to defend the "Catholic, Apostolic, and Roman faith" (*la Religion Catholique, Apostolique et Romaine*).

Calvinist militants similarly asserted their Reformed identities, referring to their comrades as the "Reformed" (*reformées*), the "defenders of the church" (*défenseurs de l'église*), or the "defenders of the true church" (*défenseurs de la vraie église*). The prince de Condé referred to an association of princes and nobles to "maintain God's honor, the peace of the kingdom, and the state and liberty of the king."[56] Huguenots also described themselves collectively as members of "the churches of God" (*les eglizes de Dieu*)[57] or the "union of the churches" (*union des églises*). And like the Catholics, Huguenot militants expressed strong religious motivations and justifications for their acts of violence.

As confessional parties articulated their own collective identities, they simultaneously constructed images of their enemies. Belligerents demonized their confessional opponents as ungodly, heretical, and monstrous enemies of God who waged war on the true faith. Armed confessional parties never envisioned their opponents as merely military and political enemies, but as dangerous heretics, atheists, or Satanic forces. Denis Crouzet has powerfully demonstrated the sacral nature of violence during the religious wars, as apocalyptic fears and providential interpretations of human events and natural phenomena shaped French people's understandings of the conflicts.[58] Some royal officials and writers challenged the confessionalized justifications of violence, as when the humanist and historian François Belleforest echoed Michel de l'Hospital's arguments that God had no need of humans to fight for his causes.[59] Religious justifications could also be mocked by opponents, as when the Parisian lawyer Étienne Pasquier commented on one armed mobilization by saying that: "they took up arms under the pretext of religion."[60] Some moderate Catholics ridiculed the Catholic Leaguers' justifications for armed violence, but many Catholics embraced the Leaguers' utter rejection of Henri de Bourbon as a heretic and illegitimate monarch. Regardless of their confessional position, the militants who waged the religious wars consistently justified their causes as godly campaigns against the forces of evil and hatred.

French Catholics called their Reformed enemies Huguenots, heretics (*hérétiques*), the "opinionated" (*opiniâtres*), or "the religious" (*les religionnaires*). Most often, Catholics dismissed their enemies as "those of the so-called Reformed religion" (*ceux de la religion prétendue réformée*), or simply "those of the religion" (*ceux de la religion*). Militant Catholic Leaguers also expressed outrage at what they regarded as the dangerous and corrupt acts of their Reformed neighbors. A manuscript *mémoire* written in 1585 by a Catholic militant, for example, alleged that the Huguenots "live off of illegal commerce and scheming, desiring the troubles for their profit and embarrassing us."[61] Another Catholic *mémoire* warned that "it is feared that the heretics are recovering during the troubles and hope to fortify their cities."[62]

Huguenot militants described Catholics as the enemies of the true churches or the enemies of God. Huguenots considered Catholics as ungodly and superstitious people who were in league with Satan. They occasionally referred to Catholics as "Papists" (*papistes*) or "those of the Roman religion." More often, they used broader terms to brand them as "the ignorant" (*ignorants*), "the evildoers" (*malicieux*), "the slanderers" (*calomniateurs*). Huguenot nobles, assemblies, and clergy declared the Catholics to be "the enemies of this state and of our religion" (*les ennemis de cet estat et de nostre relligion,*"[63] or the "enemies of God" (*les ennemis de Dieu*).[64] Huguenot soldiers and civic guards carried out iconoclastic attacks on altars, sacred artworks, and liturgical objects in Catholic churches throughout the religious wars.[65] The processes of forming armed parties and mobilizing military forces thus forged confessional identities and structured contemporaries' understandings of religious warfare.

Confessional War Aims and Strategies

Confessional parties asserted war aims and formulated strategies that were explicitly religious in nature. The Huguenot and Catholic nobles who led parties recruited confessional armies to advance specific religious agendas and political causes, so their operational and tactical practices responded to their strategic aims. Their field armies conducted military operations that aimed to control communities and transform religious practices in entire regions. When military forces seized control of cities, they instituted religious reforms by opening closed or seized religious buildings, transforming religious sites, celebrating worship services, and encouraging public devotional activity by co-religionists. The same troops targeted the religious sites of their confessional enemies, defiling their sacred spaces and engaging in acts of iconoclasm and destruction. Military governors closed offending religious spaces, expelled opposing clergy, banned "heretical" preaching, invited "correct" clergy to instruct laypeople, encouraged or forced conversions, and restricted the devotional practices of religious minorities or occupied communities.

The conduct of military campaigns displayed the confessional dimensions of violence. Huguenot armed forces aimed to "cleanse" churches through iconoclasm and sober redecoration, implementing Reformed theological aims by rejecting veneration of saints, opposing the Mass, and eliminating Catholic liturgy. Huguenot troops closed monasteries, expelled Catholic clergy, and installed consistories in cities and towns under their control. In contrast, Catholic forces aimed to restore churches, monasteries, and convents that they described as having been "defiled" by Huguenot soldiers and Reformed clergy. Meanwhile, Catholic nobles and military governors supported missionary activities by Jesuits, Capuchins, and other religious orders hoping to return the entire populace to the true faith.[66] These competing attempts to enforce religious practices and spaces, however, led many Catholic and Huguenot nobles to complain of religious persecution and repression in their correspondence, petitions, and manifestos. Arguments for freedom of conscience (*liberté de conscience*) also emerged as appeals for a redress of religious grievances, although these rarely argued for a generalized tolerance. Instead, the edicts, laws, and religious peaces that granted limited religious freedoms in France seem to have been normally presented as temporary measures rather than permanent privileges (much less inherent rights) for religious minorities.

Local histories and chronicles narrated the troubles that brought religious war to their communities, revealing the confessional nature of the fighting. The rhythms of warfare varied across the kingdom, with fighting erupting in each region at different moments. Local perspectives present the experiences of townspeople who became swept up in the religious conflicts when their communities became blockaded or besieged. For example, one manuscript chronicle provided a "discourse on the things most memorable that happened in Châteauvillain and surrounding places during the troubles."[67] This account notably begins in 1589, with the military mobilizations immediately following the assassinations of Henri of Lorraine, duc de Guise, and the cardinal de Guise by royal bodyguards at the meeting of the Estates General at Blois. Episodes of confessional violence such as massacres and assassinations framed narrative approaches to individuals' and communities' experiences of religious warfare.

Interestingly, military commanders and royal officials sometimes used the language of troubles strategically to deny the existence of confessional strife and religious warfare. Military documents and correspondence frequently avoided using terms such as religious war (*guerre de religion*) or civil war (*guerre civile*), preferring generalized descriptions of troubles and disorders. At the same time, military officers and royal officials used targeted terminology of *rebellion, révolte,* and *sédition* to brand specific opponents as illegal combatants in an unjust war. A 1628 decision (*arrêt*) of the parlement de Toulouse accused Henri II de Rohan, duc de Rohan, and Benjamin de Rohan, duc de Soubise, of rebellion, claiming that:

> in the present rebellion, which has no pretext as a war of religion—the Edicts being maintained in a saintly manner—but which is only a conspiracy against the king and his state by the duc de Rohan and Monsieur de Soubise in favor of the king of England, who has incited them to make powerful armaments on the sea, to invade the Ile de Ré, and to make themselves masters of La Rochelle by conspiracy and complicity of the inhabitants of that place and other ports, harbors and towns of the kingdom.[68]

Other royal documents, religious treatises, and polemical pamphlets similarly rejected the existence of religious war, although when authors denied the reality of confessional conflict, they seem to have done so for polemical or strategic reasons. Nobles' strategic thinking responded to confessional aims, and their operational planning advanced their parties' religious and political goals.

The Miseries of France

Religious warfare produced confessional and spiritual understandings of suffering. The military campaigns of the religious wars produced intense fighting, especially in mixed confessional regions of France, as warrior nobles and their armed parties advanced confessional goals and war aims by targeting specific communities and religious sites for attacks and sieges. Raiding parties attacked châteaux and peasant villages, devastating broad swathes of the French countryside and rural communities, while urban residents coped with blockades and formal sieges. The resulting misery of the French people can be seen in the language used throughout contemporary accounts of the impact of religious conflict on their everyday lives. Noble and administrative correspondence, for example, described the ongoing conflicts not only as generalized troubles but as catastrophes that inflicted broad pain and trauma. Louis II de Bourbon, duc de Montpensier, related news about the beleaguered Catholic city of Poitiers, which was under bombardment by a besieging Huguenot army. The duc de Montpensier remained confident that the city's defenses would hold, since Poitiers was "full of many good men," including Henri de Lorraine, duc de Guise. Montpensier indicated that he would soon be meeting with Henri III and prayed that God would "soon put an end to these miserable troubles."[69] Bishops also frequently complained of the agony of the parishioners in their dioceses, as when Charles II de Péreuse d'Escars, bishop of Langres, complained of "the misery" of his province, where "everyone is entirely ruined."[70] Provincial and diocesan assemblies similarly recorded the depredations of raiding soldiers and the devastation of crops. The deputies of the Estates of Languedoc, for example, regularly complained of the anguish of the people of their province.[71]

Municipal leaders and residents in urban communities often described their collective suffering and losses during religious conflicts in petitions and supplications to the king, royal officers, and law courts. City councils typically had firm confessional identities, except in strongly bi-confessional communities that formed a confessionally divided (*mi-partie*) city council. During each episode of religious conflict, city councils affiliated themselves with particular religious and political parties or attempted to maintain neutrality. The correspondence of municipal leaders and military officers also described the massive destruction inflicted on urban centers located in war zones. Military units extracted food, clothing, money, and material resources from communities, often taking city councilors and residents hostage to ensure the payment of military contributions. Groups of soldiers also plundered valuable belongings from homes, villages, churches, and monasteries in the countryside. Some letters issued by urban officials used the language of troubles to relate generalized violence inflicted by marauding soldiers, but often narratives designated specific enemies. The city councilors of Châlons-sur-Marne complained in 1593 that the town's workers had endured "great losses of their properties in the seizure of their horses, cows, sheep, and other animals ... by the rebels and enemies of his majesty garrisoned in Vitry-le-Francois, château de Mareval, Reims, Rethel, and other places."[72] The municipal leaders of Châlons, who embraced the royalist cause, here effectively accused their regional Catholic League enemies of rebellion and sought damages for their town's losses.[73] Cities and towns that were blockaded or besieged often experienced much greater destruction. When cities and towns resisted, confessional militancy seems to have fueled harsh punishments on besieged garrisons and townspeople.[74]

Such petitions and letters show that many urban centers were transformed by religious warfare, but chronicles and geographies also related the suffering of urban populations. The authors of chronicles recounted the devastation wrought by urban riots, religious massacres, blockades, and sieges, while geographic works detailed some of the more concrete local and regional dimensions of religious conflicts. For example, after the Huguenot-controlled town of Royan, located on the Atlantic coast south of the major Huguenot city of La Rochelle, enhanced its fortifications during the religious wars, a manuscript entitled "Plans de places fortes," described the new fortifications: "during our troubles, the Rochelais [Huguenots from La Rochelle], in order to have a free hand along the coast and a safe refuge in Guyenne when they needed it, constructed bastions, ravelins, half-bastions, and additional outworks."[75] Many other towns built new fortifications during the religious wars, often prompting nearby communities to enhance their own defenses.

French people recognized the misery of civilians during the religious wars and often responded with acts of charity. French kings also granted gifts and pensions to corporate bodies and individuals who had incurred severe losses during the fighting. Henri IV, for example, offered a pension to nuns at a convent, stating that "we desire to favorably treat the poor nuns, abbess, and convent de Canteloup, in consideration of their poverty and the inconveniences that they have suffered during the troubles and to give them means of maintaining the divine service and to relieve some of the pains that they have suffered."[76]

The immense costs of religious warfare strained the finances of the nobles who raised military forces and fought in the wars. Many warrior nobles who acted as military entrepreneurs reported massive financial and material losses that they accrued in waging religious warfare. Saladin de Montmorillon, seigneur de Montmorillon, for example, lamented his inability to pay for the marriage of his daughter or to pay for arming his son for war because of the financial losses he had suffered "since these troubles."[77] Many of these warrior nobles' own châteaux were also seriously damaged or destroyed in raiding warfare, and both noblemen and noblewomen often discussed their fears and losses in their personal and family correspondence. Many correspondents also worried about their ability to continue communicating with their family members and friends. For example, an undated letter by one nobleman stated that: "I can see that if the troubles continue in France, I will have no other way of telling you anything without the resourcefulness of this courier.[78]

Contemporaries referred to the long duration and seemingly incessant nature of the religious wars, despite truces and religious peaces. Henri III attempted to reform the letter-writing practices of the royal council, noting "the confusion that had been introduced into our council by the mischief of our times and the long continuation of the troubles."[79] A manuscript entitled "Oration on the miseries of France" lamented that it had been "such a long time that we have seen these troubles last in this kingdom."[80] French people often deplored the religious divisions in the kingdom, expressing war weariness and hoping for God's deliverance from the miseries of war.

The Pacification of Troubles

The language of troubles often included expressions of a compelling desire for peace, which could only be brought by God. Royal edicts and declarations repeatedly expressed a desire for peace and concord among the king's subjects. A 1568 pamphlet on the pacification of the troubles, for example, explained the reasons for the peace as aiming to forge religious reconciliation and concord.[81] Henri III similarly explained in a 1583 edict that God had moved him "to embrace the pacification of the troubles of his kingdom."[82] Such religious peaces aimed to end confessional conflict and civil warfare and to heal religious divisions in the kingdom, but implementing these peaces involved local negotiations, logistical planning, and military and security operations to calm religious tensions. However, even expressions of pacific motives were shaped by confessional politics and religious warfare.

Beginning with the Peace of Amboise, which ended the First War of Religion in 1563, royal edicts of pacification announced and formalized each negotiated religious peace, and were then gradually implemented by royal commissioners in every province.[83] Additional royal edicts attempted to manage demobilization processes and prevent further arming of military forces in an attempt to return the number of companies and regiments to the level that had existed before the "disorders" (*mouvements*).[84] Catherine de Médicis, queen mother and advisor to Charles IX, also organized a royal tour of the kingdom in 1563–64, conducting extensive negotiations to implement the first religious peace.[85] The outbreak of the Second War of Religion curtailed this personal approach to pacification, however. Later religious peaces instead relied on royal commissioners to conduct local negotiations in each region.

Many nobles, clergy, and municipal officials also appealed for the calling of religious conferences and assemblies to bring about peace and order. The resulting assemblies of the clergy and religious disputations sought to overcome confessional divisions and find religious compromises, but they often failed to find common ground. The Colloquy of Poissy (September to October 1561) represented the most publicized religious disputation of this type, and included direct royal involvement, but many other disputations were held in communities throughout the kingdom.[86] Public preaching also brought religious arguments and concerns about peace into churches and city squares, while religious debates proliferated in printed polemical pamphlets that brought discussions of the prospects for religious peace to broader reading and listening audiences.

Contemporaries also discussed the need for regional and general assemblies of nobles to negotiate peaces and halt religious wars. Some letters and petitions called explicitly for the assembly of the Estates General, that is, an assembly of representatives of the clergy, nobility, and commoners, to bring peace to the realm. Henri I de Montmorency, duc de Montmorency, argued in a 1574 manifesto that it was necessary to have "an assembly of the Estates General, the sole remedy to pacify the troubles."[87] The duc de Montmorency is also notable for leading a moderate Catholic party in the religiously mixed province of Languedoc, and at times seeking bi-confessional compromises. Some Catholic militants also called for assemblies of the Estates General, seeking to bring religious conformity and order to the kingdom through unified action against heresy. A letter, presumably written by an adherent of the Catholic League in the 1590s, explained that "when the great troubles and disfunction of this kingdom that have required a prompt and powerful remedy, we have always had recourse to the assembly and holding of Estates General composed of three orders."[88] Indeed, some Catholic Leaguers claimed authority to be acting in anticipation of a coming Estates General. Royal officials in Paris who supported the Catholic League also argued that they followed the decree (*lettres patentes*) issued by the duc de Mayenne, who acted as "lieutenant general of the royal state and crown of France of the general council of the Union of Catholics established in Paris, waiting for the assembly of the Estates of this kingdom."[89]

Religious peacemaking was cumulative, with each edict of pacification referring directly to previous religious peaces, and with articles confirming or replacing their individual provisions. Military commanders and town governors also periodically negotiated local and regional truces during the religious wars. For example, a regional truce was negotiated between "the seigneurs and governors of the region of Velay of one and the other party" in 1596.[90] Edicts of pacification then often referred to the terms of such regional truces, as well as to the submissions of individual military commanders and compositions of cities, and these local peace pacts and regional truces could also be partially or fully incorporated into broader religious peaces that were implemented by royal commissioners.[91]

The peace treaties themselves used the language of troubles in framing the problem of religious warfare. In the preamble of the Edict of Nantes, for example, Henri IV referred to the "frightful troubles, confusions, and disorders which prevailed at our accession to this kingdom, which was divided into so many parties and factions."[92] Many articles of the Edict of Nantes also referred to unacceptable actions made "during the troubles" and illegitimate changes that had occurred "on account of the troubles." The language of troubles then further guided the work of the pairs of royal commissioners (normally one Catholic, one Reformed) who worked to implement the religious peace in each province and region, as well as of the provincial governors and lieutenant generals who intervened in the peace implementation process, arranging local affairs between Catholics and Huguenots. Anne de Lévis, duc de Ventadour, for example, cited the Edict of Nantes in attempting to settle a dispute in Languedoc in 1600, indicating that he was "following the article 72 of the said Edict, by which all cities will enjoy the inviolable privileges, liberties, jurisdictions, and seats of justice that they had before the troubles."[93]

Religious peaces during the Wars of Religion usually called for general demobilizations of all parties' military forces across the kingdom as part of pacification. Since the processes of demobilizing field armies and garrisons required military operations involving extensive war financing and planning, however, negotiated agreements usually granted large payments to warrior nobles to reimburse them for their military expenditures and to help them pay their soldiers' wages, which were often in arrears. Peace agreements also provided for soldiers' needs while marching toward their home regions within France or beyond its borders. The conference of Loudun in 1616, for example, produced a series of articles intended for "the pacification of the troubles," and provision for the demobilization of military units.[94]

Religious peace agreements provided for general amnesties that exonerated military officers and soldiers for potentially rebellious conduct. Such general pardons provided certain exclusions for particularly heinous crimes, however, placing many officers and soldiers in ambiguous positions. As a result, monarchs issued numerous individual pardons in the wake of most religious peaces, often in response to petitions from nervous individuals. Thus, Henri IV issued a pardon in 1599 to Charles de La Verdure after he excused his support for soldiers who had pillaged the region "during the last troubles."[95]

Once religious peaces were publicized, military commanders, administrative officers, and regimental officers sought to integrate demobilized officers and soldiers into society. Army commanders and provincial officials offered recommendations for lower-ranking officers and soldiers who had been demobilized, carefully positioning their service in relationship to the "troubles," while military officers crafted histories of their subordinates' military service in civil warfare. The duc de Ventadour offered a recommendation for a nobleman "who was disabled in the king's service during the recent troubles."[96] The duc de Montmorency recommended another noble, stating: "among all the good and faithful servants that your majesty has in my government of Languedoc, must be numbered the seigneur du Brouteil, who has over the past twenty-five years and above all during the most recent troubles in your kingdom, has never missed an opportunity to testify to the zeal and ardor that he has for the good of your majesty. In addition, by his prudence and diligence, he reduced the fortress of Brescou, situated in the sea [off of Cap d'Agde], to your obedience."[97] Catholic nobles in particular were able to associate their experiences in religious warfare with royal service, while Huguenot nobles were left in precarious positions in relationship to the monarchy.

The Fires of Troubles

Religious peaces aimed to pacify the troubles by halting civil warfare and calming religious tensions between Catholic and Reformed communities. However, the implementation of religious peaces by royal commissioners was always problematic, as armed parties and local communities contested unfavorable provisions of treaties. Moreover, royal officials and Catholic clergy also expressed concerns over any innovations in religious beliefs and practices, arguing that traditions must be upheld. Catholic processions involving clergy, nobles, and civic officials sought civic unity and religious union, yet they often included armed bodyguards and civic militia companies that threatened Calvinists. Huguenots, meanwhile, worked for the restoration of what they viewed as an original and pristine church, cleansed of Papist pollution. So while contemporaries expressed hopes of healing religious divisions and ending heresy and impieties through a religious union and restoration of a unified Christian church, they could not agree on what form a unified Christian church should take. Many French people worried that the fires of recent troubles would reignite.

Military commanders and municipal leaders routinely complained of violations of peace agreements. An undated manuscript *mémoire,* presumably written by a moderate Catholic, argued that the so-called Reformed heresy and the religious divisions it produced were the source of all the kingdom's woes. The anonymous author of this *mémoire* insisted that order needed to be re-established, "and by this means, we and they can live in peace and union, without having to fear that every three years new troubles and civil wars will recommence."[98] Even as Catherine de Médicis worked to implement a religious peace in 1579, she observed that Huguenots "seem to be searching only to recommence the troubles."[99]

Religious peaces attempted to restrict Reformed preaching generally, while permitting certain locations for Calvinist worship. Royal administrative officers and judges periodically enforced specific restrictions on Reformed practice in regional and local contexts. For example, in 1569 the parlement de Paris issued injunctions to halt Calvinist preaching and prevent the instruction of Reformed beliefs.[100] Predominantly Reformed cities and towns, however, were usually able to establish consistories to enforce Reformed beliefs and morals in their communities, and Huguenot military governors and town councils supported

Reformed ministers and sought to block Catholic clergy from preaching in their urban spaces. Some religious peaces, moreover, confirmed certain communities as Reformed-controlled security towns (*places de sûreté*) as concessions to Huguenot military commanders and municipal authorities.

The Edict of Nantes (1598) is certainly the best-known religious peace, but disputes over the Edict's implementation continued for years, especially in mixed-confessional regions and communities in southern France.[101] Bi-confessional teams of royal commissioners struggled to negotiate local settlements in accordance with the Edict of Nantes's complex provisions.[102] Cities and towns that had been designated *places de sûreté* and that had Reformed majority populations resisted restoring Catholic worship and delayed the implementation of other pro-Catholic articles of the Edict of Nantes. In 1606, six years after the promulgation of the Edict of Nantes, royal commissioners in Montauban reported on their continuing efforts to implement the religious peace there: "We have come to this city of Montauban following the command that it has pleased your majesty to give us to complete the implementation of the Edict of Nantes and the responses made by your majesty in his council on the articles which were presented to him by the syndic of the clergy of the diocese, and after having heard the said syndic of the clergy and the consuls and inhabitants [of Montauban]."[103] Montauban would continue to experience religious tensions between its Reformed residents, who represented a significant majority, and its Catholic minority.

Religious tensions in urban communities often hindered the permanence of peaces, with disagreements reigniting confessional conflicts and religious riots. Royal administrative officers and military commanders recognized the threat of disorder and worried about the potential for renewed religious warfare. For example, when armed Huguenots surprised the town of Fiac and seized its château in 1600, the duc de Ventadour reported to the duc de Montmorency, who was the provincial governor of Languedoc, that the Huguenots "had razed and demolished [the château], chased out the priests and all the Catholic inhabitants, and mortally wounded the first consul." The duc de Ventadour further noted that "if a similar attack had been committed by Catholics on those of the said religion, the fire of past troubles would have been reignited in all four corners of your government [of Languedoc]." He claimed that local Catholics were not alarmed, however, because they had confidence that their provincial governor would deliver justice.[104]

Continuing confessional tensions sparked urban riots, regional conflicts, and religious wars well into the seventeenth century. Henri IV, for example, intervened in 1602 to protect Catholics in the countryside around the Huguenot stronghold of Nîmes from seizing and destroying their grain.[105] Royal officials also frequently reported on the alarms and rumors that continued to disturb the peace. A presiding judge of the parlement de Toulouse wrote in 1604 of the need "to appease the troubles, disorders, acts of defiance, quarrels, and divisions of the city of Le Puy [en Velay]."[106] Following the shocking assassination of Henri IV in May 1610, the duc de Ventadour stressed the importance of his presence in the province of Languedoc, claiming that "without that [presence], great troubles and disorders will occur."[107] No major fighting erupted that year, but successive confessional conflicts and religious wars continued until the Huguenot military forces collapsed in the 1620s, finally bringing the troubles to an end.[108]

Conclusion

The descriptions of "troubles" in French society reflected the religious divisions that had emerged in France as Latin Christianity progressively splintered across Europe during the sixteenth century. Reformed preaching, conversions, and clandestine worship created deep fears and tensions in French communities, which were generally affiliated with Catholic reforms and confessional identities. As political instability grew, religious disputes and animosities fueled religious riots and confessional conflicts across the kingdom. A language of troubles increasingly described warfare waged in the name of God between armed groups with distinct confessional affiliations. Political, economic, social, and cultural factors certainly figured in the religious wars but were generally articulated through a confessional lens. The language of troubles could be deployed in many different contexts, but almost always alluded to religious disputes and confessional conflicts. Contemporaries often used the language of troubles polemically against their opponents, ensuring that the concept of religious warfare was hotly contested and often problematic.

Huguenot and Catholic militants engaged in warfare that they understood as fundamentally religious in nature and described through this language of troubles. Nobles formed religious and political parties, constructing collective identities as defenders of God's honor and of the true Christian faith. Various parties mobilized armed forces and pursued confessional agendas and war aims, using specific strategies and military operations to target urban centers and sacred sites for religious transformation. Confessionalized parties branded their opponents as heretics and enemies of God, creating the conditions for massacres and atrocities. The practices of religious warfare thus responded to the polemical language of troubles but also enlarged it with each episode of confessional violence.

Although French people continued to articulate hopes for religious peace and dreams of Christian unity well into the seventeenth century, few Catholics or Huguenots seem to have had any doubts that their wars were indeed fought substantially over religious issues. Even optimistic texts that expressed desires to restore Christian unity referred to God's religion and the true faith in the singular. Both Catholic and Huguenot militants claimed to be fighting to uphold God's honor and to protect His Church. Religious pluralism and toleration remained negative concepts that were acceptable for most people only as temporary states until an imagined Christian unity could be restored.

Despite the rich evidence of the language of troubles and the practices of religious warfare, some historians continue to minimize the confessional dimensions of the conflicts in sixteenth-century France. William T. Cavanaugh notably dismisses the entire notion of "wars of religion," claiming that the concept is a statist historiographical construction that forged a "myth of religious violence."[109] Barbara Diefendorf has refuted Cavanaugh's arguments, offering a forceful restatement that the Wars of Religion were indeed fundamentally waged over religious issues and controversies.[110] Examining the language of troubles and the associated practices of organizing confessional parties demonstrates that the concept of religious warfare indeed emerged during the conflicts that contemporaries referred to as "troubles" and that we call the French Wars of Religion. If we look beyond the battlefield, we can see how the language of troubles effectively described the practices and experiences of religious warfare in early modern France. Historians of war, culture, and society need to be attentive to the roles that religion may have played in other periods of armed conflict as well.

Notes

1 Research for this chapter was made possible by generous residential fellowships at the Institut d'Études Avancées de Paris and the IMéRA in Marseille, as well as a sabbatical from Northern Illinois University. All translations are mine, unless otherwise noted.

2 *Edict et declaration faicte par le roy Charles IX. de ce nom sur la pacification des troubles de ce Royaume : le xix iour de Mars, mil cinq cens soixante deux. Publié en la Cour de Parlement à Paris, le vingtseptiesme iour dudict mois* (Paris: Robert Estienne, 1563).

3 James B. Wood, *The King's Army: Warfare, Soldiers, and Society during the Wars of Religion in France, 1562–1576* (Cambridge: Cambridge University Press, 1996), 184–204.

4 Mack Holt, *The French Wars of Religion, 1562–1629,* 2nd ed. (Cambridge: Cambridge University Press, 2005); Mack Holt, "Putting Religion Back into the Wars of Religion," *French Historical Studies* 18 (Fall 1993): 524–51; Barbara B. Diefendorf, *Beneath the Cross: Catholics and Huguenots in Sixteenth-Century Paris* (Oxford: Oxford University Press, 1991); Denis Crouzet, *Les guerriers de Dieu. La violence au temps des troubles de religion (vers 1525–vers 1610),* 2 vols. (Seyssel: Champ Vallon, 1990); Philip Benedict, *Rouen during the Wars of Religion* (Cambridge: Cambridge University Press, 1981); Natalie Zemon Davis, "The Rites of Violence," in *Society and Culture in Early Modern France* (Stanford, CA: Stanford University Press, 1975), 152–87.

5 Jérémie Foa, *Le tombeau de la paix. Une histoire des édits de pacification (1560–1572)* (Limoges: Presses universitaires de Limoges, 2015); Penny Roberts, *Peace and Authority during the French Wars of Religion, c. 1500–1600* (Houndmills: Palgrave Macmillan, 2013); Brian Sandberg, *Warrior Pursuits: Noble Culture and Civil Conflict in Early Modern France* (Baltimore: Johns Hopkins University Press, 2010);

Allan A. Tulchin, *That Men Would Praise the Lord: The Triumph of Protestantism in Nîmes, 1530–1570* (Oxford: Oxford University Press, 2010); Nicolas Le Roux, *Les Guerres de religion, 1559–1629* (Paris: Belin, 2009); Mark Greengrass, *Governing Passions: Peace and Reform in the French Kingdom, 1576–1585* (Oxford: Oxford University Press, 2007); Mark Konnert, *Local Politics in the French Wars of Religion: The Towns of Champagne, the Duc De Guise, and the Catholic League, 1560–95* (Aldershot: Ashgate, 2006); Ariane Boltanski, *Les ducs de Nevers et l'état royal. Genèse d'un compromis (ca 1550–ca 1600)* (Geneva: Librairie Droz, 2006); Raymond A. Mentzer and Andrew Spicer, eds., *Society and Culture in the Huguenot World, 1559–1685* (Cambridge: Cambridge University Press, 2002); Nicolas Le Roux, *La faveur du roi. Mignons et courtisans au temps des derniers Valois (vers 1547–vers 1589)* (Seyssel: Champ Vallon, 2000); Olivier Christin, *La paix de religion. L'autonomisation de la raison politique au XVIᵉ siècle* (Paris: Seuil, 1997).

6   This chapter utilizes diverse manuscript and printed collections from the Bibliothèque Nationale de France [hereafter, BNF] and several Archives départementales [hereafter, AD], and Archives municipales [hereafter, AM] located in cities that were within the early modern provinces of Guyenne, Languedoc, and Provence, which had heavily mixed Catholic and Reformed populations.

7   John Q. Lynn, *Battle: A History of Combat and Culture* (Boulder, CO: Westview, 2003); Geoffrey Parker, *The Military Revolution: Military Innovation and the Rise of the West, 1500–1800*, 2nd ed. (Cambridge: Cambridge University Press, 2003); J. R. Hale, *War and Society in Renaissance Europe, 1450–1620* (Baltimore: Johns Hopkins University Press, 1985); William H. McNeill, *The Pursuit of Power* (Chicago, IL: University of Chicago Press, 1982); John Keegan, *The Face of Battle: A Study of Agincourt, Waterloo, and the Somme* (London: Penguin, 1976).

8   Ilya Berkovich, *Motivation in War: The Experience of Common Soldiers in Old-Regime Europe* (Cambridge: Cambridge University Press, 2017); Michael J. Hughes, *Forging Napoleon's Grande Armée : Motivation, Military Culture, and Masculinity in the French Army, 1800–1808* (New York: New York University Press, 2012); Alan Forrest, *Napoleon's Men: The Soldiers of the Revolution and Empire* (London: Bloomsbury, 2006); James M. McPherson, *For Cause and Comrades: Why Men Fought in the Civil War* (Oxford: Oxford University Press, 1997); Alan R. Millett and Williamson Murray, eds., *Military Effectiveness*, 3 vols. (Boston: Allen and Unwin, 1988); John A. Lynn, *The Bayonets of the Republic: Motivation and Tactics in the Army of Revolutionary France, 1791–94* (Urbana, IL: University of Illinois Press, 1984).

9   Sigrun Haude, *Coping with Life during the Thirty Years' War (1618–1648)* (Leiden: Brill, 2021); Barbara Donagan, *War in England 1642–1649* (Oxford: Oxford University Press, 2010); Sandberg, *Warrior Pursuits*; Peter H. Wilson, *The Thirty Years War: Europe's Tragedy* (London: Penguin, 2009); Peter J. Arnade, *Beggars, Iconoclasts, and Civic Patriots: The Political Culture of the Dutch Revolt* (Ithaca, NY: Cornell University Press, 2008).

10   See, for example, Andrew Cunningham and Ole Peter Grell, *The Four Horsemen of the Apocalypse: Religion, War, Famine, and Death in Reformation Europe* (Cambridge: Cambridge University Press, 2000); Wood, *The King's Army*; Charles Carlton, *Going to the Wars: The Experience of the British Civil Wars 1638–1651* (London: Routledge, 1994).

11   Penny Roberts, "The Most Crucial Battle of the Wars of Religion? The Conflict over Sites for Reformed Worship in Sixteenth-Century France," *Archiv für Reformationsgeschichte* 89 (1998): 247–67.

12   Arlette Jouanna, *La France du XVIe siècle, 1483–1598* (Paris: Presses Universitaires de France, 1996), 325–40; Jacques Dupâquier, ed., *De la Renaissance à 1789*, vol. 2

of *Histoire de la population française* (Paris: Presses Universitaires de France, 1988), 81–94.

13 Religious polemics employed the imagery of the world turned upside down. Luc Racaut, *Hatred in Print: Catholic Propaganda and Protestant Identity during the French Wars of Religion* (Aldershot: Ashgate, 2002), 81–98.

14 Davis, "The Rites of Violence," 152–87.

15 Jérémie Foa, *Tous ceux qui tombent. Visages du massacre de la Saint-Barthélemy* (Paris: La Découverte, 2021); Tulchin, *That Men Would Praise the Lord;* Diefendorf, *Beneath the Cross,* 49–63, 84–88.

16 *Le Resveil-matin des françois. Touchant les troubles & mouvuements de ce temps* (1622).

17 Susan Broomhall and Sarah Finn, eds., *Violence and Emotions in Early Modern Europe* (London: Routledge, 2019).

18 Pierre Viret, *De la vraye et favsse religion* ... (n.p.: Jean Rivery, 1560).

19 Gabriella Scarlatta and Lidia Radi, eds., *Representing Heresy in Early Modern France* (Toronto: University of Toronto Press, 2017).

20 See Anonymous, *Henri IV, vainqueur de la Ligue*, painting, c. 1600; Anonymous, *Henri IV pourchassant l'Hérésie*, print, 17th c.; *Ordre des villes dostage...*, c. 1622; Anonymous, *La Vraye Femme*, print, 17th c., BNF, Estampes, Marolles, Réserve TF-1-fol.

21 Racaut, *Hatred in Print*, 33–35.

22 Crouzet, *Les guerriers de Dieu*, 1: 163–318.

23 Racaut, *Hatred in Print;* Jeffrey K. Sawyer, *Printed Poison: Pamphlet Propaganda, Faction Politics, and the Public Sphere in Early Seventeenth-Century France* (Berkeley, CA: University of California Press, 1990). Roger Chartier and Henri-Jean Martin, eds., *Histoire de l'édition française. 1. Le Livre conquérant. Du Moyen Age au milieu du XVIIe siècle*, 2ᵉ éd. (Paris: Fayard, 1989).

24 Arlette Jouanna, *Le Devoir de révolte. La noblesse française et la gestation de l'État moderne, 1559–1661* (Paris: Fayard, 1989), 281–312.

25 Andrew Pettegree, *The Invention of News: How the World Came to Know About Itself* (New Haven: Yale University Press, 2014); Brendan Dooley, ed., *The Dissemination of News and the Emergence of Contemporaneity in Early Modern Europe* (Farnham: Ashgate, 2010).

26 Diefendorf, *Beneath the Cross*, 50–63.

27 Henri Lancelot Voisin de La Popelinière, *La Vraye, & entiere histoire des trovbles et choses memorables adduenues, tant en France qu'en Flandres, & pays circonuoisins, depuis l'an mil cinq cents soixante & deux* ... (Basel: Barthelemy Germain, 1578); *Original des troubles de ce temps* ... (Nantes: Nicolas des Marestz and François Fauerye, 1592).

28 Louis de Bourbon, prince de Condé, to Antoine de Bourbon, roi de Navarre, Orléans, 13 June 1562, copy, BNF, Mss. fr. 4737, fᵒ 102–4.

29 Louis d'Orléans, *Apologie ov defence des Catholiqves vnis les vns auec les autres, contre les impostures des Catholicques associez à ceux de la pretenduë Religion* (n.p.: n.p., 1586), 5.

30 "Procession pour la pacification des troubles du dimanche 21 juin 1562," BNF, Mss. fr. 5221, fᵒ 114–17.

31 Henri III to Jacques de Savoie, duc de Nemours, Paris, 28 August 1578, BNF, Mss. fr. 3397.

32 Greengrass, *Governing Passions*, 370.

33 Diefendorf, *Beneath the Cross*, 64–92; Penny Roberts, "Emotion, Exclusion, Exile: The Huguenot Experience during the French Religious Wars," in *Feeling Exclusion: Religious Conflict, Exile and Emotions in Early Modern Europe*, eds. Giovanni Tarantino and Charles Zika (London: Routledge, 2019), 11–26.

34 Mémoire, BNF, Mss. fr. 3363, f° 156–57.
35 Claude de Vergy, comte de Champlitte, to Louis de Gonzague, duc de Nevers, Gray, 21 May 1593, BNF, Mss. fr. 3631, f° 52.
36 Jouanna, *Le Devoir de révolte*, 368–90.
37 Ibid.
38 Brian Sandberg, "'Accompanied by a Great Number of Their Friends': Warrior Nobles and *Amitié* during the French Wars of Religion," in *Friendship and Sociability in Early Modern Europe: Contexts, Concepts, and Expressions,* eds. Amyrose McCue Gill and Sarah Rolfe (Toronto: Centre for Reformation and Renaissance Studies, 2014), 171–91.
39 *Sommaire declaration et confession de foy, Faitcte par Monseigneur le Prince de Condé, contre les calomnies & impostures des ennemis de Dieu, du Roy, & de luy* (n.p. [Orléans]: n.p., 1562).
40 *Traicté d'association faicte par Monseigneur le Prince de Condé auec les Princes, Cheualiers de l'ordre, Seigneurs, Capitaines, Gentilshommes & autres de tous estats, qui sont entrez, ou entreront cy apres, en ladicte association, pour maintenir l'honneur de Dieu, le repos de ce Royaume, & l'estat & liberté du Roy soubs le gouuernement de la Royne sa mere* (n.p. [Orléans]: n.p., 1562).
41 *Prieres ordinaires des soldatz de l'armee conduite par monsieur le Prince de Condé: accommodees selon l'occurrence du temps. Zvvey schone gebet vvelche die soldaten die unter dem herrn Printzen von Conde ligen beide abends und morgens vvenn sie auf und von der vvacht zihen sprechen* (n.p. [Orléans]: n.p., 1562).
42 *La Juste et saincte defense de la ville de Lyon* (Lyon, 1563).
43 Philip Benedict, *Christ's Churches Purely Reformed: A Social History of Calvinism* (New Haven: Yale University Press, 2002), 144–45.
44 Benedict, *Rouen during the Wars of Religion*, 95–97.
45 "L^{res} de comission du gouuernem^t de Normandye po^r mons^r dAumalle durant les troubles aduenues en France en lannée g v^C lxii," BNF, Mss. fr. 4682, f° 7–8.
46 Ibid.
47 Benedict, *Rouen during the Wars of Religion*, 99–103.
48 "L^{res} de comission du gouuernem^t de Normandye po^r mons^r dAumalle durant les troubles aduenues en France en lannée g v^C lxii," BNF, Mss. fr. 4682, f° 7–8.
49 Wood, *The King's Army*, 184–204.
50 "Copp^e de la cõmission du roy aux elluz de Paris po^r f.e leuer deux cens pionniers et cent cheuaulx dartillerie en le^r ellõn, du iie octobre 1567," BNF, Mss. fr. 4682, f° 131–32.
51 "Lettre missive d'un gentilhomme," 1567, in *The French Wars of Religion: Selected Documents,* ed. David Potter (Houndmills: Macmillan, 1997), 100–2. The translation is Potter's.
52 Louis I de Bourbon, prince de Condé, "La requeste du prince de Condé et gentilshommes de France," c. October 1567, SP70/95, f° 96–97, in *The French Wars of Religion*, 102–3. The translation is Potter's.
53 Benedict, *Christ's Churches*, 144–8.
54 Robert M. Kingdon, "Calvinism and Resistance Theory, 1550–1580," in T*he Cambridge History of Political Thought, 1450–1700*, ed. J. H. Burns (Cambridge: Cambridge University Press, 1991), 200–14.
55 Nikki Shepardson, *Burning Zeal: The Rhetoric of Martyrdom and the Protestant Community in Reformation France, 1520–1570* (Bethlehem, PA: Lehigh University Press, 2007).
56 *Traicté d'association...* (See note 40).
57 Pasteur Segond to M. de Galerende and M. de Basin, deputés généraux des eglises, Saint-Paragoire, 12 January 1629, BNF, Dupuy 100, f° 306–7.

58 Crouzet, *Les guerriers de Dieu*, 1: 131–62 and 2: 361–403.

59 François Belleforest, *L'histoire des nevf roys Charles de France* (Paris: Olivier de P. L'Huiller, 1568).

60 Pasquier quoted in Jean-Baptiste de La Curne de Sainte-Palaye, *Dictionnaire historique de l'ancien langage françois* ou *Glossaire de la langue françoise depuis son origine jusqu'au siècle de Louis XIV* (Paris: L. Favre, Honoré Champion, 1875–1882).

61 "Sur la resolution s^te et treschrestiene que voz ma^tez ont faict de ne voulloir plus qu'une religion en France," BNF, Mss. fr. 3363, f° 160.

62 Mémoire, n.d., BNF, Mss. fr. 3362, f° 89.

63 "Acte de ceux de la religion pretendue refformée assemblez a Anduze par lequel ilz promettent de vouloir demeurer fermes en leur union et en la jonction de leurs armes avec celle du roy d'Ang^re & des s^rs de Rohan & de Soubize," BNF, Dupuy 100, f° 296–97.

64 Pasteur Segond to M. de Galerende and M. de Basin, deputés généraux des eglises, Saint-Paragoire, 12 January 1629, BNF, Dupuy 100, f° 306–7.

65 Olivier Christin, *Une révolution symbolique: l'iconoclasme huguenot et la reconstruction catholique* (Paris: Éditions de Minuit, 1991); Solange Deyon and Alain Lottin, *Les 'Casseurs' de l'été 1566: L'iconoclasme dans le Nord de la France* (Paris: Hachette, 1981); Robert Sauzet, "L'iconoclasme dans le diocèse de Nîmes au XVIe et au début du XVIIe siècle," *Revue d'histoire de l'église de France* 66 (1980): 5–15.

66 Brian Sandberg, "The Means to Rebuild the Church: Noble Networks, Piety, and Religious Patronage in Southern France and Tuscany," *Sixteenth Century Journal* 52 (Fall 2021): 667–95.

67 "Discours des choses plus memorables aduenues a Chasteauuilain & lieux circonuoisins ez derniers troubles depuis l'an 1589 iusques a l'an 1593," Mss. fr. 20153, f° 371–81.

68 *Arrest de la cour de Parlement contre les rebelles* (Toulouse: Raymond Colomiez, 1628), AD Tarn, C 207.

69 Louis II de Bourbon, duc de Montpensier, to Jacques de Savoie, duc de Nemours, camp de Beaulieu près Loches, 5 August 1577, BNF, Mss. fr. 3338, f° 95–96.

70 Charles II de Péreuse d'Escars, évêque de Langres, to Louis de Gonzague, duc de Nevers, Mussy, 14 March 1593, BNF, Mss. fr. 3624, f° 38.

71 For representative examples, see the deliberations of the Estates of Languedoc in September and October 1621, AD Hérault, C 7059, f° 18–72.

72 Échevins de Châlons to Louis de Gonzague, duc de Nevers, Châlons, 6 July 1593, BNF, Mss. fr. 3625, f° 63.

73 On local rivalries and the complex politics of the province of Champagne during the wars of the Catholic League, see: Konnert, *Local Politics in the French Wars of Religion*, 209–55, 260–64.

74 Brian Sandberg, "'The Enterprises and Surprises that They Would Like to Perform': Fear, Urban Identities, and Siege Culture during the French Wars of Religion," in *The World of the Siege: Representations of Early Modern Positional Warfare*, eds. Anke Fischer-Kattner and Jamel Ostwald (Leiden: Brill, 2019), 265–87; Brian Sandberg, "'Only the Sack and the Noose for its Citizens': Atrocities against Civilians in the Wars of Religion in Early Seventeenth-Century France," in *Inventing Collateral Damage: Civilian Casualties, War and Empire*, eds. Stephen J. Rockel and Rick Halpern (Toronto: Between the Lines Press, 2009), 97–114.

75 "Plans de places fortes," BNF, Mss. fr. 15380, f° 96.

76 "Don a des religieuses d'une rente quilz donnent au roy," copy, BNF, Mss fr. 4014, f° 193.

77   Saladin de Montmorillon, seigneur de Montmorillon, to Louis de Gonzague, duc de Nevers, Vésigneux, 8 May 1593, BNF, Mss. fr. 4710, f° 126. On Saladin de Montmorillon and the château de Vésigneux, see: comte de Chastellux, "Vésigneux," *Bulletin de la Société nivernaise des lettres, sciences et arts* (1890): 152–73.

78   [?] to duchesse de Nemours, n.p., n.d., BNF, Mss. fr, 3397, f° 67.

79   "Ensuit la forme des lettres de con^er au conseil destat de sa ma^te," BNF, Mss. fr. 4805, f° 875–77.

80   "Oraison sur les miseres de la France et pour lestat dicelle recommandee," BNF, Mss. fr. 4681, f° 1–28.

81   *Discovrs svr la pacification des trovbles de l'An 1567* (Anvers: Conras Thetieu, 1568).

82   "Deffence de f.e aucun amas de gens de guerre," copy, BNF, Mss. fr. 4805, f° 722–26.

83   Foa, *Le tombeau de la paix*; Christin, *La paix de religion.*

84   "Estat des gens de guerre que le roy veult entretenir doresnavant," BNF, Mss. fr. 15582, f° 198–99.

85   Jean Boutier, Alain Dewerpe, and Daniel Nordman, *Un tour de France royal. Le voyage de Charles IX (1564–1566)* (Paris: Aubier, 1984).

86   Stuart Carroll, *Martyrs and Murderers: The Guise Family and the Making of Europe* (Oxford: Oxford University Press, 2009), 128–59; Michael Wolfe, "Exegesis as Public Performance: Controversialist Debate and Politics at the Conference at Fontainebleau (1600)," in *Politics and Religion in Early Bourbon France,* eds. Alison Forrestal and Eric Nelson (Houndmills: Palgrave Macmillan, 2009), 65–85.

87   BNF, Mss. fr. 3392, f° 44–47.

88   [?] to Louis de Gonzague, duc de Nevers, n.d., BNF, Mss. fr. 3616, f° 14.

89   Lettres patentes of the Catholic League, Paris, 9 November 1589, BNF, Mss. fr. 26477, f° 35–36.

90   "Articles accordees entre les seigneurs gouuerneurs du pays de Vellay de lung & lau.e partye pour le repos & soulagem^t dicelluy," BNF, Mss. fr. 3584, f° 34–36.

91   Jérémie Foa, "Making Peace: The Commissions for Enforcing the Pacification Edicts in the Reign of Charles IX (1560–1574)," *French History* 18, 3 (2004): 256–74.

92   "The Edict of Nantes with Its Secret Articles and Brevets," translated by Jotham Parsons, 41.

93   Anne de Lévis, duc de Ventadour, to Henri I de Montmorency, duc de Montmorency, Montpellier, 9 December 1600, BNF, Mss. fr. 3589, f° 37–38.

94   "Dernier Prolongation de la suspension darmes jusques au vi^e may 1616," BNF, Mss. Fr. 3807, f° 83.

95   "Pardon," copy, BNF, Mss. 4014, f° 193–94.

96   Anne de Lévis, duc de Ventadour, to Henri I de Montmorency, duc de Montmorency, Paris, 23 March 1608, BNF, Mss. fr. 3602, f° 36.

97   Henri I de Montmorency, duc de Montmorency, to Henri IV, Pézenas, 13 October 1606, BNF, Mss. fr. 23198, f° 476–77.

98   Mémoire, n.d., BNF, Mss. fr. 3386, f° 14–15.

99   Catherine de Médicis to Henri I de Montmorency, duc de Montmorency, Lyon, 30 September 1579, BNF, Mss. fr. 3330, f° 19–20.

100   *Arrest et ordonnance de la covrt de parlement ...* (Paris: Jean Caniurt and Jean Dallier, 1569).

101   Brian Sandberg, "'Re-establishing the True Worship of God': Divinity and Religious Violence in France after the Edict of Nantes," *Renaissance & Reformation / Renaissance et Réforme* 29 (2005): 139–82.

102 Claude de Vic and Joseph Vaissete, *Histoire générale de Languedoc*, 15 vols. (1730–1745; reprint, Toulouse: Privat, 1872–1892), vol. 14: 880–900; Raymond A. Mentzer, "L'édit de Nantes et l'établissement de la paix en Languedoc," in *Paix des armes, paix des âmes*, eds. Paul Mironneau and Isabelle Pébay-Clottes (Paris: Imprimerie Nationale, 2000), 295–301.

103 Forest, Issac[?], and Carlencas to Henri IV, Montauban, 19 October 1606, BNF, Mss. fr. 23198, f 472–75.

104 Anne de Lévis, duc de Ventadour, to Henri I de Montmorency, duc de Montmorency, Montpellier, 13 December 1600, BNF, Mss. fr. 3589, f 42–43.

105 Lettres patentes de Henri IV, Saint-Germain-en-Laye, 13 August 1602, BNF, Mss. fr. 3564, f 65–66.

106 de Verdun to monsieur de Fresne, Toulouse, 11 March 1604, BNF, Mss. fr. 23198, f 84–87.

107 Anne de Lévis, duc de Ventadour, to Henri I de Montmorency, duc de Montmorency, Montpellier, 28 May 1610, BNF, Mss. fr. 3602, f 90.

108 Sandberg, *Warrior Pursuits,* 253–92; Sandberg, "Only the Sack and the Noose for its Citizens," 97–114.

109 William T. Cavanaugh, *The Myth of Religious Violence: Secular Ideology and the Roots of Modern Conflict* (Oxford: Oxford University Press, 2009), 3–14, 123–80.

110 Barbara Diefendorf, "Were the Wars of Religion about Religion?" *Political Theology* 15, 6 (2014): 552–63.

Further Reading

Carroll, Stuart. *Martyrs and Murderers: The Guise Family and the Making of Europe.* Oxford: Oxford University Press, 2009.

Davis, Natalie Zemon. "The Rites of Violence." In *Society and Culture in Early Modern France*, 152–87. Stanford, CA: Stanford University Press, 1975.

Diefendorf, Barbara B. *Beneath the Cross: Catholics and Huguenots in Sixteenth-Century Paris.* Oxford: Oxford University Press, 1991.

Foa, Jérémie. "Making Peace: The Commissions for Enforcing the Pacification Edicts in the Reign of Charles IX (1560–1574)." *French History* 18, 3 (2004): 256–74.

Holt, Mack P. *The French Wars of Religion, 1562–1629*, 2nd ed. Cambridge: Cambridge University Press, 2005.

Roberts, Penny. *Peace and Authority during the French Wars of Religion, c. 1500–1600.* Houndmills: Palgrave Macmillan, 2013.

Wood, James B. *The King's Army: Warfare, Soldiers, and Society during the Wars of Religion in France, 1562–1576.* Cambridge: Cambridge University Press, 1996.

9 Why Serve in Wars in Seventeenth-Century Europe?

The Case of Soldiers in Poland-Lithuania

Dariusz Kołodziejczyk

In 1588, Alessandro Valignano, a Jesuit father appointed "Visitor" to the province of India, prepared a list of gifts that the pope should send to the Ming emperor to pave the road to China for Catholic missionaries. Among the most desirable objects Valignano listed were beautifully illustrated books; yet, curiously, he stressed that the pictures they contained should *not* depict war or martyrdom.[1] Matteo Ricci further elaborated on this implied Chinese distaste of warfare, over twenty years later observing that "just as among us it is a fine thing to see an armed man, among them it is considered bad, and they are fearful to see such a horrible thing."[2] At the beginning of his stay in China, Ricci in fact ridiculed the femininity of Chinese men and their lack of masculine virtues; yet with time, he learned to praise his hosts' aversion to violence. Though his image of China may have been idealized, there was surely a grain of truth in his comparison between violence-prone Europe and violence-eschewing China.

Academic textbooks covering the early modern history of Europe are dominated by grand narratives that stress the role of interstate competition and interreligious (or interconfessional) conflict. I admit that I have myself frequently invoked easy schemes in my classroom to simplify complicated issues and expose larger geopolitical connections to my students. How better to explain the seemingly exotic rapprochement between the French king Francis I and the Ottoman sultan Suleyman in the 1530s than by referring to the fear of "France," ruled by the Valois dynasty, of being surrounded by Habsburg-ruled "Spain" and "Germany"? Similarly, a tacit anti-Habsburg cooperation between the Polish and Ottoman courts, visible at the same period in Hungary, makes perfect sense if one assumes that Polish statesmen already then nourished the paranoia of a German-Russian encirclement that was to haunt their descendants through the partition era until the twentieth century, when this nightmare came true in the form of the Molotov-Ribbentrop Pact.[3]

Similarly, the narrative of the Thirty Years War gains in clarity and attractiveness if one mentally constructs a broad anti-Habsburg coalition consisting not only of Protestants but also Orthodox Christians as well as Sunni Muslims. Close contacts between Kyrillos Loukaris, the Greek patriarch of

DOI: 10.4324/9781003157700-12

Constantinople, and Cornelis Haga, the Dutch envoy to the Porte, embodied the idea of a larger anti-Catholic coalition that was conceived in some Protestant and Orthodox minds of the era.[4] Zbigniew Wójcik, a leading Polish specialist in the early modern political history of Eastern Europe, long ago posed the question as to whether the wars fought by Poland-Lithuania against Sweden and Muscovy in the first half of the seventeenth century should not be regarded as integral components of the Thirty Years War.[5] Another Polish historian, Franciszek Suwara, in his monograph devoted to the Polish-Ottoman campaign of 1620, boldly posits that Ottoman engagement against Poland in the fall of 1620 proved decisive for the Habsburg victory over Bohemian Protestants, secured on 8 November of the same year at the battle of White Mountain. Busy fighting Polish troops in Moldavia, the Ottomans were allegedly unable to effectively assist their vassal Gábor Bethlen, the Calvinist prince of Transylvania, who was too weak alone to aid his religious brethren in Bohemia. Hence, by engaging and distracting the Porte, the Poles "saved" the Habsburgs by enabling them to quell the Protestant rebellion.[6] Although challenged by other historians and perhaps oversimplified, Suwara's thesis merits our attention as it stresses the interconnectedness of events, not just within Christian Europe, but on a larger Eurasian scale.

In this context, it is also worth invoking the intensification of contacts between Catholic Europe and Shiite Iran, partly motivated by plans of an anti-Ottoman league. Already at the turn of the sixteenth century, Shah 'Abbas invited the pope, the emperor, and the kings of Spain and Poland to jointly attack the Ottoman sultan, who after all was a Sunni Muslim. The idea of a military cooperation between Europe and Safavid Iran might have been a fantasy, yet some curious connections should not be ignored. Murtaza Pasha, the Ottoman governor of Buda and then commander-in-chief on the Polish border, who in 1634 negotiated the renewal of peace with his Polish peer, Hetman Stanisław Koniecpolski, perished in 1636 as the commander of Yerevan, killed in the Safavid assault on the fortress. Incidentally, Shah Safi, who commanded the siege army, then hosted in his retinue Giovanni da Lucca, a Dominican friar, sent shortly before as a special envoy of the same Polish hetman with whom Murtaza had negotiated just seventeen months earlier. When in 1637 King Władysław IV of Poland resolved to strengthen his relations with Vienna by marrying Cecilia Renata, a sister of Emperor Ferdinand III Habsburg, he almost simultaneously sent an embassy to Isfahan with the proposal of a triple anti-Ottoman alliance.[7]

A thesis about the predominance of confessional factors in military confrontations in which Poland-Lithuania took part in the seventeenth century finds apparent support if one draws the list of its major adversaries and calculates the years of combat. These were Sweden (thirty-four years[8]), Russia (thirty-three years[9]), and the Ottoman Empire (twenty-four years[10]), and it is worth noting that the wars against non-Catholic Christian adversaries lasted longer than the wars against a Muslim neighbor, notwithstanding the label *antemurale*

Christianitatis ("bulwark of Christianity") by which Polish propagandists presented their country at the courts of Europe.

Yet no matter how attractive such grand narratives, reminiscent of Zbigniew Brzezinski's *The Grand Chessboard* or Samuel Huntington's *The Clash of Civilizations*, appear to a popular audience today, it is worth asking *whether*—or better *to what extent*—they appealed to early modern humans, especially those who resolved to choose a military career in seventeenth-century Europe. Whereas it is doubtful whether an ordinary soldier or even a career officer, shared his superiors' broader geopolitical views, he might have nurtured confessional as well as proto-national prejudices, especially against those neighbors against whom Polish-Lithuanian troops crossed their sabers most often. Yet such a hypothesis should not be taken on faith and requires verification.

In a recent book focused on the everyday life of the soldiers serving in Polish-Lithuanian armies in the seventeenth century, Tadeusz Srogosz lists five incentives that motivated young men to enter military service:

1 Material enrichment, whereby it is worth noting that the hope for spoils appeared more important than the expectation of normal salaries that were paid irregularly and with rising delays throughout the century. Material motives played the greatest role among commoners and petty nobles, especially the younger sons who could not count on a landed inheritance that warranted the preservation of noble status.
2 Social promotion through royal benefices and grants—and even ennoblement in the case of foreigners and commoners who distinguished themselves in the royal service.
3 Prestige and fame that a young man could gain through his service in elite cavalry formations. The curiosity and excitement that accompanies exotic travels and service in foreign lands (including service overseas in the European Indian companies) can also be included in this category.
4 The duty call to defend one's fatherland, instilled especially in young nobles through the Jesuit school system that extolled ancient Roman virtues in the curriculum.
5 Religious motives.[11]

It is symptomatic that in the above list, religious and "patriotic" motives (whatever this last term may refer to) are placed at the end, though this does not imply that they were meaningless. In the present chapter, I compose a ranked list of enemies with whom Polish-Lithuanian soldiers preferred to cross sabers and discuss the reasons for these preferences. Fortunately, we have at our disposal dozens of written memoirs left by soldiers who served in the Polish-Lithuanian armies of the seventeenth century. Among the authors, we find both "natives"— born, educated, and trained in Poland-Lithuania—and "foreigners," including such personages as the German Hieronymus Christian von Holsten, the

Frenchman Philippe Dupont, and the Scotsman Patrick Gordon. This is hardly a representative sample. Literacy among soldiers was far from common in early modern Europe and even those who could write rarely nourished literary ambitions. So most of our writers were nobles, typically with some school training, most often belonging to the lower and middle rather than to the upper nobility. Needless to say, one must be wary of the authors' efforts at self-presentation, and their decisions to expose certain aspects while shrouding others with silence, but the interrogation of sources is a standard problem with which every historian must engage.

By far the richest in our sample are the memoirs of Jan Chryzostom Pasek, written in the 1690s and covering the years between 1656 and 1688. In Polish historiography, Pasek is sometimes compared to the famous English diarist Samuel Pepys. The author, a petty Mazovian nobleman born in 1636 and educated at the Jesuit College at Rawa, entered military service in the hope of raising his status, which he eventually accomplished by becoming a landed possessor—partly due to his war savings, and partly due to his marriage with a rich widow, effected in 1667. Pasek participated in campaigns against Sweden, Transylvania, and Muscovy and even fought abroad in the Polish auxiliary corps sent to allied Denmark. Throughout his career he served under the command of Stefan Czarniecki, the palatine of Rus' (today a region split between southeastern Poland and western Ukraine) and later the Crown field hetman, and in his memoirs he expressed deep personal loyalty to the commander who guaranteed frequent victories to his soldiers.[12] In 1658, when a number of his fellows left the army out of the fear of departing to distant Denmark, Pasek received a letter from his father, who advised him to follow his military commander wherever the latter ordered him to go.[13] Family considerations and military traditions played a significant role in Pasek's decisions. When he resolved to enroll in a "Hungarian" campaign in 1657, he did so because his cousin, Filip Piekarski, had already done so.[14] Another relative, Adrian Piekarski, served as the chaplain with the Polish troops in Denmark. Although Pasek did not participate in the Turkish wars of the 1670s and 1680s, his two nephews did, so his memoirs also contain first-hand relations regarding the latter campaigns. It is to Pasek that we owe the famous praise of the Turkish wars that he penned in his report on the Battle of Vienna of 1683: "A war against the Turk is pleasant and desirable for everyone because it is no pity to risk one's skin if one knows that after a victory there will be enough means to buy a plaster and to bandage a wound." Describing Ottoman luxury, the author can barely conceal his envy in noting that even Turkish horses were kept under tents, covered with silken blankets from cold and rain. At the same time, however, he criticizes Turkish effeminacy, much as Matteo Ricci had once ridiculed the Chinese. At the end of his statement, Pasek adds that by fighting against the Turk one also served God, but it is clear to the reader that religious concerns played merely an ornamental role in his argument.[15] In the aftermath of the battle, Pasek is also dismayed by the cruelty with which the Germans tortured and killed Ottoman prisoners of war[16] and mutilated their bodies after death, describing an event in which a German soldier killed a Turkish captive who had been taken prisoner by Pasek's nephew. For the Polish noble, an Ottoman captive was a financial asset, to be redeemed by his relatives, typically through the mediation of Armenian merchants. Pasek nonetheless tries to rationalize and explain the allies' behavior by invoking the great hatred Germans harbored for the Turks for having taken so many of their lands and humiliating them so frequently in combat.[17]

Curiously, Pasek's anthropological observations found an analogy in distant Scandinavia, when he comments that the king of Denmark had an inherent hatred of the Swedes (*mając innatum przeciwko Szwedom odium*).[18] He also recalls that after the conquest of Sønderborg by the Poles, the king of Denmark had asked for its Swedish commander to be delivered to him alive because "he nourished some great rancor against him, [so] I do not know, how he was received there."[19] For anyone familiar with Orientalist stereotypes that attributed fanaticism and irrational behavior to Eastern Europe,[20] it is quite amusing to read this Catholic Polish noble who calmly explained the behavior of Germans and Danes as motivated by their immoderate passions.

Religious and confessional aspects are present in Pasek's narrative, yet his attitude was far from the ardent fanaticism of which baroque literature has been often accused. While in Denmark he attended prayers at Lutheran churches, remarking that their interiors, decorated with images and altars, far surpassed in richness the churches of Polish Calvinists. He admitted that many Polish soldiers visited Lutheran churches in Denmark merely to see "beautiful girls and their customs."[21] He labeled the Danes as Lutherans (*lutrzy*), yet at the same time he referred to them as allies (*foederati*).[22] Most interestingly, he almost married a Danish girl and was willing to remain in Denmark but was persuaded to abstain from his decision by his uncle, the Jesuit chaplain Adrian Piekarski, who warned him that he would gain a wife but lose his soul by possibly becoming a Protestant.[23]

Referring to the Russians, Pasek repeated the stereotype according to which the tsar was more venerated than God in Muscovy. Yet he also recorded his own feelings of intimidation at the appearance of Muscovite boyars, who impressed him with their long, majestic patriarchal beards. On one occasion, during a battle fought in 1660, Pasek spared and liberated a young Russian prisoner with the words "go back to your mother" (*utikaj do matery*); although, admittedly, the compassion of our memoirist was bought with an ornamented cross worth twenty ducats.[24]

Even those Poles who had sided with the Swedish king in 1655—and were thus branded as traitors in the subsequent collective memory and nationalist historiography—earned a rather dispassionate remark in Pasek's memoirs. Under the year 1656, we read that "there were still many of our Poles at the side of the Swedish king; the Unitarians[25] and Lutherans had their own banners, under which many Catholics [also] served, some *per nexum sanguinis* (through ties of blood), others for the sake of spoils and wantonness."[26]

The relaxed attitude toward confessional differences should not come as a surprise. According to Marek Wagner, in the second half of the seventeenth century, only around eighty percent of officers in the Polish Crown army were Catholics, while the remaining twenty percent were almost equally divided between Protestants and Orthodox Christians.[27] This estimation does not take into account the Tatar Muslims, who served under separate banners, especially in Lithuania. When Pasek deplored that in 1672 the Lithuanian Tatars—"our foster children" (*nasi wychowańcy*)—had betrayed the Commonwealth and sided with the Ottoman invaders, his vocabulary reveals an emotional attachment to the ancient comrades in arms.[28] Another soldier and memoirist, Stanisław Zygmunt Druszkiewicz, records that when in 1654 he obtained a royal patent allowing him to recruit 120 Tatar horsemen, volunteers flocked not only from Lithuania but also from Crimea; and apparently he had no inhibitions to enroll the subjects of the Crimean khan, especially as Poland-Lithuania and the Crimean Khanate formed an alliance against Muscovy between 1654 and 1666.[29] Druszkiewicz was so fond of Tatar fashion and armor that in 1676 he was almost killed in a skirmish by the Poles, who mistook him for a Crimean Tatar.[30]

Cavalier behavior was less visible in domestic wars. During the Cossack uprisings both sides excelled in cruelty; impaling or putting to the sword was a standard mode of dealing with captured enemies.[31] Further, the Polish intervention in Muscovy in the years 1604 to 1612 bore traits of a civil war as the Poles supported pretenders to the Russian throne—first the two False Dmitrys, and then Władysław Vasa. A Polish participant in the intervention, Samuel Maskiewicz, recalls in his memoirs that after a battle near Moscow in 1612, the Polish troops impaled all Russian captives except skilled artisans, who were taken back to Poland.[32] During the civil war in Poland in 1665–66, noble rebels massacred thousands of royal supporters taken captive at the battle of Mątwy (1666), while a member of the rebel troops, Jakub Łoś, recalls that the royalists used to raid the rebels' houses and rape their wives and daughters with no regard to their noble status.[33]

To understand why soldiers fought in these brutal wars of the seventeenth century, it may help to turn to a hypothetical young Polish noble who enlisted in the royal or a private magnate army. How might this individual have seen the world around him? What motives did he have, and what preconceptions of those against whom he fought? In all likelihood, he would have ranked his enemies in the following fashion.

The Ottomans and the Tatars

Fighting against the Ottomans and Crimean Tatars promised material benefits through potentially rich spoils—especially horses, arms, and textiles—and the redemption of captives that was facilitated by the established custom of mediation by Armenian merchants and the institution of "brotherhoods" (*pobratymstwa*) often concluded between Polish and Tatar noblemen.[34] If captured, a Polish noble could similarly count on redemption or prisoner exchange, typically negotiated and declared upon the renewals of peace treaties between the Polish kings, the sultans, and the khans. It must be noted, however, that this opportunity was usually limited to nobles and professional soldiers, especially officers, whereas rank-and-file soldiers recruited from among commoners often ended up on Ottoman galleys. A war against "the Turk" also brought prestige,[35] social recognition, and—in the case of death—salvation for those fighting "the enemy of Christ."

The Muscovites

Notwithstanding numerous similarities, a war against Muscovy seemed less attractive. Potential spoils were less valuable and potential danger in the case of defeat seemed greater. The Muscovite government eagerly used captives, including Polish-Lithuanian nobles, in its eastward expansion, so it often preferred to use them as settlers in Siberia than to exchange them for Muscovite prisoners held in Poland-Lithuania.[36] The mutual cruelty, too, was far greater. This might be explained by the fact that the Muscovite army was largely composed of peasants who did not share the chivalric values of the European (and Middle Eastern) nobility, and vice-versa. Peasants were not protected by codes of honor shared by the Polish-Lithuanian nobles. Pasek, who narrated at length his mercy displayed to a young Russian noble (see above) and his efforts to capture two boyars that ended less happily for the captives, matter-of-fact comments that the Muscovite infantrymen who tried to defend themselves in a birch wood and then surrendered were put to the sword (*w pień wycięto*) so that their bodies lay one upon another and the blood ran like rainwater.[37] Christian von Holsten, who had served in the Swedish army but, after being captured, entered the Polish service, recalls grim scenes that followed an order issued in 1660 by the Polish military command, according to which all Russian prisoners who surrendered in the camp at Chudniv—apart from the highest dignitaries—were to be put to the sword or delivered to the Tatars, while Polish soldiers who tried to hide Russian captives were to be punished by death. Holsten proudly reported that at least Germans in Muscovite service were saved by other Germans serving the Polish king.[38]

Other European Armies

The attractiveness of encounters with Swedes and other European soldiers was limited by the lesser value of potential spoils, given the rising proportion of peasant infantrymen in Western European armies, and by the rising effectiveness of firearms throughout the seventeenth century, which left the

Polish-Lithuanian cavalry at a disadvantage, still valued for their maneuverability on the immense war theater of Eastern Europe, yet less effective in siege warfare and pitched battles.³⁹ Polish hussars, who annihilated the Swedish infantry at the battle of Kircholm in 1605, were no longer able to repeat such a victory a hundred years later.

The Cossacks and Other Rebel Subjects

Although the Ukrainian Cossacks perceived themselves as free warriors and were indeed regarded as such by royal and state authorities during the periods of cooperation with regular Polish-Lithuanian troops against the Ottomans, Tatars, and Muscovites, whenever they revolted against the Commonwealth, they were immediately labeled as rebellious serfs and treated accordingly. Small wonder that they repaid with the same coin. In 1652, when the army of the Polish Crown was crushed by a Cossack-Tatar coalition at Batoh, the Cossacks infamously insisted that all Polish captives be slaughtered, and they largely succeeded, even though their Tatar allies managed to keep some prisoners alive by dressing them in Tatar clothes.⁴⁰ In an analogous situation, the Cossacks allied with Polish troops during a skirmish in 1671 and refused to keep Tatar captives alive, and the few Tatars who survived owed their life to the Poles.⁴¹ Christian von Holsten recalls that in 1660, during a Polish-Cossack encounter, "the Cossacks fought like desperadoes because neither side could count on pardon."⁴² Social tensions hidden behind the Polish-Cossack wars were clearly perceived by Khan Islam III Giray, who in the midst of peace negotiations of 1653 sent a conciliatory letter to King Jan II Kazimierz: "I am the lord and monarch of the great Crimean hordes, and Your Royal Majesty is also the mighty lord and monarch of these lands [here]. Almighty God sees that I do not wish Your Royal Majesty anything wrong, [especially as you are] in such dire straits, because I ought to respect you, as one monarch should [respect] another monarch."⁴³ Although the khan was then in a formal alliance with the Cossacks who defied Polish suzerainty, he hinted to the king that as a monarch he hardly sympathized with rebels who rioted against another monarch. If the Cossacks were not treated by Polish nobles as peer rivals, rebellious commoners could count on mercy even less. For a Ukrainian peasant, the entry of Polish troops into his village did not differ much from the entry of Tatars or Turks. Likewise, the Lithuanian nobleman Filip Kazimierz Obuchowicz calmly observed that when, in 1649, the town of Mazyr (Mozyrz) in present-day Belarus was retaken by Lithuanian troops, all Cossacks that had captured the town earlier were put to the sword along with local commoners.⁴⁴ Druszkiewicz's memoirs contain a short unemotional note recording "the capture of Łęczyca and putting [of its] Jews to the sword" (*Łęczycę [...] wzięto i Żydów wycięto*), a reference to the wholesale massacre of the town's Jewish community, which was accused of cooperation with its Swedish garrison on the eve of the Polish attack.[45] The only moderating factor to the soldiers' fury and wantonness was the concern of patrons and owners of the localities in question who worried about a potential decrease in their incomes.

Conclusion

To conclude, the picture that emerges in this analysis suggests a Marxist rather than a Huntingtonian pattern. Seventeenth-century Europe might have become confessionalized, but this designation did not yet include its soldiers—notwithstanding the chaplains, sermons, and holy images that often accompanied soldiers on campaigns and the harsh punishments meted out for sacrileges.[46] Nobles and professional soldiers had much in common despite ethnic, confessional, and even religious differences.[47] In regard to shared values, lifestyle, and taste, a Catholic Polish hussar might have had more in common with a Muslim Ottoman *sipahi* (cavalryman) or a Protestant Swedish Reiter than with a Catholic peasant in Mazovia. One may also argue that in the case of a Polish noble soldier, fighting the Tatars and Turks enabled him access to luxurious Oriental objects that his contemporaries in Portugal, England, or the Netherlands could only obtain through (not always peaceful) commercial activity in the Indian Ocean. Christian-Muslim controversy certainly played a role in motivating military service, but this role should not be overemphasized.

While looking for a transnational framework of early modern warfare, current scholarship focuses on fiscal-military hubs that functioned as centers of expertise, resource accumulation, and production and that were partly independent of state actors, thus shaping interstate political order across Europe.[48] Still, when we consider such issues as the availability of external credit or the opportunity of purchasing war materials from outside, we typically think of a European state surrounded by European neighbors. This is no longer the case when we think of individual soldiers. The Germans and the Swiss served in various European armies and, to take one famous example, the Scotsman Patrick Gordon worked for the rulers of Sweden, Poland, and Russia.[49] The Tatars, Nogays, Circassians, and even Buddhist Kalmyks served alternatively the khan, the tsar, and the Polish king. The French are another example of "transreligious soldiers of fortune" who were in the employ of Muslim rulers long before the Enlightenment era. When, in 1687, Philippe Dupont, a French military engineer in the Polish royal service, was sent to the Ottoman border fortress of Kamieniec Podolski (Kamaniçe) to redeem war prisoners, he discovered that several of his countrymen served in the local "Turkish" garrison.[50] In 1678, a Polish envoy to Constantinople took great pains to secure the redemption of a French infantry major, Louis de Thévenin, who had been taken prisoner during the Ottoman conquest of Pidhaitsi (Podhajce) in 1675. The reason was not merely humanitarian. Thévenin had revealed to his captors his intimate knowledge of military plans of Polish border fortresses, so the Polish side had justifiable reason to suspect that he might be tempted to enter the sultan's service.[51] Thus tracing the career paths of individual soldiers, regarded as non-state actors, we can perhaps move beyond the conventional division between "Europe" and "Asia" and highlight the complex networks of loyalties and allegiances of ordinary soldiers in the complicated landscape of seventeenth-century East Central Europe.

Notes

1 Mary Laven, *Mission to China: Matteo Ricci and the Jesuit Encounter with the East* (London: Faber & Faber, 2012), 93.
2 Ibid., 181.
3 In 1514, Emperor Maximilian I entered an alliance with Muscovy, directed against Poland-Lithuania. In order to dissuade the Habsburg from this cooperation, King Sigismund I of Poland consented in 1515 to withdraw his eventual pretensions to the thrones of Bohemia and Hungary in the case of an extinction of the local branch of the Jagiellonian dynasty. When it happened indeed in 1526, and the last Jagiellonian king of Bohemia and Hungary perished at the battle of Mohács, Sigismund willy-nilly consented to the Habsburg inheritance, thus opening the way to their predominance in Central Europe that was to last until 1918, yet he secretly began to support John Zapolya, an anti-Habsburg pretender to the throne of Hungary who was also supported by Istanbul. On the notion of France and Poland as the two "pillars" of Ottoman policy in Europe during the reign of Suleyman the Magnificent, cf. Dariusz Kołodziejczyk, "The Ottoman Porte, Poland and Central Europe from the 15th until the early 17th century," in *The Sultan's World: The Ottoman Orient in Renaissance Art*, eds. R. Born, M. Dziewulski, and G. Messling (Brussels: Hatje Cantz, 2015), 24. On the documents of "eternal peace" concluded by Sultan Suleyman with Poland and France in 1533 and 1536, respectively, see idem, *Ottoman-Polish Diplomatic Relations (15th–18th Century): An Annotated Edition of 'Ahdnames and Other Documents* (Leiden: Brill, 2000), 48 and 117–18.
4 The most substantial study on Loukaris remains Gunnar Herring, *Ökumenisches Patriarchat und europäische Politik 1620–1638* (Wiesbaden: Franz Steiner Verlag, 1968). On Haga, see Alexander de Groot, *The Ottoman Empire and the Dutch Republic: A History of the Earliest Diplomatic Relations 1610–1630* (Leiden-Istanbul: Nederlands Historisch-Archaeologisch Instituut, 1978). For a more recent publication, forwarded by Halil Inalcík and containing a publication of relevant sources, see B. Arí and L. Kírval, eds., *Four Centuries of Diplomatic and Economic Relations between Turkey and The Netherlands (1612–2012)* (Istanbul: Panteia Press, 2014).
5 Zbigniew Wójcik, *Historia powszechna XVI–XVII wieku* (Warsaw: Państwowe Wydawn. Nauk., 1973), 345–47.
6 Franciszek Suwara, *Przyczyny i skutki klęski cecorskiej 1620 r.* (Cracow: Gebethner and Wolf, 1930), 133. The campaign of 1620 was named after Țuțora (Cecora), a Moldavian village on the Prut river where a military camp was set by the Polish troops, yet their actual defeat took place further to the north on 7 October during their withdrawal to Poland. The Polish commander-in-chief, Hetman Stanisław Żółkiewski, was killed in combat.
7 Dariusz Kołodziejczyk, "Historical Introduction," in *Stosunki dawnej Rzeczypospolitej z Persją Safawidów i katolikosatem w Eczmiadzynie w świetle dokumentów archiwalnych* [*The Relations of the Polish-Lithuanian Commonwealth with Safavid Iran and the Catholicosate of Etchmiadzin in the light of archival documents*], eds.,

204   *Dariusz Kołodziejczyk*

Stanisław Jaśkowski, Dariusz Kołodziejczyk, and Piruz Mnatsakanyan (Warsaw: Archiwum Główne Akt Dawnych, 2017), 70–83. See also idem, "The relations between the Polish-Lithuanian Commonwealth and Safavid Iran. Some comments on their character and intensity," in *Eastern Europe, Safavid Persia and the Iberian World: Frontiers and Circulations at the Edge of Empires*, eds. J. Cutillas Ferrer and Ó. Recio Morales (Valencia: Albatros Ediciones, 2019), 37–40.

8   With the years of combat 1600–22, 1625–29, and 1655–60.
9   With the years of combat 1604–19, 1632–34, and 1654–67, including five years of Polish-Lithuanian semi-official military involvement at the beginning of the Time of Troubles.
10 With the years of combat 1620–21, 1633, 1672–76, and 1683–99.
11 Tadeusz Srogosz, *Życie codzienne żołnierzy armii koronnej i litewskiej w XVII wieku* (Oświęcim: Napoleon V, 2018), 25–32.
12 [T]rzymałem się Czarnieckiego i z nim zażywał czasem okrutnej biedy, czasem też i rozkoszy; gdyż właśnie był wódz maniery owych wielkich wojenników i szczęśliwy; sufficit, że po wszystek czas mojej służby w jego dywizyi nie uciekałem, tylko raz, a goniłem—mógłby razy tysiącami rachować; see Jan Chryzostom Pasek, *Pamiętniki*, ed. R. Pollak (Warsaw: Państwowy Instytut Wydawniczy, 1971), 7. For an English translation, see *Memoirs of the Polish Baroque. The Writings of Jan Chryzostom Pasek, a Squire of the Commonwealth of Poland and Lithuania*, ed. C. Leach (Berkeley: University of California Press, 1976).
13 Ibid., 11.
14 Ibid., 8.
15 Ibid., 355.
16 *Niemcy zaś żywcem nic nie brali, ale zabijali crudelissime*; ibid., 365.
17 Ibid., 366.
18 Ibid., 10.
19 Ibid., 28.
20 These stereotypes are most persuasively presented and provided with a historical context in Larry Wolff, *Inventing Eastern Europe: The Map of Civilization on the Mind of the Enlightenment* (Stanford: Stanford University Press, 1994).
21 Ibid., 16.
22 Ibid., 33 and 38.
23 *[U]łowią cię, że tu mieszkać będziesz—z jakim przestajesz, takim się stajesz—zostaniesz lutrem. Aż tu będzie piękna: żony dobrać, a duszę stracić*; see ibid., 58. Admittedly, Piekarski's admonition also contained common-sense arguments that have retained validity in regard to international marriages until the present day, namely that Pasek's parents would not receive any help from their son in their old age if their future contacts were to be limited to the exchange of letters.
24 Ibid., 96, 110, and 115.
25 In the memoir referred to as "Arians" (*aryjani*).
26 Ibid., 5.
27 Marek Wagner, *Korpus oficerski wojska polskiego w drugiej połowie XVII wieku* (Oświęcim: Napoleon V, 2015), 112.
28 Pasek, *Pamiętniki*, 309. On the Lithuanian Tatars, their military service, and legal status in Poland-Lithuania, see Zygmunt Abrahamowicz, "Lipka," in *The Encyclopaedia of Islam*, 2nd ed., vol. 5 (Leiden: Brill, 1986), 765–67; Jacek Sobczak, *Położenie prawne ludności tatarskiej w Wielkim Księstwie Litewskim* (Warsaw: Państwowe Wydawn. Nauk., 1984).
29 Stanisław Zygmunt Druszkiewicz, *Pamiętniki 1648–1697*, ed. M. Wagner (Siedlce: Wydaw. AP, 2001), 91.

30 Ibid., 100.
31 Renata Gałaj, "Okrucieństwa wojenne w świetle pamiętników staropolskich," in *Studia z dziejów polskiej historiografii wojskowej*, vol. 5, ed. B. Miśkiewicz (Poznań: UAM, 2001), 125–39; cf. Srogosz, *Życie codzienne*, 237–38.
32 Samuel Maskiewicz, "Dyjariusz," in *Moskwa w rękach Polaków. Pamiętniki dowódców i oficerów garnizonu polskiego w Moskwie w latach 1610–1612*, eds. M. Kubala and T. Ściężor (Kryspinów: Wydawn. DiG, 1995), 213; cf. Srogosz, *Życie codzienne*, 237.
33 Jakub Łoś, *Pamiętnik towarzysza chorągwi pancernej*, ed. R. Śreniawa-Szypiowski (Warsaw: Wydawn. DiG, 2000), 122–24; cf. Srogosz, *Życie codzienne*, 180.
34 Srogosz, *Życie codzienne*, 237; Dariusz Kołodziejczyk, "Permeable Frontiers: Contacts between Polish and Turkish-Tatar Elites in the Early Modern Era," in *Foreign Drums Beating: Transnational Experiences in Early Modern Europe*, eds. B. Forsén and M. Hakkarainen (Helsinki: Finnish Society for Byzantine Studies, 2017), 161.
35 Not only Ottoman tents but also Ottoman camels were eagerly imported as exotic symbols of prestige; Pasek witnessed the massive import of camels that followed the 1673 victory at Hotin (Pasek, *Pamiętniki*, 325), and Druszkiewicz himself presented a few camels to his patron Marcin Zamoyski in 1684 (Druszkiewicz, *Pamiętniki*, 137—a letter from 9 May 1684 enclosed in the edition).
36 During the negotiations on the reestablishment of peace between Poland-Lithuania and Muscovy in 1664, the Russian head plenipotentiary Afanasij Ordin-Naščokin warned the tsar that a massive departure of Polish-Lithuanian captives might affect the security and wellbeing of Russia's Siberian provinces; see Sergej Solov'jov, *Istorija Rossii s drevnejšix vremen* (St. Petersburg, 1851–79), book 3 (vols. 11–15), 161. For an edition of two memoirs of Polish soldiers exiled to Siberia in the seventeenth and eighteenth century, see A. Kuczyński, ed., *Dwa polskie pamiętniki z Syberii* (Warsaw: Polskie Towarzystwo Ludoznawce, 1996); both authors of memoirs, Adam Kamieński Dłużyk (born ca. 1635) and Ludwik Sienicki (1677–1757), managed to return home from their captivity; yet most of the prisoners, especially common soldiers, never returned, prevented from doing so either by the Russian authorities or by having consciously embarked on a career in the Russian Empire. Among the latter was the grandfather of Ivan Kozyrevskij, the future explorer of the Kuril Islands; cf. Antoni Kuczyński, Borys Polewoj, Zbigniew J. Wójcik, "Adam Kamieński Dłużyk i jego *Diariusz więzienia moskiewskiego, miast i miejsc* z około 1672 roku," in ibid., 62–66.
37 Pasek, *Pamiętniki*, 99; the same event is recorded by another eyewitness, Jakub Łoś, who recalled that the surrounded Muscovites "asked for peace and left [their refuge] entering an open field, like sheep, when the Palatine [i.e., Stefan Czarniecki, the palatine of Rus'] ordered everybody to attack them, and they were killed to the last one" (*miru wołali i tak wyszli w pole czyste jako owce, kędy jm. pan wojewoda rozkazał wszystkim skoczyć na nich, i wybici są co do jednego*); Łoś, *Pamiętniki*, 97.
38 H. Lahrkamp, ed., *Kriegsabenteuer des Rittmeisters Hieronymus Christian von Holsten, 1655–1666* (Wiesbaden: Steiner, 1971), 36; for a Polish translation, see Hieronim Chrystian Holsten, *Przygody wojenne 1655–1666*, ed. T. Wasilewski (Warsaw: Instytut Wydawniczy Pax, 1980), 64. The composite character of the Muscovite army, which consisted of both local peasants and hired infantrymen from Western Europe, is fairly reflected in a Polish chancery diary from the Smolensk campaign of 1633–34; a Muscovite captive reported to his Polish captors that the tsar had sent 12,000 additional troops to the main army corps besieged near Smolensk, yet these reinforcements consisted merely of common peasants; in addition the tsar was to send 40,000 rubles to the Netherlands to recruit 4,000 infantrymen, to be used

against the Polish-Lithuanian enemy (*powiedział, iż car za prośbę wielką Sejna [...]* *posyła mu 12 000 ludzi duchownych albo monasterskich, których duchowieństwo* *ruskie wyprawiło, ale wszystko chłopstwa a prostego [...]. Tenże język powiadał, iż* *Moskwa 40 000 rubli posyła do Holandyjej na zaciągnienie piechoty 4000 na nas*); see M. Nagielski, ed., *Diariusz kampanii smoleńskiej Władysława IV 1633–1634* (Warsaw: Wydawnictwa Uniwersytetu Warszawskiego, 2006), 131–32.

39　Cf. Dariusz Kołodziejczyk, "Az oszmán 'katonai lemaradás' problémája és a kelet-európai hadszíntér" [The problem of Ottoman "military backwardness" and the East European theater of war], *AETAS* 3, no. 4 (1999): 142–48.

40　According to a contemporary Polish chronicler and soldier: "There were few who thanks to Tatar compassion were hidden from his [i.e., the Cossack leader Bohdan Khmelnytsky's] tyrannical fury" (*Mało było takich, których miłosierdzie tatarskie od jego tyrańskiej złości ukryło*); Mikołaj Jemiołowski, *Pamiętnik dzieje Polski zawierający (1648–1679)*, ed. J. Dzięgielewski (Warsaw: Wydawn. DiG, 2000), 98. One of the prisoners killed in this massacre was Marek Sobieski, an elder brother of the future king; cf. Dariusz Kołodziejczyk, *The Crimean Khanate and Poland-Lithuania: International Diplomacy on the European Periphery (15th–18th Century). A Study of Peace Treaties Followed by Annotated Documents* (Leiden: Brill, 2011), 162.

41　Srogosz, *Życie codzienne*, 237; S. Cramer, ed., *Das Reisejournal des Ulrich von Werdum (1670–1677)* (Frankfurt: Peter Lang International Academic Publishers, 1990), 538–39; for a Polish translation, see Ulryk Werdum, *Dziennik podróży 1670–1672. Dziennik wyprawy polowej 1671*, ed. D. Milewski (Warsaw: Folia Toruniensia, 2012), 248.

42　*Kriegsabenteuer des Rittmeisters Hieronymus Christian von Holsten*, 32; cf. Holsten, *Przygody wojenne*, 59.

43　Kołodziejczyk, *The Crimean Khanate and Poland-Lithuania*, 309.

44　Dobyci przez szturm, wszyscy [Kozacy] z gminem miejskim pod szablę poszli; see H. Lulewicz and A. Rachuba, eds., *Pamiętniki Filipa, Michała i Teodora Obuchowiczów (1630–1707)* (Warsaw: Wydawn. DiG, 2003), 234.

45　Druszkiewicz, *Pamiętniki*, 113; an equally dispassionate mention can be found in the memoirs of Jakub Łoś, another participant of the massacre, who recalled that "eight thousand Jews were put to the sword" (Żydów wyścinano na osiem tysięcy); see Łoś, *Pamiętniki*, 72; the number of slaughtered Jews given by Łoś is surely exaggerated; the whole Jewish community in Łęczyca likely did not exceed one thousand.

46　Interestingly, at the initial period of the Polish occupation of Moscow, Polish-Lithuanian commanders tried hard not to offend local Orthodox sensibilities; in 1610 a drunken Polish soldier who shot at an icon of the Holy Virgin was condemned to death and burned alive while his hands were cut off; see Maskiewicz, "Dyjariusz," 177. Robert Frost maintains that confessionalization took place in the Polish-Lithuanian army around the mid-seventeenth century, although—unlike in Western Europe—it did not result in strengthening the state. The author also invokes examples displaying the role of religion in a soldier's mental world, yet he does not prove that confession determined individual decisions regarding a soldier's enrollment in the army or his participation in a given campaign; idem, "Konfesjonalizacja a wojsko w Rzeczypospolitej 1558–1668," in *Rzeczpospolita wielu wyznań. Materiały z międzynarodowej konferencji, Kraków, 18–20 listopada 2002*, eds. A. Kaźmierczyk, A. Link-Lenczowski, M. Markiewicz, and K. Matwijowski (Cracow: Księg. Akademicka, 2004), 89–98. On his part, Mirosław Nagielski argues that at least by the 1670s confession was of secondary importance in the Commonwealth's troops, invoking their adherence to the camp of Carl X Gustaf, the Protestant king of Sweden, in the initial phase of the Second Northern War or the continuous careers of Protestants in the Commonwealth's army despite the antidissident laws that had been passed by the Diet in 1638 and 1658; idem, "Z problematyki wyznaniowej armii Rzeczypospolitej połowy XVII wieku," in ibid., 99–118. On the soldiers' religious life and religiosity, see also Srogosz, *Życie codzienne*, 159–61.

47 On the mutual perception of the Polish nobles and the members of Ottoman "military class," see Kołodziejczyk, "Permeable frontiers," 158–60. In 1676, after the Polish-Lithuanian and Ottoman negotiators concluded an armistice at Žuravno (Żurawno), the soldiers of both sides spontaneously left their trenches, mutually declaring brotherhood and engaging in trade, thus causing much confusion to their commanders who took pains to force them back to their positions; see Dariusz Kołodziejczyk, *Podole pod panowaniem tureckim. Ejalet kamieniecki 1672–1699* (Warsaw: Polczek, 1994), 83.

48 See the website of a project entitled "The European Fiscal-Military System 1530–1870," headed by Peter Wilson from the University of Oxford and funded by the European Research Council: https://fiscalmilitary.history.ox.ac.uk/home (accessed 23 Apr. 2020).

49 In reference to seventeenth-century Eastern Europe, Andrzej Rachuba labels Germany, Switzerland, and Scotland "the reservoirs of mercenary armies"; see idem, "Oficerowie armii litewskiej z armii szwedzkiej i oficerowie armii szwedzkiej z armii litewskiej w latach 1655–1660," in *Wojny północne w XVI–XVIII wieku. W czterechsetlecie bitwy pod Kircholmem*, eds. B. Dybaś with A. Ziemlewska (Toruń: Tow. Nauk. w Toruniu, 2007), 151–63, esp. 151.

50 Philippe Dupont, *Mémoires pour servir à l'histoire de la vie et des actions de Jean Sobieski III du nom rois de Pologne*, ed. J. Janicki (Warsaw: J. Berger, 1885), 223.

51 Dariusz Kołodziejczyk, "Turecki spis jeńców z ziem Rzeczypospolitej wziętych do niewoli podczas kampanii Sziszman Ibrahima paszy w 1675 roku," *Biblioteka Epoko Nowożytnej* 5 (2016): 567–82; Konrad Bobiatyński, et. al, eds., *Hortus bellicus. Studia z dziejów wojskowości nowożytnej* (Warsaw: Neriton, 2017), 573.

Further Reading

Kołodziejczyk, Dariusz. *The Crimean Khanate and Poland-Lithuania: International Diplomacy on the European Periphery (15th–18th Century). A Study of Peace Treaties Followed by Annotated Documents*. Leiden: Brill, 2011.

Kołodziejczyk, Dariusz. "The Ottoman Porte, Poland and Central Europe from the 15th until the early 17th century." In *The Sultan's World: The Ottoman Orient in Renaissance Art*, edited by R. Born, M. Dziewulski, and G. Messling, 223–27. Brussels: Hatje Cantz, 2015.

Kołodziejczyk, Dariusz. "Permeable Frontiers. Contacts between Polish and Turkish-Tatar Elites in the Early Modern Era." In *Foreign Drums Beating: Transnational Experiences in Early Modern Europe*, edited by B. Forsén and M. Hakkarainen, 153–78. Helsinki: Finnish Society for Byzantine Studies, 2017.

Leach, C., ed. *Memoirs of the Polish Baroque. The Writings of Jan Chryzostom Pasek, a Squire of Poland and Lithuania*. Berkeley, CA: University of California Press, 1976.

University of Oxford. "The European Fiscal Military System 1530–1870." Accessed April 23, 2020. https://fiscalmilitary.history.ox.ac.uk/home.

Part III
The Costs of War

10 At Home and Away

The Impact of Warfare upon Officers' Wives in Seventeenth-Century Sweden

Mary Elizabeth Ailes

Throughout her marriage to Lennart Torstensson, Brita De la Gardie's life focused on warfare. Torstensson was a Swedish nobleman who had a highly successful career in the Swedish military. Throughout the late 1620s and early 1630s, he rapidly rose through the ranks to become a general. During the late 1630s, he acted as second in command to Johan Banér who had been appointed as the Swedish army's commander after Gustavus Adolphus's death at the Battle of Lützen in 1632. After Banér's death in 1641, Torstensson became commander of the Swedish forces in the Holy Roman Empire, a responsibility that he fulfilled until his retirement in 1651. During her husband's career, De la Gardie followed him wherever his responsibilities took him, and made a life for their family whether they were living in an occupied city, on the march, or periodically at home in Sweden. She even remained with her husband through her pregnancies. Their oldest son, Gustaf Adolf, was born in Sweden in 1634 during a period when Torstensson was at home recovering from an earlier period as a prisoner of war.[1] Their remaining children, however, would be born during their father's military campaigns in the Holy Roman Empire. Their next child Märta Elisabet was born in a military camp during the Swedish attack on Dewitz in January 1639, their second son Anders was born in the Swedish-occupied town of Stralsund in 1641, and their third son Johan Gabriel was born in the Swedish military's winter quarters at Saltzwedel in 1642. The birthplace of their youngest child, Margareta Catharina, is not recorded. However, she probably was born in the Holy Roman Empire as her birth occurred in the spring of 1643 a few months after her mother was present at the siege of Leipzig.[2] In a manner similar to many officers' wives of this era, De la Gardie gave birth at home in Sweden, in occupied towns, in military camps, and on the march depending upon where the circumstances took her family.

De la Gardie's experiences were not unusual. From the mid-1500s through the 1650s, the Swedish crown continuously waged wars against its neighbors as it sought to expand and protect its growing empire around the Baltic Sea. The core of Sweden's army consisted of peasants conscripted throughout Swedish territory. However, because the Swedish empire's manpower was too small to

DOI: 10.4324/9781003157700-14

fulfill the kingdom's military needs, the crown also frequently employed foreign soldiers and regiments.[3] Regardless of where the soldiers and officers originated, however, all armies during this era had a multitude of women who traveled with the forces to provide for the soldiers' and officers' domestic and emotional needs.

Warfare during this era created many opportunities, but also hardships for women associated with soldiers and officers. Women who remained at home gained opportunities to manage families' properties and financial affairs while women on campaign managed households and ran small businesses within the campaign community. While constant warfare opened up new possibilities for women, it also created great difficulties as some experienced problems managing their families' affairs and as they encountered the dangers of participating in campaigns. In investigating women's experiences of warfare, scholars have addressed a diversity of subjects including women's changing involvement in campaigns, their participation in sieges, as well as their involvement in specific conflicts such as the Civil Wars in the British Isles and the Thirty Years War.[4] In analyzing women's experiences of war, these works allude to the disruptive impact of the era's constant wars upon family life, but this is an issue that has not been directly investigated. This article's purpose is to examine the impact of warfare upon the family lives of officers' wives by focusing particularly on seventeenth-century Sweden. It will investigate the impact that following their husbands on campaign had upon officers' wives and their children, the strategies these women used to create stable homes for their families in the midst of conflicts, and the social networks that they called upon to help them achieve these goals.

Women and Their Families on Campaign

Throughout the seventeenth century, every European army entered the field followed by many camp followers consisting of soldiers' and officers' families and servants. No state possessed the administrative structures or resources to consistently supply its armies in the field. Instead, commanders relied upon camp followers and sutlers to care for soldiers' and officers' domestic needs.[5] According to the Swedish Articles of War, men were allowed to bring their wives and other family members with them on campaign. Although the regulations did not prevent conscripts from having their families accompany them, they most likely did not possess the financial resources to pay for their relatives' passages to the war zones abroad and thus left them at home.[6] Instead, many soldiers picked up women during campaigns to take care of domestic issues such as maintaining a household and acquiring supplies.[7] It is impossible, however, to know exactly how many women were involved in this process. The main source for studying the Swedish army is the *rullor*, which are lists of soldiers' names in every regiment used as a basis for payment.[8] These lists allow soldiers' careers to be followed on a month-to-month basis. However, no other information is given such as other individuals or family members who accompanied particular regiments in the field. With regard to the officers' families, they were more likely to bring their wives and children on campaign. Officers, who usually were of noble status, possessed greater wealth and thus could afford to finance their families' trips overseas. While the *rullor* do not reveal the numbers of officers' wives following the army, other sources provide insight into their experiences. Materials such as officers' correspondence and family papers usually record the economic, military, and political affairs of the family's male members.[9] Women's activities, however, can be glimpsed in the background of eyewitness accounts of campaigns and family records detailing individual officers' military service. Because of the richer available source material, this article focuses upon the officers' wives.

On campaign, officers' wives carried out many responsibilities. Some activities, such as cooking, acquiring food, and managing households, they shared in common with all women who followed armies throughout Europe during this period. Others were more exclusive to their higher societal status such as entertaining their husbands' guests and colleagues and acting as intermediaries between their husbands and petitioners. Despite the useful roles that women played in supporting the army, officers usually were skeptical about whether the women's valuable services outweighed the potential problems their presence could pose. In particular, officers wanted to ensure that women were not distractions to the troops, were not sources of discipline problems, and did not place burdens upon the military personnel or resources.[10] Robert Monro, a Scottish officer who served in the Swedish army during the Thirty Years War, commented on this concern when he urged soldiers to leave their wives at home or settle them away from the fighting. In his book *Monro, His Expedition with the worthy Scots Regiment called Mac-Keys*, Monro stated that men who had their wives follow them on campaign faced difficulties because their attention would be split between fulfilling their military duties and safeguarding their families.[11] During his time in the Swedish military, Monro followed his own advice. Even though he brought his wife and children with him on campaign, he settled them in the Swedish-occupied city of Stettin (Szczecin). While Monro did not record what his family experienced while living in Stettin, they would have been part of a community of other families associated with the Swedish military. This city on the Baltic came under Swedish occupation in 1630 and remained a Swedish possession for the entire seventeenth century.[12] Throughout the Thirty Years War, it served as a place where women and children connected to the Swedish army took refuge from the ravages of campaigns. During the time that Monro's family lived in the city, the Swedish commander Gustav Horn sent his wife and children to live there after his wife had contracted the plague while following her husband on campaign.[13] Anna Margareta Bielke, the wife of General Nils Brahe, also traveled to Stettin in

1632 where she gave birth to her daughter Elsa Elisabet. Bielke had followed her husband on campaign throughout the Swedish army's first few years of involvement in the Thirty Years War but chose to temporarily leave the campaign community to give birth in safer surroundings.[14] The Swedish queen Maria Eleonora also spent time in Stettin with her court in 1633 while bringing her husband's body back to Sweden after his death at the Battle of Lützen.[15] The experiences that Monro's wife had while living in an occupied city probably were very similar to those of other military wives who had been left at home or settled away from the fighting. Most likely her daily life centered on running the household and raising their children. She remained in Stettin for at least three years while her husband took part in the Swedish campaigns. Monro in his book commented that at one point he was able to take a furlough and travel to the city to visit his family. However, he only could visit them for one night, and then as he stated, "I was not suffered in three years time to returne, so long as his Majesty [Gustavus Adolphus] lived, which was much to my prejudice."[16] Living away from her husband during this period would have allowed Monro's wife to act temporarily as the head of the household and oversee the family's social, domestic, and financial welfare.

Officers were not only concerned that wives and children could be a distraction but also that being on campaign could bring them into dangerous situations. Although commanders did their best to safeguard camp followers both on the march and during battles, sometimes their efforts did not work. The women who accompanied the Scottish forces that marched across Norway on their way to join the Swedish army in 1612 experienced the full force of a military attack. The local peasants, who regarded the Scots as an enemy force, lured the troops into a mountain valley and buried many of them in a rock slide. Those who did not die in the initial attack were either killed in the ensuing fight or imprisoned by the peasants.[17] According to stories preserved amongst the peasants' descendants, the wife of George Sinclair, who was the Scottish force's commander, was part of the group caught in the valley. Her fate is unclear as there are differing accounts. According to one story her infant son, who had been born earlier on the journey, was shot during the attack. While trying to protect him, she stabbed a Norwegian peasant who tried to help her. In response, other peasants shot and killed her. Another story stated that the peasants threw her into a river after her son was shot and that they both drowned. There is also evidence to suggest that she may have been among the prisoners whom the peasants held.[18] Battles were not the only danger that women could experience while in a war zone. Traveling also could bring individuals into threatening situations. In June 1646, Christina Mörner, the wife of Major-General Axel Lillie, was traveling from Leipzig to Pomerania. On her journey, Colonel Otto Schulmann and Philipp Herlin, a treasurer for the military, accompanied her. When they neared Brandenburg an der Havel, they surprised a group of Imperial cavalrymen who were also traveling in the area. When the two groups encountered each other, they exchanged gunfire.

In the ensuing fight, Schulmann was fatally shot in the neck while Mörner escaped with a minor gunshot injury to her arm.[19]

Because of the inherent danger and difficulties associated with military campaigns, officers and their wives developed a variety of strategies to safeguard their families. A life event when such issues became particularly concerning was when a woman became pregnant. For married couples, pregnancy was both a joyous event and one that raised fears for the woman and the unborn child. The birth of children was important for maintaining familial lineage and for ensuring family control over property and wealth. For women, successful pregnancies also demonstrated that they could fulfill expected societal roles as mothers and caregivers. At the same time, the lack of medical care to handle difficult births and the potential medical problems that accompanied them meant that many women and babies died as a result of childbirth.[20] Giving birth in the unsanitary conditions of a military camp or out in the open on the march, compounded the potential dangers that all women faced during childbirth. Some officers' wives settled into occupied cities late in their pregnancies to ensure that they had a safe place to give birth. In 1656, Hedvig Mörner, the wife of General Robert Douglas, took advantage of such an opportunity while following her husband on the Swedish campaign in Poland during the Second Northern War (1655–60). Douglas took part in the Swedish invasion of Poland in 1655, and throughout 1656 participated in campaigns in eastern Prussia and central Poland.[21] It appears that Mörner accompanied her husband in the field throughout 1655 and into 1656. It is not known when she left the army, but sometime in the spring or summer of 1656, she traveled to the Swedish-occupied city of Toruń (Thorn) where she gave birth to the couple's youngest child, Carl, on August 2, 1656.[22] While Toruń provided a safe haven for families associated with the Swedish military in the war's early stages, as the conflict progressed it became significantly more dangerous. In 1657, Denmark declared war on Sweden, and King Charles X Gustavus withdrew many troops from Poland to deal with this new threat. The Poles, with the aid of other allies, began to successfully fight back against the Swedes and laid siege to cities that the Swedes had occupied including Toruń.[23] During the siege of Toruń in 1658, the wife of Rudolf Bruce, a lieutenant in the Swedish military, gave birth. However, neither she nor the baby survived the delivery. This was not the only tragedy that Bruce experienced. During the attack, he was severely wounded and captured by Polish forces. He would spend two years imprisoned until a peace treaty was signed in 1660.[24]

While some women voluntarily traveled to occupied cities to give birth, others had to be convinced that it was in their best interest to follow such a course of action. During the Thirty Years War, Agneta Horn, the wife of Colonel Lars Cruus, followed her husband on campaign. In 1648 soon after their wedding took place, Horn and Cruus left Sweden, and throughout the fall and winter, Horn accompanied her husband's troops. Horn wrote an autobiography of her experiences and frequently commented on being sick and uncomfortable throughout these travels because she was pregnant. She also wrote that both her father and her grandfather sent letters to Cruus urging him to leave her in an occupied city so that she would be safe.[25] Both men had first-hand experience of warfare and understood the dangers that Horn potentially could encounter. Horn's father, Gustav Horn, was a field marshal in the Swedish army. He had commanded the Swedish army in the Holy Roman Empire after Gustavus Adolphus's death. In 1634, he was captured during the Battle of Nördlingen and spent years as a prisoner of war until he was exchanged in 1642.[26] Horn's grandfather, Axel Oxenstierna, was the head of the Swedish regency government for Queen Christina and oversaw the Swedish war effort during the Thirty Years War after the king's death.[27] Despite their efforts to encourage Cruus to leave Horn in an occupied city, Horn repeatedly refused stating that she wanted to accompany her husband. It was only when she was about to give birth that she agreed to travel to Wismar where her son Gustav was born on April 8, 1649.[28]

Besides taking up residence in an occupied city another strategy that some women used to ensure a safe location for giving birth was to travel home to Sweden. Maria Euphrosyna, the wife of Magnus Gabriel De la Gardie, made this decision after she became pregnant while visiting her husband overseas. Her husband was the governor-general of Livonia, a Swedish province in the eastern Baltic. When her husband took up this position in 1655, Maria Euphrosyna and the couple's children remained in Sweden because the plague was prevalent throughout the region where her husband was traveling, and Maria Euphrosyna was planning to help her sister-in-law, Queen Hedvig Eleonora, in her upcoming childbirth. In 1656, Maria Euphrosyna traveled to Riga to visit her husband but was forced to flee after only fourteen days as the city came under attack.[29] She would return to Riga for another visit in 1658 but would not stay there long as her husband had been ordered to travel to Copenhagen to participate in negotiations with the Danish crown. During this visit, Maria Euphrosyna became pregnant and after a period of time in Denmark, she received permission from her brother, King Charles X Gustavus, to leave the war zone as her pregnancy was advancing. She then traveled to Stockholm with a female servant and after her arrival gave birth to her daughter Hedvig Ebba.[30]

While many women sought refuge in occupied cities or visited friends and family in Sweden to give birth in safe surroundings, such opportunities were not available to all women. Many officers' wives in a similar manner to Beata De la Gardie followed their husbands throughout their military careers and gave birth during battles, sieges, in military camps, or on the march. Although the sources often reveal little information about individuals' decisions on where to give birth, the circumstances in which their deliveries occurred sometimes suggest why particular women gave birth in the midst of a campaign. For some women, as their delivery approached, the possibility of traveling to a safer location was impossible. Anna Margareta von Haugwitz, followed her husband, Carl Gustaf Wrangel, throughout his military career and gave birth to most of their children while on campaign. The birth of their daughter, Margareta Juliana, particularly illustrates why some women remained within a military camp to give birth. In the autumn of 1642, Wrangel participated in the siege of Leipzig and the Second Battle of Breitenfeld. Throughout this period, his pregnant wife and one-year-old son, Hannibal Gustavus, stayed in the Swedish siege camp. On November 4, two days after the Second Battle of Breitenfeld, Haugwitz gave birth to a daughter, Margareta Juliana.[31] In the midst of a siege and in the aftermath of a major battle, both of which occurred deep within hostile territory, the opportunity to travel safely away from the army most likely was not possible.

Though women and their families used a variety of strategies to attempt to ensure that children could be born in safe surroundings, once the children were born, they also employed different tactics to try to shield them from the dangers of campaigning. One choice that some families made, particularly when they had very small children, was to leave their children at home with other relatives. The practice of children from noble families being raised by other relatives in Sweden was not unusual during this period. Children could be sent to the homes of relatives for better education, new training or skill sets, and to create social connections that would be useful to them as adults.[32] High death rates due to the extended wars that the Swedish crown engaged in throughout this period and the Swedish empire's rapid expansion that led many noblemen to serve overseas both in administrative and military capacities also contributed to the practice of children being raised on the home front by other relatives.[33] Robert Douglas and Hedvig Mörner reached this decision when Douglas was sent to Poland in the 1650s. Although Mörner followed her husband on campaign, once she safely gave birth in Toruń, she returned to Sweden to be with the couple's other children who had been left at home. This decision led to a long separation of the parents from their older children and from each other. When Mörner returned home, she had not seen her other children in about a year, while Douglas saw their children after a two-year separation.[34]

Other couples made the decision to stay together and kept their children with them throughout a campaign or a war. Very little evidence exists of children's experiences within the campaign community. They were present in baggage trains that followed every European army throughout this period and were subject to all the dangers that everyone on campaigns potentially experienced. Opportunities for more normal experiences of family life arose when armies took control over fortresses or towns, which allowed families associated with the troops to have a more secure living situation. In such situations, parents took advantage of opportunities to improve their children's lives that would only be available in the setting of an occupied town or fortress. One such concern was children's education. While families were constantly on the move during a campaign, the opportunity for their sons to receive formal education would be severely limited. However, when families spent extended periods of time in an occupied city, the possibility existed for their sons to attend local educational institutions. During the Swedish occupation of Erfurt in the 1630s, several officers including General Johan Banér sent their sons to the local Jesuit school.[35] During his time as the commander in Leipzig (1642–45), Axel Lillie sent his sons, Axel and Gustaf Helmer, to local schools. They had followed him on campaign throughout their childhoods and with the family staying in Leipzig for three years, they were able to take advantage of local educational opportunities. In the 1650s, Axel, Gustaf Helmer, and their younger brother Leonard, who had been born in Swedish-occupied Stettin in 1642, continued their education in Greifswald while their father was serving as governor-general of Swedish-controlled Pomerania.[36] Education could extend not only to attending formal schools but also to the acquisition of skills that would be useful in social settings or in court. In 1658, when he became governor-general of Livonia, Robert Douglas hired a dance master who traveled with the family from Sweden to their residency in Riga.[37] In the 1600s, ballet became very popular at the Swedish court and was used as a means both for entertainment and propaganda.[38] The teacher's efforts were put on display when the Douglas children performed a ballet at the royal court in Stockholm in 1659.[39]

Staying in a fortified city or fortress also allowed for officers and their families to enjoy a more normal family life and activities associated with more peaceful times. Events centered around children, such as baptisms, were celebrated in these places. In 1641, Lennart Torstensson wrote to Axel Oxenstierna informing him about the birth of his son, Anders, in the Swedish-occupied city of Stralsund. In his letter, he asked if Oxenstierna would serve as Anders's godfather and told him that the baptism would take place in Stralsund on March 2, 1641.[40] In 1649, Carl Gustaf, the son of Colonel Johan Nern was baptized in Swedish-occupied Leipzig. Among the godparents at the ceremony were Maria Euphrosyne, the wife of General Magnus Gabriel De la Gardie. Other godparents included the child's namesake Field Marshal Carl Gustaf Wrangel and Wrangel's seven-year-old daughter Margareta Juliana.[41] The practice of choosing godparents served as a means to strengthen social ties between families, friends, and colleagues. Parents often asked individuals from a higher social status to serve as godparents as this practice helped to cement relationships that could be beneficial both to their children and themselves.[42] This was a practice that Nern followed with his other children. In 1652, he asked his commanding officer Duke Adolf Johan to serve as the godfather for his daughter who had been born on May 22, 1652, and invited him to attend her baptism.[43]

Networks of Support

Following regiments on campaign held both benefits and disadvantages to officers' families. Many women undoubtedly accompanied their husbands to maintain their relationships, keep their families together, and provide support and care for their husbands. These experiences were not, however, wholly positive as life on campaign could be difficult, frightening, lonely, and deadly.

When problems arose, officers and their wives often reached out to and relied upon networks of individuals within the military community to help them overcome these difficulties. They called upon family members, friends, and colleagues to provide aid and in turn gave support when it was requested of them. In this manner, families associated with the Swedish military, and in particular the women within it, bound themselves together into a tight-knit group that helped each other overcome the challenges they all encountered as a result of their service to the crown.

A time when the military community particularly came together to provide aid arose when an individual experienced the death of a spouse. For men whose wives died while they were on campaign, the immediate issue that arose was who would care for their children. Men whose families followed them into the field relied upon their wives to maintain a household and raise their children. Often their military responsibilities meant that they had to be away from their families, if they had been settled in an occupied town or region, or that they could not be continually present to help care for young children and infants if their family was in a nearby military camp. In such situations, men turned to other female relatives to step into the role of caregiver. Colonel Gustav Evertson Horn experienced this situation in 1643 when his wife Maria Mörner died during childbirth at the family's estate in Finland. Since their marriage in 1638, Mörner had followed her husband on campaign and had given birth to a daughter in 1639 who died at a young age, their son Evert who was born in the Swedish-occupied town of Stettin in 1640, their daughter Margaretha who was born in Swedish-occupied Erfurt in 1641, and their daughter Maria who was born in Finland in 1642.[44] When Mörner died, Horn was campaigning in the Holy Roman Empire while his young children were in Finland without his wife to oversee their upbringing. A family genealogy reveals that his daughter Margaretha went to live with Horn's mother Margaretha Fincke and then later lived with a relative of his second wife Barbro Kurck. Where the other children lived and who cared for them is not recorded though presumably these same relatives would have taken them in as well. Margaretha later served as a lady-in-waiting to Queen Christina and then went to live with her aunt, Hedvig Mörner, who took her on campaign when she followed her husband, Robert Douglas, to the battlefields of Poland in the late 1650s.[45] Margaretha was in Toruń with Mörner when she gave birth to her son Carl in 1656.[46] In this manner, Mörner was carrying on a family tradition as her older half-sister Christina raised Hedvig and her siblings after their mother died in 1634. Christina was married to Major-General Axel Lillie who during the Thirty Years War served as the commander of the Swedish forces at Stralsund and later became governor of Pomerania and commander of Leipzig during the Swedish occupation. Throughout the 1630s and 1640s, Hedvig and her siblings lived in the Lillie household and followed the couple on campaign in the Holy Roman Empire.[47]

Hedvig Mörner not only raised her niece, but she also brought the children of her husband's colleagues into her household when their wives passed away. In 1660, Major-General Bengt Horn's wife died. He had been governor of the Swedish province of Estonia but had recently resigned when he became a member of the Council of the Realm, which was the highest political appointment that a Swedish nobleman could hold. At the time of his wife's death, he had four children who were seven years of age and younger including his infant son Bengt who had been born earlier in 1660.[48] During this same period, Horn's older brother Christer Horn also became a widower. He had been governor-general of the Swedish provinces of Ingria and Kexholm in the eastern Baltic but had also resigned when he became a member of the Council of the Realm in 1660. When his wife died in 1659, he also had four children who were five years of age and younger.[49] Douglas knew the brothers from having served in the military with them in various capacities, and in 1658 he had become Bengt Horn's commanding officer when the crown appointed him to be the overall commander in Estonia and Livonia. When Douglas learned that Bengt Horn's wife had passed away, he suggested that both brothers send their children to Douglas's home in Riga where his wife could take care of them. In 1661, Douglas sent a request to the Swedish College of War asking that he be transferred back to Sweden. One reason he stated for this request was his wife's workload in taking care of many small children.[50] The practice of extended families raising children was not unusual among the Swedish nobility given the crown's constant need for noblemen to serve overseas in administrative and military capacities, and because of high death rates due to military service. Other military commanders who grew up in such situations included Field Marshal Jakob De la Gardie who was raised by his grandmother and older sister after his mother died giving birth to him and after his father drowned during a military campaign.[51] Lennart Torstensson was also raised by his grandmother and after her death by his maternal aunt's husband.[52]

For women, when their husbands died, an immediate concern was to secure the financial future for themselves and their children. Without their husbands' income, women faced very uncertain futures. One means to avoid financial disaster was to gain control over land that the crown had donated to their husbands. As the crown was chronically short of cash, it donated land to military officers and government officials to compensate them for back pay, unpaid pensions, and expenses incurred while serving the crown. Although only men were supposed to have access to land donations, the crown recognized that maintaining control over these grants would be important to the holder's immediate family. To ensure that officers and government officials' wives were not left without financial support, the crown allowed widows to maintain control over land donations after their husbands' deaths.[53] Barbara Kinnaird, a Scottish woman who followed her husband William Ogilwie to Sweden sometime during the early 1600s, discovered for herself the importance of these grants once she was widowed. Ogilwie, who had served as a colonel in the Swedish army, died sometime before 1613.[54] In 1604 and 1607, the Swedish king Charles IX granted him two land donations in Sweden.[55] After his death, the crown gave Kinnaird two land donations in Finland in exchange for the land her husband had received in Sweden, which the crown had re-claimed. Even though she married another Scottish officer in Swedish service, Colonel Samuel Coburn, Kinnaird maintained control over her first husband's land. Additionally, after her second husband's death in 1621, Kinnaird kept control over a portion of the land the crown had donated to him in Finland as well.[56] Some of the land that she had acquired after her first husband's death later passed on to her stepson, Patrick Ogilwie, whom Kinnaird raised first by herself and later with the help of her second husband.[57] Land donations thus were important to widows such as Kinnaird because they provided a means of support not only for themselves but also for their children and other family members.

Despite gaining access to land donations, some women continued to experience financial difficulties after their husbands' deaths. Another tactic they used to ameliorate this hardship was to turn to their friends and husbands' colleagues for financial assistance. Isabella Spens, the wife of Colonel James Ramsay, found herself experiencing financial difficulties after her husband died as a prisoner of war in 1638.[58] After her husband's death, she wrote two letters to Queen Christina requesting help in reacquiring control over land that the crown had donated to her husband and for reimbursement of money that the crown owed to her husband.[59] In addition to petitioning the crown for aid, Isabella Spens also turned to her female friends within the military community in Sweden for help. In the later 1640s, she borrowed money from the wives of Robert Douglas and Frans Sinclair, officers who had served with her husband in the Swedish army.[60]

A different issue that caused officers' wives to step in to offer aid was the birth of a child. Due to the difficulties and dangers of life on campaign, some women chose to stay in Sweden with friends or relatives when their husbands traveled to the war zones. Ebba Brahe, the wife of Field Marshal Jakob De la Gardie, provided a stable home for her friends and her daughters-in-law while their husbands fulfilled their military responsibilities. In 1640, Brahe's friend Margareta Boije, the wife of Colonel Arvid Forbus, decided to stay in Sweden with Brahe and De la Gardie when her husband returned to the battlefields of the Thirty Years War. Boije and Forbus had married in 1639 while he was on a temporary leave of absence. When he had to resume his military responsibilities, Boije decided to remain with her friends because she was pregnant. Later in 1640, she gave birth at their home. In the summer of 1641, Forbus returned to Sweden to visit his family. He had been named commander of western Pomerania and the city of Stralsund. When her husband traveled back to the Holy Roman Empire to take up his command, Boije accompanied him and for the next ten years made a home for the family in various Swedish-occupied cities in the Holy Roman Empire.[61] In the 1650s, Ebba Brahe also opened up her home to her daughters-in-law while her sons were participating in the Swedish campaigns in Poland and Denmark. During this period, Beata Elisabet von Königsmarck, the wife of Pontus Fredrik De la Gardie, and Ebba Sparre, the wife of Jacob Kasimir De la Gardie, both stayed in Stockholm with Brahe and gave birth while in residence there.[62]

Aid and support provided between military families helped create social obligations that bound individuals into networks of friends, colleagues, and relatives who could rely upon each other. As individuals requested and received help, they in turn responded to others providing help when needed. Such ties could extend over very long periods of time with individuals being called upon years after they had served together. Ebba Brahe experienced the benefits of such ties when she moved to Sweden as her husband transitioned from a military career to serving in the royal administration. In the 1620s, Brahe's husband, Jakob De la Gardie, had served as the governor-general of the Swedish province of Livonia. In 1628, the crown moved him into government administration where his work focused upon leading the College of War and serving as a member of the Council of the Realm. In 1629, he accompanied the king to the Holy Roman Empire to take part in the negotiations leading up to Sweden's entrance into the Thirty Years War. While De la Gardie was in the Holy Roman Empire, his family traveled to Sweden and began to look for a place to live. Ebba Brahe, who had been overseas with her husband for the last nine years, had hoped to find a house to rent in Stockholm while the family looked for a house to buy. However, several attempts to rent a house fell through and she eventually turned to Gabriel Oxenstierna for help. During the 1610s, Oxenstierna had served in the Swedish military in the eastern Baltic provinces where Brahe's husband was the commander. At the time that Brahe and her children returned to Sweden, Oxenstierna was serving in the Swedish royal administration as a member of the Council of the Realm. He arranged for the family to stay in the house of his brother Axel Oxenstierna who was overseas at that time.[63]

Conclusion

Throughout the seventeenth century, warfare took a dramatic toll on Swedish military personnel and their families. Continuous campaigns abroad could mean exposing women and children to the dangers of warfare in order to keep families together, or to extended periods of separation if women and children remained at home. Even when families traveled together across the battlefields of Europe, military dependents were often settled into occupied towns or fortified sites thus helping to ensure their safety, but also bringing about periods of separation. High death rates for all members of the campaign community also meant that families were frequently shattered as individuals succumbed to the diseases, depravation, and injuries that stalked every early modern European army. Despite these difficulties, families tried to stay together and developed varied strategies to protect themselves and their dependents. Additionally, they developed networks among their extended families, friends, and colleagues to give each other aid during times of difficulty. Expressions of love and support illustrate that women were instrumental in fostering these ties and that the men in their lives had great respect for the sacrifices that their wives and female relatives made in order to preserve their families.

In 1672 Alexander Forbes, Eleventh Lord of Forbes, another Scottish immigrant to Sweden, expressed his devotion to his wife and recognized her instrumental role in holding their family together when he named her as the executor of his estate. To justify this decision, Forbes highlighted his close relationship with his wife stating:

> I have had a most sweet bedfellow and Companion of her in every turn of fortune for these forty-six years and the current year. She is the parent of thirteen most sweet and beloved children, ... She was the solace and comfort of my youth and I have found by experience that she is the most careful nurse of my old age and excellently versed in all Oeconomy and disciplines.[64]

As Alexander Forbes explained, he and Elizabeth had shared a long, happy marriage full of many different experiences. Beginning in 1630, Alexander Forbes pursued a varied career that ultimately resulted in his family leaving Scotland, and settling in Sweden. His connection to Sweden began in the 1630s when he recruited and commanded a Scottish regiment in the Swedish army, and served as a diplomatic representative of Charles I to the Scandinavian courts. After returning to the British Isles in the early 1640s, Forbes traveled to Sweden in the late 1640s to petition the crown to provide military aid to Charles I during the Civil Wars. In 1652, Queen Christina granted him a commission to raise sunken ships in Swedish waters to supply the ships' guns to Charles II.[65] In the end, this scheme proved impractical, and Forbes gave up the plan and settled into his new home within the Swedish empire.[66] Through these differing experiences, Elizabeth followed in her husband's wake. A family genealogy reveals that in the early 1630s, she probably remained in Scotland with their young children while Forbes served in the Swedish army because she did not give birth to any children during this period. When he began his diplomatic career, however, Elizabeth was constantly by his side and gave birth to ten more children as the family moved between the European continent and the British Isles before they eventually settled in Sweden.[67] Throughout Forbes's career, his wife maintained a stable home for the family and supported her husband, which allowed him to fulfill his responsibilities whether overseas or closer to home.

Per Brahe, who had served as an officer in the Swedish army during the Thirty Years War and later in the royal administration as a member of the Council of the Realm expressed similar sentiments about his first wife, Kristina Katarina Stenbock, after her death in 1650. In his diary, Brahe described their twenty-two-year marriage as one that had been filled with great harmony and love. During his military service, his wife followed him twice to the Holy Roman Empire, and remained steadfastly by his side. He believed that her death had brought great sorrow to all who knew her.[68]

Throughout this era, military officers depended upon their wives and other female relatives to fulfill domestic responsibilities and to provide them with emotional support. As Forbes and Brahe stated, officers recognized that their wives' actions to maintain strong familial ties and ensure their family members' welfare contributed to their abilities to fulfill their military responsibilities. Additionally, the aid their wives provided to their extended family, friends, and colleagues fostered ties within the larger military community. Women's activities such as running households, raising children, and providing help during times of difficulty, thus played a role in supporting the Swedish military and contributed to the Swedish crown's military success.

Notes

1 Gustaf Elgenstierna, *Den introducerade svenska adelns ättartavlor med tillägg och rättelser* vol. 8 (Stockholm: P.A. Norstedt & Söners Förlag, 1925–36), 331; Peter H. Wilson, *The Thirty Years War: Europe's Tragedy* (Cambridge, MA: The Belknap Press of Harvard University Press, 2011), 514.
2 Elgenstierna, 331.
3 Mary Elizabeth Ailes, *Military Migration and State Formation: The British Military Community in Seventeenth-Century Sweden* (Lincoln: University of Nebraska Press, 2002), 7.
4 Works focusing on women's roles in the campaign community include Barton C. Hacker, "Women and Military Institutions in Early Modern Europe: A Reconnaissance," *Signs: Journal of Women in Culture and Society* 6, no. 4 (August, 1981): 643–71; John A. Lynn, *Women, Armies, and Warfare in Early Modern Europe* (Cambridge, UK: Cambridge University Press, 2008); Maria Sjöberg, *Kvinnor i fält* (Möklinta: Gidlunds Förlag, 2008); Mary Elizabeth Ailes, "Camp Followers, Sutlers, and Soldiers' Wives: Women in Early Modern Armies (c.1450–c. 1650)," in *A Companion to Women's Military History*, eds. Barton C. Hacker and Margaret Vining (Leiden: Brill, 2012), 61–91. For women's involvement in sieges see Ulinka Rublack and Pamela Selwyn, "Wench and Maiden: Women, War and the Pictorial Function of the Feminine in German Cities in the Early Modern Period," *History Workshop Journal* 44, (1997): 1–21; Brian Sandberg, "'Generous amazons came to the breach': Besieged Women, Agency and Subjectivity during the French Wars of Religion," *Gender and History* 16, no. 3 (November, 2004): 654–88. Works investigating women's experiences during the Civil Wars in the British Isles include Alison Plowden, *Women all on Fire: The Women of the English Civil War* (Stroud: Sutton, 2000); Michelle A. White, *Henrietta Maria and the English Civil Wars* (Burlington, VT: Ashgate, 2006); Ann

Hughes, *Gender and the English Revolution* (London: Routledge, 2012); Fiona McCall, "Women's Experience of Violence and Suffering as Represented in Loyalist Accounts of the English Civil War," *Women's History Review* 28, no. 7 (2019): 1136–56. Works analyzing women's involvement in the Thirty Years War include Tryntje Helfferich, *The Iron Princess: Amalia Elisabeth and the Thirty Years War* (Cambridge, MA: Harvard University Press, 2013); Mary Elizabeth Ailes, *Courage and Grief: Women and Sweden's Thirty Years' War* (Lincoln: University of Nebraska Press, 2018).

5 Lynn, 19–24.

6 Ailes, *Courage and Grief*, 20.

7 Sjöberg, 70.

8 Rullor 1620–1723, SE/KrA/0022, Krigsarkivet, Stockholm.

9 Many of the family papers of officers who served in the Swedish army have been archived in the Biographica Collection, SE/RA/756/756.1, located at Riksarkivet, Stockholm.

10 Ailes, "Camp Followers, Sutlers, and Soldiers' Wives," 80–84.

11 Robert Monro, *Monro, His Expedition with the worthy Scots Regiment called Mac-Keys*, ed. William S. Brockington, Jr. (Westport, CT: Praeger, 1999), 292.

12 Wilson, 463, 717.

13 Agneta Horn, *Leverne* (Stockholm: Albert Bonniers Förlag, 1961), 12–16.

14 Elgenstierna, vol. 1, 558.

15 Horn, 23.

16 Monro, 151.

17 Thomas Mitchell, *History of the Scottish Expedition to Norway in 1612* (London: T. Nelson and Sons, 1886), 53–54.

18 Mitchell, 77, 108–9.

19 Alexander Zirr, *Die Schweden in Leipzig. Die Besetzung der Stadt im Dreißigjährigen Krieg (1642–1650)* (Leipzig: Leipziger Universitätsverlag, 2017), 315.

20 Owen Hufton, *The Prospect Before Her: A History of Women in Western Europe, 1500–1800* (New York: Alfred A. Knopf, 1996), 177, 185.

21 Archibald Douglas, *Robert Douglas. En krigargestalt från vår storhetstid* (Stockholm: Bonniers, 1957), 180–99.

22 Douglas, 192.

23 Paul Douglas Lockhart, *Sweden in the Seventeenth Century* (New York: Palgrave Macmillian, 2004), 95–97; Robert Frost, *After the Deluge: Poland-Lithuania and the Second Northern War 1655–1660* (Cambridge, UK: Cambridge University Press, 1993), 1–2.

24 Rudolf Bruce to Charles XI, Biografica, SE/RA/756/756.1/B/B51. Also see Elgenstierna, vol. 1, 628.

25 Horn, 115–23.

26 Elgenstierna, vol. 3, 667.

27 For an overview of Oxenstierna's career after Gustavus Adolphus's death see Sven A. Nilsson and Margareta Revera, "Axel Oxenstierna," in *Svenskt biografiskt lexikon*, vol. 28 (1992–94): 504; available from https://sok.riksarkivet.se/sbl/artikel/7882; Internet; accessed February 18, 2021.

28 Horn, 132.

29 Maria Euphrosyna, *Riks-Canzlerens, Gr. Magni Gabr. De la Gardies Frus, Pfalz-Grefvinnan Maria Euphrosynas egenhändiga Lefvernes Beskrifning. Vånnegarn, 1682, Nov.* in *Handlingar til uplysning af Svenska historien*, ed. Eric Michael Fant (Uppsala: Johan Edman, 1789), 42–44.

30 Maria Euphrosyna, 46–47.

31 Arne Losman, *Carl Gustaf Wrangel och Europa. Studier i kulturförbindelser kring en 1600-talsmagnat* (Stockholm: Almqvist & Wiksell International, 1980), 25–26.

32 Liisa Lagerstam, *A Noble Life: The Cultural Biography of Gabriel Kurck (1630–1712)*, trans. Susan Sinisalo (Helsinki: Academia Scientiarum Fennica, 2007), 43.

33 Svante Norrhem, *Kvinnor vid maktens sida 1632–1772* (Lund: Nordic Academic Press, 2007), 53–54.

34 Douglas, 208.

35 Holger Berg, *Military Occupation under the Eyes of the Lord: Studies in Erfurt during the Thirty Years' War* (Göttingen: Vandenhoeck & Ruprecht, 2010), 298–99.

36 Elgenstierna, vol. 4, 657–58.

37 Hans Werner, "Elbsvaneordenens Candorin, Conrad von Hövelen og Danmark," *Personhistorisk tidskrift* no. 3–4 (1940): 229.

38 Gunilla Dahlberg, "The Theatre around Queen Christina," *Renaissance Studies* 23, no. 2 (2009): 168–70.

39 Werner, 229.

40 "Lennart Torstensson to Axel Oxenstierna, Stralsund, February 26, 1641" in *Rikskansleren Axel Oxenstiernas skrifter och brefvexling*, part 2, vol. 8 (Stockholm: P.A. Norstedt & Söner, 1897), 347.

41 Zirr, 319–20.

42 Solveig Fagerlund, "Women and Men as Godparents in an Early Modern Swedish Town," *The History of the Family* 5, no. 3 (2000): 356.

43 Johan Nern to General-Governor Adolf Johan, Gothenburg, May 26, 1652, Stegeborg samlingen, vol. E382, SE/RA/720810/003/01, Riksarkivet, Stockholm.

44 Gabriel Anrep, *Svenska adelns ättartavlor*, vol. 2 (Stockholm: P.A. Norstedt & Söner, 1864), 296.

45 Anrep, vol. 2, 296.

46 Douglas, 245.

47 Ailes, *Courage and Grief*, 38–39.

48 Anrep, vol. 2, 305.

49 Anrep, vol. 2, 304–5.

50 Douglas, 315.

51 Anrep, vol. 1, 560.

52 Elgenstierna, vol. 8, 331.

53 Ailes, *Military Migration*, 72–75, 85–86.

54 Elgenstierna, vol. 5, 537.

55 Sven A. Nilsson, *På väg mot militärstaten. Krigsbefälets etablering i den äldre Vasatidens Sverige* (Uppsala: Historiska Institutionen, 1989), 38, 96.

56 B. Boëthius, "Samuel Cobron eller Cockburn," in *Svenskt biografiskt lexicon*, vol. 8 (1929): 678; available from http://sok.riksarkivet.se/sbl/artikel/14907; Internet; accessed October 27, 2015.

57 Elgenstierna, vol. 5, 538.

58 Erik Spens, "A Scots Lady in Alghult," *The Scottish Genealogist* 31, no. 2 (June, 1984): 44–46.

59 Isabella Spens to Queen Christina, Biographica, SE/RA/756/756.1/R/1/R3.

60 For documents about these loans see Jakob Sinclair to Charles X Gustavus, Biographica, SE/RA/756/756.1/S/S33 and Charles XI to the Douglas family, July 13, 1681, Biographica, SE/RA/756/756.1/S/S33.

61 Bengt Hildebrand, "Arvid Forbus," in *Svenskt biografiskt lexikon*, vol. 16 (1964–66): 272; available from https://sok.riksarkivet.se/sbl/artikel/14314; Internet; accessed February 9, 2021.

62 Ailes, *Courage and Grief*, 122–23.

63 Svante Norrhem, *Ebba Brahe. Makt och kärlek under stormaktstiden* (Lund: Historiska Media, 2007), 57. For information about Gabriel Oxenstierna's career see Robert Sandberg, "Gabriel Oxenstierna Gustafsson," in *Svenskt biografiskt lexikon*, vol. 28 (1992–94): 524; available from https://sok.riksarkivet.se/sbl/artikel/7932; Internet; accessed February 16, 2021.

64 "Copy of Testament of Alexander, Lord Forbes, made in Latin in Sweden, and Translated into English," in *The House of Forbes*, eds. Alistair Tayler and Henrietta Tayler (Aberdeen: Aberdeen University Press, 1937), 199.

65 "Memorial of Alexander Forbes, 11th Lord Forbes present to King Charles II, Westminster, August 26, 1661," in *The House of Forbes*, 185–87. Also see the documents about the controversy surrounding the raising of the ships in Biographica, SE/RA/756/756.1/B/F15.

66 Alastair Tayler and Henrietta Tayler, "Introduction," in *The House of Forbes*, 184–89.

67 "List of the Family of Alexander Forbes, 11th Lord of Forbes" in *The House of Forbes*, 188.

68 Per Brahe, *Svea Rikes Drotset Grevfe Per Brahes tänkebok* (Stockholm: Carl Delén, 1806), 75–76.

Further Reading

Ailes, Mary Elizabeth. *Courage and Grief: Women and Sweden's Thirty Years' War*. Lincoln: University of Nebraska Press, 2018.

Hacker, Barton, and Margaret Vining, eds. *A Companion to Women's Military History*. Leiden: Brill, 2012.

Hacker, Barton. "Women and Military Institutions in Early Modern Europe: A Reconnaissance." *Signs: Journal of Women in Culture and Society* 6, no. 4 (Summer, 1981): 643–71.

Lynn, John A. *Women, Armies, and Warfare in Early Modern Europe*. Cambridge: Cambridge University Press, 2008.

Sjöberg, Maria. "Women in Campaigns 1550–1850 Households and Homosociality in the Swedish Army." *The History of the Family* 16, no. 3 (2011): 204–16.

11 Fighting War Pestilence

Habsburg Strategies of Disease Management during the Ottoman War (1737–39)

Sabine Jesner

The global spread of COVID-19 beginning in Spring 2020 demonstrated that the arrival of a previously unknown virus poses an enormous challenge for societies and governments, and that an effective containment strategy is connected with the need for coordinated governmental action as well as targeted measures to protect vulnerable groups from deadly infections. This modern pandemic has thus served as a reminder that fighting infectious diseases on a global scale is not solely a phenomenon of past ages, but has also renewed scholarly interest in how past societies managed such challenges. In particular, the bubonic plague, caused by the bacterium *Yersenia pestis*, was recognized by European contemporaries of the late medieval and early modern era as one of the most fearsome threats. Today, even a casual stroll through many Central European towns reveals that the cultural testaments of the fight against this disease are reflected not just in documentary sources, but in the very urban architecture. For example, plague columns on central squares of former urban Habsburg centers, such as the *Piața Unirii* in Timișoara (Romania) or the *Alte Platz* in Klagenfurt (Austria), are silent witnesses of this menace and aspects of a monumental culture of memory. In Vienna, the *Karlskirche* is the most impressive symbol of this type. In 1713, the construction of the church was initiated by Emperor Charles VI (1685–1740) after an epidemic of bubonic plague that touched the territory of present-day Austria. He named it after his patron, St Charles Borromeo (1538–84), who as archbishop of Milan was revered as a healer of those who suffered from the plague.[1]

During the eighteenth century, managing epidemics became an important field of duty for early modern states. As they recognized that their populations were a valuable economic commodity (a view encouraged by mercantilism), they sought more effective ways to protect them. The French philosopher Michel Foucault has thus described the eighteenth-century state as an initiator, organizer, and controller of health politics.[2] Indeed, in the Habsburg monarchy the improved organization of health care under the political program of "medical policing" was introduced by the Viennese court in the eighteenth century, at a time when the state developed internal and external strategies to

DOI: 10.4324/9781003157700-15

take care of its population.[3] The spread of communicable diseases was perceived by the Habsburgs as an existential threat and was also clearly recognized as a constant companion of war since warfare fostered the spread of diseases in both the military and civilian populations.

In this article, I examine the nexus of what Friedrich Prinzing has termed "war pestilences"[4] and early modern state strategies for containing the spread of communicable diseases in wartime. The focus is on a bubonic plague outbreak during the Habsburg-Ottoman War of 1737–39. The disease was brought from the Ottoman Empire to the Habsburg Principality of Transylvania in 1737 and was finally stopped in 1740. Based on archival research of selected case studies, primarily in the unpublished files and protocols of the Sanitary Court Commission (*Sanitätshofcommission*), the Aulic War Council (*Hofkriegsrat*), and the so-called *Feldakten* (files from the battlefield) that are stored at the Austrian State Archives, I explore how the Habsburg civil and military apparatuses tried to curb infectious disease during wartime and which methods, old and new, were used in this endeavor. The region under study comprises the southeastern regions of the Habsburg monarchy; it includes the historical provinces of Serbia, Slavonia, Syrmia, Transylvania, Oltenia, the Banat of Temeswar, parts of Hungary, and the Habsburg Military Border.

Bound by a 1726 agreement for military assistance, the Habsburg monarchy was obligated to enter the Russo-Ottoman War (1736–39) as a Russian ally in 1737, when Russia cited the occasional raids of Crimean Tatars on the Russian border zones as *casus belli*. The "weak" Ottoman Empire was not as near collapse as expected by its opponents, however, which led the Habsburgs to reach a separate peace with the Ottomans in 1739. By this Treaty of Belgrade, the Habsburgs ceded all territorial gains made since the peace treaty of Passarowitz in 1718, including Oltenia, the Kingdom of Serbia, and a small strip of Bosnia, but excluding the Banat of Temeswar. Later in 1739 the Russians and Ottomans also reached a peace through the Treaty of Niš, by which Russia gained the demolished fortress of Azov (in modern Russia) and the city of Zaporizhzhia (in modern Ukraine), but Russian merchants lost economic rights on the Black Sea. The road to these peace treaties was paved with pain and suffering for ordinary soldiers, both due to combat and bubonic plague, which touched the hostile armies and surely influenced the war's outcome. The unfavorable outcome for the Viennese court has led historian Claudia Reichl-Ham to question, in her study of this Ottoman War, whether it had been purged from memory to become a "forgotten war" (*vergessene Krieg*).[5] Indeed, German-language military historiography has particularly neglected this Ottoman War,[6] a lacuna that has also resulted in a dearth of studies on the war's entanglement with epidemics.[7] The intertwining of war and medicine, however, is no longer terra incognita for historians. Well-known scholars have investigated the subject across a full range of historical periods, providing it now with a robust theoretical framework.[8]

Epidemic Threats and Early Modern Warfare

When it came to the first major confrontation between Ottoman and Habsburg troops in the summer of 1737, following a nearly twenty-year period of peace, familiar opponents came into conflict. Though they had been political and military adversaries for centuries, however, they also had a more fearsome enemy in common: the bubonic plague. Indeed, the Ottoman Empire—including "the Levant"—had been so frequently touched by the bubonic plague that it was seen by many contemporaries as a dangerously unhealthy part of the world, with certain provinces under Ottoman suzerainty, such as the Danubian principalities of Moldova and Walachia (Muntenia, Oltenia) which were adjacent to Habsburg territory, being perceived as locales posing an especially increased health risk. It is from this perception that the long-standing designation of the Balkan peninsula as "dirty" derives.[9]

The spread of the disease within the Ottoman Empire has often been portrayed as something encouraged by the prevailing perception within the Muslim empire of illness as divine purpose, coupled with an official neglect of precautions in the case of bubonic plague.[10] Thus Daniel Panzac has shown that during the eighteenth century, the city of Istanbul was afflicted by the malady for a full sixty-eight years, Bosnia for forty-one years, and Bulgaria for eighteen years.[11] Recently, however, the assumption of "fatalistic" Ottoman responses to contagious disease has been questioned by historians. In particular, Birsen Bulmuş has highlighted the efforts of learned Ottomans to implement preventive measures.[12] Furthermore, faced with such widespread disease, people in the Islamic world often turned to various types of religious protection, just as in Habsburg-dominated Catholic areas, where activities such as venerating plague saints (Sebastian, Rochus, or Charles) were very popular. A short work by Eleazar Birnbaum concerning the mystical handling of *veba* (plague) among educated people in the Ottoman Empire describes one of the methods used in fighting it: "If the plague occurs in a town, he should write these seven verses [of the Koran], obliterate them with water and put [this water] out in the cold night, and drink some of the water every morning; and [thus] he will be safe from the pestilence and the plague."[13]

Disease and plague were also concerns for the Habsburgs, and with the transformation of the Habsburg military forces from a mercenary army into a standing army in the seventeenth century, proper care for troops became a major concern for the state. Having a potent army on standby obliged the early modern state to provide soldiers not only with weapons, clothes, and nutrition but also with medicine in war and peacetime—a very expensive undertaking.[14] The problem of soldiers' health became especially pressing after the conclusion of the Commerce and Shipping Treaty between the Viennese court and the Ottoman High Porte in Passarowitz (1718) since this facilitated both commercial contacts and the transmission of infectious diseases between the empires.[15] In addition, the Habsburg campaigns against the Ottomans in Southeast Europe were scheduled as yearly campaigns for the spring and summer months, followed by a recuperative phase during which the troops stayed in "winter quarters" in the surrounding areas of the region in conflict,[16] and between 1737 and 1740 the Habsburg regiments rested in Hungary, Syrmia, Transylvania, or the Banat of Temeswar—unfamiliar climates that also exposed the soldiers to disease. These included not just plague, but also *morbus hungaricus* (or Hungarian Disease), that is, typhus fever, which had spread fear among Habsburg soldiers since the seventeenth century.[17] The disease resulted from poor sanitation and hygiene, circumstances that are known in general as a catalyst for maladies. For the bubonic plague, the more frequent occurrence of rats in the warm summer months coincided with the military campaigns and engendered higher rates of infections among the troops and the local civil population of the combat areas in Southeast Europe. During this period, however, neither the presence of rats nor the wearing of dirty clothes contaminated by fleas were seen as extraordinary problems of hygiene.[18]

Habsburg Health Management

In autumn 1737, shortly after the Habsburg army had begun a campaign in the region, merchants from Wallachia conveyed the plague to the Habsburg province of Transylvania. Local authorities, however, were unsure whether the disease was the "real" bubonic plague or another malady, so the Transylvanian government (*Gubernium*) in Sibiu intensified its communication with the central institutions in Vienna to seek clarity and advice. Soon the lethal disease was recognized as the bubonic plague,[19] an outbreak that revealed sanitation and hygiene deficits throughout the southeastern provinces of the monarchy. In the following years, the Sanitary Court Commission in Vienna then served as a discussion platform for sanitary questions,[20] but the plague outbreak also induced the installation of subordinated Sanitary Commissions (*Sanitätscommissionen*) in the most important urban regional centers. In particular, the commissions in Sibiu, Timişoara, Belgrade, Osijek, Szeged, and Buda became important regional institutions in the coordination of the health strategies designed and promoted by the authorities in Vienna.[21]

While the Sanitary Court Commission, which included both medical professionals and assisting staff (clerks), played an important advisory role, disease management was most closely directed by the provincial commissions, which were composed of medical, civil, and military individuals who were more familiar with the local circumstances and often held highly ranked military or administrative posts in these Habsburg provinces.[22] The local expert commissions were authorized to implement the enacted regulations in compliance with the imperial center in Vienna, and provided the Sanitary Court Commission with regularly written reports about the health status in the far-off provinces.

These reports, recommendations, and evaluations, as well as the elaborated instructions by the central Sanitary Court Commission, then functioned as a basis for additional reports directed to Emperor Charles VI. As another sign of the ties between war and disease management, these reports were sent to the emperor via the Aulic War Council, with which the Sanitary Court Commission worked closely. The Aulic War Council also gathered additional information from the subordinated general commands in the provinces or directly from the battlefield.[23] Simultaneously, the Court Chamber (*Hofkammer*) was responsible for relevant financial questions, while the General War Commission (*Generalkriegscommissariat*), which was subordinated to both central authorities, the Court Chamber and the Aulic War Council, also played a role in disease management and containment. Of particular importance in this regard were the actions of the war commissioners (*Kriegscommissare*) assigned to provincial authorities or to the army.[24] During wars, and supported by the Provision Office (*Provisionsamt*), the commissioners were responsible for the transport of infected or wounded soldiers directly from the battlefield or from the mobile field camps to the hospitals. Moreover, they had to monitor the financial budget for those hospitals.[25]

Numerous actors and institutions had to operate seamlessly to avoid or mitigate disease outbreaks among the soldiers. To be sure, however, curing war pestilence was not solely a military fight; the civil administration in the provinces was also of importance.[26] An episode that occurred in the Serbian town of Sremski Karlovci in 1738 illustrates well the dual engagement of the civil and military apparatuses to control a particular outbreak of bubonic plague. In September of that year, the field surgeon (*Feldscherer)* Franz Läßer, appointed and stationed at the civil administrative center (*Cameral-Provisoriat-Ambt*) in Petrovaradin, and staff surgeon (*Stabschirurg)* Stang from the garrison of its fortress, discovered that 66 families were infected by the plague in the town of Sremski Karlovci. This community had become part of the Habsburg monarchy with the Treaty of Karlowitz (1699)[27] and since 1713 had served as residence for the metropolitan of the Serbian Orthodox Church. After Stang informed his superior, the commander of the fortress Freiherr von Pfefferkorn,[28] about the outbreak, and the commander then forwarded the information to the main administrative center of the province, the *Landes-Deputation* of Slavonia in Osijek (Esseg), and to the commander-in-chief of the Habsburg army, Joseph Lothar Dominik Graf von Königsegg-Rothenfels (1673–1751).[29] Similarly, the head of the *Cameral-Provisoriats-Ambt*, Casper Joseph Pertsch, informed the *Landes-Deputation*.[30] Within a few days of detection by the two surgeons, the *Landes-Deputation* of Slavonia informed the Aulic War Council that the bubonic plague raged in Sremski Karlovici—and that this fact had been concealed by three regional officeholders.[31] At this moment, the second year of the war was coming to a close; the Habsburg troops encamped in Zemun, near Belgrade, and other parts of the army were already infected by the plague, which made the information concerning the outbreak of the disease extraordinarily important to Commander-in-chief Königsegg. This was particularly pressing as the march into safe winter quarters lay ahead.[32]

Containment Measures: Sanitary Cordon and Quarantine

Rural areas were generally not equipped with an adequate military medical infrastructure. The first printed field hospital regulation (*Feldspitalsordnung*) with a constitutional character for all Habsburg troops was issued on 1 May 1738. After considering the local conditions, the installation of three hospitals was decreed: a main hospital in Belgrade (*Haupt-Spital*) and two subsidiary hospitals (*Filial-Spital*) in Pančevo and Banatska Palanka.[33] The decree, however, did not provide practical guidance for managing the plague, though a separate set of instructions for military physicians, *Instruction für einen Kayserlichen Feld-Staabs-Medicum*, did respond to the threat of the epidemic. Military physicians were instructed under point eleven of these instructions rigorously to isolate soldiers suffering from contagious diseases from those who were merely wounded.[34]

In January 1738, four months before this regulation on field hospitals was published, Habsburg communication channels were dominated by an alarming suspicion. One battalion of the Infantry Regiment Grünne stationed in the Banat of Temeswar was rumored to have been infected by the bubonic plague. The battalion had been transferred via Wallachia and Transylvania, where the plague had broken out, before entering into the neighboring region of the Banat. There they would indeed spread the disease, but also provide an interesting example of Habsburg disease containment measures. In Transylvania itself, the outbreak had been contained to an extremely low infection rate, so once the plague came to Banat, the central authorities in Vienna first moved to adopt the same successful and tested modes of pestilence control. Enacted decrees included, among other measures, the closure of schools and churches, a ban on public meetings, the strict segregation of infected houses— even whole districts—as well as the shooting of pets. The measures were supported by the local garrisons. The Viennese court also sent additional epidemiologists (*Contagionschirurgen* or *Pestchirurgen*) to the newly infected province, where they were instructed by the Sanitary Commission in Sibiu to carry out medical checks and treatment.[35]

In the case of Banat, however, this ambitious catalog of measures was insufficient. It soon became obvious that a stricter containment strategy was needed, one which had to be based on more effective modes to interrupt the chains of infection. Ultimately, the systematic procedure chosen was built on separation and isolation. The practice of sanitary cordoning was the most important of these measures. For centuries, restrictions of movement via militarized cordons had served as effective tools of containment,[36] and such practices were now put to use in Banat on both a large and small scale. On a large scale, sanitary cordons were geographically adapted and relocated according to the imminent threat, and the natural landscape was utilized in the establishment of sizeable barriers. For example, a militarized cordon, rigorously guarded, was soon installed along the Mureș, Tisza, and Drava rivers.[37]

Cordoning on a small scale can be seen in the treatment of the infectious battalion Grünne. Until the infection was discovered, the battalion had been stationed in the fortress situated in the center of the settlement of Timișoara. Then, in February 1738, staff physician Tobias Dolfin, staff surgeon Marianus Caunes, and a second surgeon named Delabarre issued simplified reports regarding an epidemic disease with buboes that had appeared among the soldiers. Yet the physicians, underestimating the dangerous nature of this deadly disease, soon thereafter informed the authorities that outbreaks had subsided.[38] A report covering the period between 19 February and 1 March 1738, however, reveals that more and more soldiers were exhibiting typical symptoms of bubonic plague: high fever, headaches, limb pain, a general if intense feeling of malaise, and—the most striking medical evidence—the formation of buboes in the groin and armpit area, which were caused by an infection of the lymph nodes.[39]

Given the clear evidence of an outbreak, an imperial decree dated 24 March 1738 instructed the entire battalion Grünne to leave the fortress and camp in the area known as the "suburban town" (*Vorstadt*). There those soldiers suspected of suffering from the malady were separated from the other troops, and were isolated in shanties and in thirteen Armenian houses that were rented by the Court Chamber. They were then strictly guarded by the local military garrison and a cluster of selected civilians.[40] The battalion Grünne as a whole consisted of 140 soldiers. On 15 March 1738, doctors Dolfin and Caunes reported twenty-eight infected; four days later the number had risen to thirty-one, with two ill women and one child among them.[41] There is little information concerning the medical care for this sequestered cluster of people, but it is documented that the isolated area included two hospitals or lazarets. The first offered space for twenty individuals and was situated circa 900 meters from the fortress. The surgeon, who oversaw daily medical care, was accommodated nearby. The second hospital was located about one kilometer to the north of the first lazaret and was divided into two sections: one exclusively for infected persons (twenty-one individuals) and one for generally sick people (fifty individuals).[42] The following month (April), the entire settlement of Timișoara was deemed a *locus infectus* and locked down.[43] In May, there was a decline in infection rates among the battalion Grünne, though by this point it was apparent that the disease had spread beyond the city to the rural population of the Banat.[44] The spread of the malady among the locals evoked fear, leading to flight from the region and thereby jeopardizing the supply chain of the army, as the local population produced significant amounts of grain and facilitated its transportation.[45]

Coinciding with the order for the battalion Grünne to leave the fortress, Timişoara had been ordered to place all potentially infected individuals—all those who had been in touch with the suspected soldiers—into quarantine for forty-two days. The abandoned living quarters were ordered to be cleaned and aired by specially appointed staff, who were afterward disallowed from contact with healthy persons. Moreover, the beds, sheets, and clothing of infected soldiers had to be burned.[46]

Other purification methods—washing, airing, and fumigating—were also in use in the official temporary quarantine stations (*Contumazen*). These stations were first established in 1738 and then formed a new and important type of sanitary institution for the fight against the bubonic plague within the Habsburg monarchy. During the second half of the eighteenth century, such quarantine stations became permanent institutions and would play a key role in the medical control of border-crossing between the Habsburg and Ottoman empires over the next hundred years. For centuries, therefore, quarantining was the most significant strategy in preventing the spread of disease that had already arrived in or on the border of the monarchy. In contrast to medical isolation, which is applied to confirmed cases of infection, quarantining is marked by a prophylactic separation of those who may have been exposed to a contagious disease. For the Habsburgs, a militarized sanitary cordon along the rivers and a series of quarantine stations[47] installed along—or at least around—the sanitary cordons served to maintain a strict surveillance of movement. The quarantine stations were built in such a manner that allowed the rapid transfer of the facilities to other places as the need arose. This mobility was an important factor given the shifting placement of the cordons to correspond to the intensity and spread of the plague. In particular, the protection of the imperial center was a major concern for the Viennese court. As a result, the routes from the war region toward Vienna were subject to advanced controls, and the crossing of the active sanitary cordon was solely allowed via quarantine stations. In the quarantine stations, all kinds of travelers, merchants, and especially soldiers were put into medical seclusion. The stations' staff (*Contumazmeister, Reinigungskommissar, Contumazchirurg*) was appointed by the Viennese court and was tasked with the medical surveillance of the people in the stations. The timespan for the medical screening was set at forty-two days.[48]

These strict guidelines, however, gave rise to various forms of abuse. For the winter break, military officers often wanted to travel to Vienna or toward their homeland, and they frequently undertook these journeys without going through the compulsory quarantine.[49] Efforts to loosen restrictions officially, however, were unsuccessful. In the field camp in Vinča (near Belgrade) in September 1738, for example, the inquiry of Commander-in-chief Königsegg as to whether all army members were still required to submit to a forty-two-day quarantine was crisply confirmed.[50] The travel issue also sparked fresh debate as to how the army was mandated to handle epidemic disease during warfare, and the strategically important fortresses of Belgrade and Petrovaradin, which were known as infected areas, also demanded a more stringent set of measures. In October 1738, the Sanitary Court Commission thus prepared a thematic catalog of the current problems it was facing, which was presented and discussed at a specially convened "Sanitary Conference" and finally reported to Charles VI. In its report, the Sanitary Court Commission outlined two connecting points. On the one hand, the medical security of Inner Austria and Upper Austria should be guaranteed by the closure of the borders with Hungary. On the other hand, the deliberations were oriented toward the infected army and, more specifically, toward the policy to contain the spread of disease. Under these premises new regulations were compiled; officers were only allowed to leave the army if they were in charge of a military mission—and then with only simple baggage. This was linked with a mandatory travel route and concrete instructions as to the location of the quarantine station at which they had to endure quarantine. Private journeys required the formal permission of the Aulic War Council. The Sanitary Court Commission deemed the movement of troops and the cohabitation of soldiers in so-called *Kameradschaften* as high-risk factors in the rapid rise of infection rates among the military. The commission insisted on the strict segregation of coresidents after an individual among the *Kameradschaft* was diagnosed as infected. Additionally, all clothes and bed sheets had to be burned, and the isolated had to be examined daily by a field surgeon and an epidemiologist. The route to winter quarters had to be far from potentially infected villages, and particular care had to be taken to ensure that the locals who assisted in transportation should not come from those villages. Such measures gave rise in the Sanitary Court Commission to the claim that rural communities spread plague rumors to avoid having to assist the army.[51]

Logistics became a challenge in other ways as well. The waterways, especially the Danube, were critical transportation routes in the military supply chain, and due to the epidemic threat, it was deemed necessary to adjust shipping procedures. In March 1739, therefore, a comprehensive instruction on the handling of water transport was published. From Vienna to Belgrade, five locations on the Danube were now designated as stations: Buda, Balaton-földvár, Baja, Futog, and Belgrade. At each station, the ship's crew was to be replaced, and crew members were ordered to take care to avoid contact with potentially infected individuals. Moreover, to reduce contact among crew members and encourage social distancing, a type of sutler (*Marquetanter*) was appointed in the last three stations (Baja, Futog, and Belgrade) to oversee the crew's food supply and provisioning. If a suspected illness nevertheless occurred, the individual in question would be required to travel to the nearest lazaret, and the remaining crew is forced to quarantine. Since at the time, the new instruction was issued, Belgrade was still infected, the rules required that ropes and oars be sanitized before a ship from there entered the healthy station of Futog. The disinfection process would be organized shortly before arrival in Futog and be conducted by thirty so-called sanitation assistants (*Reinigungsknechte*).[52]

Conclusion

By 1738, the emperor and his counselors in Vienna had become disappointed about the course of the war, which had so far not brought the envisioned outcome, and the army's commander-in-chief, Königsegg, faced criticism for his hesitant engagement against the Ottomans on the battlefield. In his defense, Königsegg penned emotional reports to Vienna, in which he described the alarming sanitary circumstances and the permanent presence of the epidemic threat, factors that significantly reduced the impact of the Habsburg army and provoked human suffering among soldiers and civilians.[53] Nevertheless, his personal efforts to protect the army from the plague remained unappreciated, and after the summer campaign, Königsegg was replaced as commander-in-chief of the Habsburg army by Georg Olivier Graf von Wallis-Carighmain (1673–1744).[54]

The conflict between the Habsburg and Ottoman empires was not alone responsible for the calamitous state of affairs in Southeast Europe, but the war along with the outbreak of plague plunged the region into crisis. This crisis also highlighted how in the 1730s the Viennese court did not have a ready masterplan to deal efficiently with the problem of pestilence. Border closure and military cordoning had been previously practiced in the Habsburg monarchy, as were the strategies of isolation and separation, but these older containment measures, together with newer adaptions of the local infrastructure and the consolidation of the measures introduced during the war by Emperor Charles VI, would now pave the way for a more complex plague prevention system on the external border of the monarchy. In particular, the formation of a militarized cordon and the installation of temporary quarantine stations during the Habsburg-Ottoman war provided the basis for this new "medicalized border." In response to the threat of epidemic, border closures were initiated, and medical screening and mandatory quarantine became entry conditions for all those arriving from the Ottoman Empire.[55] In the decades following the war, soldiers on the Habsburg Military Border effectively monitored what came to be known as the "Habsburg Cordon Sanitaire," receiving land to cultivate as a type of military fief in exchange for their surveillance.[56] Inspired by the Venetian sea quarantines, this prophylactic concept at the Habsburg border would become one of the most important plague prevention strategies until the late nineteenth century. It would also be adopted both by the Ottoman Empire and Russia in the 1830s.[57] As a testament to its effectiveness, the southeastern margins of the monarchy were virtually free from the disease by 1740.

During the fight against the bubonic plague, the Viennese court also began to intensify its control over the movements of people and goods, and the Habsburg government created new institutions—such as the Sanitary Commissions in single provinces and quarantine stations—and introduced new sanitary regulations for medical staff and procedures in warfare. In the long run, these measures had a positive effect on the quality of medical service in the region, but they were only as good as those enforcing them. Deliberate concealment of plague outbreaks in rural villages and the half-hearted execution of instructions facilitated the spread of the malady. Johann Anton de Jean von Hansen (1686–1760), for example, a high-ranking administrative officer (*Administrationsrat*) and member of the Sanitary Commission in Timişoara, was one of many officials accused of laxness during the implementation of the new sanitary requirements.[58] Perhaps it was a guilty conscience that led him to commission the impressive Holy Trinity Column on the *Piaţa Unirii* in Timişoara, which today reminds us of the lethal epidemic that plagued the southeastern margin of the Habsburg monarchy during the Habsburg-Ottoman War.

Notes

1 Christine M. Boeckl, "Vienna's Imperial Plague Monument: Its Symbolism and Functions. The Bubonic Plague as a Power Symbol," in *Plague between Prague & Vienna: Medicine and Infectious Diseases in Early Modern Central Europe*, eds. Karel Černý and Sonia Hornová (Prague: Academia, 2018), 145–67; Bruno Atalić, "Religion against Plague: The Most Holy Trinity Monument in the City of Osijek," in Černý and Horn, *Plague between Prague & Vienna*, 168–88.
2 Michel Foucault, "Die Politik der Gesundheit im 18. Jahrhundert," *Österreichische Zeitschrift für Geschichtswissenschaft* 7, no. 3 (1996): 311–13.
3 George Rosen, *From Medical Police to Social Medicine: Essays on the History of Health Care* (New York: Science History Publications, 1974).
4 Cholera, dysentery, plague, smallpox, typhoid fever, and (louse-borne) typhus fever. See Friedrich Prinzing, *Epidemics Resulting from Wars* (Oxford: Clarendon Press, 1916), 4.
5 Claudia Reichl-Ham, "Der vergessene Krieg? Wahrnehmungen zum 2. Türkenkrieg Karls VI. von 1737 bis 1739," in *Die Türkenkriege des 18. Jahrhunderts. Wahrnehmen—Wissen—Erinnern*, eds. Wolfgang Zimmermann and Josef Wolf (Regensburg: Schnell & Steiner, 2017), 73–100.
6 Moriz von Angeli, "Der Krieg mit der Pforte 1736 bis 1739," *Mittheilungen des K. K. Kriegs-Archivs* (1881): 247–338, 409–79; Karl Roider, *The Reluctant Ally: Austria's Policy in the Austro-Turkish War 1737–39* (Baton Rouge: Louisiana State University Press, 1972); Theodor Tupetz, "Der Türkenfeldzug von 1739 und der Friede von Belgrad," *Historische Zeitschrift* 40 (1878): 1–51; Selim Güngörürler, *The repercussions of the Austro-Russian-Turkish War (1736–1739) on the diplomacy and the international status of the Ottoman Empire* (Istanbul: Libra Kitapçılık ve Yayıncılık, 2014); Virginia H. Aksan, *Ottoman Wars 1700–1870: An Empire Besieged* (London: Longman/Pearson, 2007).
7 Concerning the development of medicine in relation to medical care, public health, and as scholarly discipline in the Habsburg monarchy during the eighteenth century see Erna Lesky, "Österreichisches Gesundheitswesen im Zeitalter des aufgeklärten Absolutismus," *Archiv für österreichische Geschichte* vol. 122, (1959);

Erna Lesky, "Die josephinische Reform der Seuchengesetzgebung," *Sudhoffs Archiv für Geschichte der Medizin der Naturwissenschaften* 40, 1 (1956): 78–88; Johannes Wimmer, *Gesundheit, Krankheit und Tod im Zeitalter der Aufklärung: Fallstudien aus den habsburgischen Erbländern* (Vienna: Böhlau, 1991); Teodora Daniela Sechel, ed., *Medicine Within and Between the Habsburg and Ottoman Empires: 18th–19th Centuries* (Bochum: Verlag Dieter Winkler, 2011); for military medicine see Joachim Moerchel, *Das österreichische Militärsanitätswesen im Zeitalter des aufgeklärten Absolutismus* (Frankfurt/Main: Peter Lang, 1984); Matthias König, *Blutiges Handwerk. Die Entwicklung der österreichischen Feldsanität zwischen 1748 und 1785* (Vienna, Bundesministerium für Landesverteidigung und Sport, 2011); Salomon Kirchenberger, *Geschichte des k. und k. österreichisch-ungarischen Militär-Sanitätswesens* (Vienna: Josef Šafář, 1895); Salomon Kirchenberger, "Zur Geschichte des österreichischen Militär-Sanitäts-Wesens während des Zeitraumes vom 14. bis 18. Jahrhundert," *Streffleur's Österreichische Militärische Zeitschrift* 22, no. 3–4 (1881): 55–73.

8 Fielding H. Garrison, *Notes on the History of Military Medicine* (Hildesheim, New York: G. Olms Verlag, 1970); Roger Cooter, "Medicine and the Goodness of War," *Canadian Bulletin of the History of Medicine* 7, no. 2 (1990): 147–59; Mark Harrison, "The Medicalization of War—The Militarization of Medicine," *Social History of Medicine* 9, no. 2 (1996): 267–76; Mark Harrison, "Medicine and the Management of Modern Warfare," *History of Science* 34 (1996): 380–409; Sebastian Pranghofer, "The Early Modern Medical-Military Complex: The Wider Context of the Relationship Between Military, Medicine, and the State," *Canadian Journal of History* 51, no. 3 (2016): 451–72; Erica Charters, *Disease, War and the Imperial State: The Welfare of the British Armed Forces during the Seven Years' War* (Chicago: University of Chicago Press, 2014); Christopher Storrs, "Health, Sickness and Medical Services in Spain's Armed Forces c.1665–1700," *Medical History* 50, no. 3 (2006): 325–50; Hans Zinsser, *Rats, Lice and History: Being a Study in Biography, Which, After Twelve Preliminary Chapters Indispensable for the Preparation of the Lay Reader, Deals With the Life History of Typhus Fever* (Boston: Atlantic Monthly Press by Little, Brown, and Company, 1935); Roger Cooter, "Of War and Epidemics: Unnatural Couplings, Problematic Conceptions," *Social History of Medicine* 16, no. 1 (2003): 283–302; Matthew Smallman-Raynor and Andrew Cliff, *War Epidemics: An Historical Geography of Infectious Diseases in Military Conflict and Civil Strife, 1850–2000* (Oxford: Oxford University Press 2004); or Rebecca Seaman, *Epidemics and War: The Impact of Disease on Major Conflicts in History* (Santa Barbara: ABC-CLIO, 2018).

9 Marie-Janine Calic, *Südosteuropa. Weltgeschichte einer Region* (Munich: C.H. Beck Verlag, 2016), 265–69.

10 Nükhet Varlık, *Plague and Empire in the Early Modern Mediterranean World: The Ottoman Experience, 1347–1600* (Cambridge: Cambridge University Press, 2015).

11 Daniel Panzac, "plague (veba; waba)," in Gabor Agoston and Bruce Masters, eds., *Encyclopedia of the Ottoman Empire* (New York: Facts on File, 2009), 462–63.

12 Birsen Bulmuş, *Plague, Quarantines and Geopolitics in the Ottoman Empire* (Edinburgh: Edinburgh University Press, 2012).

13 Found, translated, and dated to the eighteenth or early nineteenth century by Eleazar Birnbaum. See Eleazar Birnbaum, "A Cure for the Plague, and Other Prescriptions," in *Ottoman War and Peace: Studies in Honor of Virginia H. Aksan*, eds. Frank Castiglione, Ethan Menchinger, and Veysel Şimşek (Leiden: Brill, 2019), 172.

14 Michael Hochedlinger, *Austria's Wars of Emergence, 1683–1797* (London: Routledge, 2013), 98–150.

15 Sabine Jesner, "Grenzschutz im und gegenüber dem Südosten," in *Wir und Passarowitz. 300 Jahre Auswirkungen auf Europa*, eds. Betina Habsburg-Lothringen and Harald Heppner (Graz: Universalmuseum Joanneum, 2018), 56–61.

16 For Habsburg warfare see Michael Hochedlinger, "Onus Militare": Zum Problem der Kriegsfinanzierung in der frühneuzeitlichen Habsburgermonarchie 1500–1750," in *Kriegsführung und Staatsfinanzen. Die Habsburgermonarchie und das Heilige Römische Reich vom Dreißigjährigen Krieg bis zum Ende des habsburgischen Kaisertums 1740*, ed. Peter Rauscher (Münster: Aschendorff, 2010), 81–136.

17 Johann Georg Heinrich Kramer, *Medicina Castrensis d.i. Bewährte Artzney wider die im Feld und Guarnisons unter Soldaten grassirenden Krankheiten* (Vienna: Peter Conrad Monath, 1745); Heinz Flamm, "Das Fleckfieber und die Erfindung seiner Serodiagnose und Impfung bei der k. u. k. Armee im Ersten Weltkrieg," *Wiener Medizinische Wochenschrift* 165 (2015): 152–53; Zinsser, *Rats*, 268.

18 Stefan Winkle, *Geißeln der Menschheit: Kulturgeschichte der Seuchen* (Düsseldorf, Zürich: Artemis & Winkler, 1997), 422–27.

19 Report of the Sanitary Commission in Sibiu from 16 July 1738. See OeStA [Österreichisches Staatsarchiv] KA [Kriegsarchiv] ZSt [Zentralstellen] MilKom [(Militär-) Hofkommissionen] Sanitätshofkommission Akten [Files] 1, 1738-Julius-31. For the protocol of the Sanitary Court Commission see OeStA KA ZSt MilKom Sanitätshofkommission Bücher [Books] (1606–1752), fol. 34–40. For a list of infected villages, including data concerning infected and recovered patients see the letter of Commander Lobkovic from Sibiu on 28 Dec. 1737 see OeStA KA ZSt HKR [Hofkriegsrat] HR [Hauptreihe] Akten, No. 205, 1738-February-98.

20 From 1719, the commission was named *commissio sanitatis aulica*. See Lesky, "Österreichisches Gesundheitswesen," 12.

21 The members of the local commissions were not paid for their work. An application of the commission in Sibiu for financial compensation for their work was defeated in July 1738. See the note of the Sanitary Court Commission to the Sanitary Commission in Sibiu from 28 July 1738: OeStA KA ZSt MilKom Sanitätshofkommission Bücher (1606–1752), fol. 93–6.

22 "[E]ine mixtin ex Militari, Camerali, Politico-Civili und Medico zusammen gesetzte Commission." See the report of the Sanitary Commission in Sibiu from 19 July 1739: OeStA KA ZSt MilKom Sanitätshofkommission Akten 1, 1738–July-31.

23 See the protocol of the Sanitary Court Commission: OeStA KA ZSt MilKom Sanitätshofkommission Bücher (1606–1752).

24 For the holdings of the General War Commission see OeStA KA ZSt GKK [Generalkriegskommissariat].

25 Solomon Kirchenberger, "Die älteste selbstständige 'gedruckte' Feld-Spitalsordnung der österreichischen Armee," *Der Militaerarzt. Zeitschrift für das gesammte Sanitätswesen der Armeen* XXXV (1901): 38–9, 68.

26 The military medical staff was subordinated to the Aulic War Council and as a part of the army they held military ranks and were assigned to different military units. See Kirchenberger, *Geschichte des k. und k.*, 1–7. The 1785 formation of the *Josephinum* established a unique military-medicine education. See Brigitte Lohff, *Die Josephs-Akademie im Wiener Josephinum: Die medizinisch-chirurgische Militärakademie im Spannungsfeld von Wissenschaft und Politik 1785–1874* (Vienna, Cologne, Weimar: Böhlau, 2019).

27 Sremski Karlovci is named after the same locality where the treaty had been concluded in historical Syrmia. Today, Sremski Karlovci is a city in Serbia.

28 Stang had interviewed the local populace. His description of the symptoms identified the malady as the bubonic plague: "[...] an allen sowohl alten als Kindern

böse Beulen an den dreissigsten Theilen als an der Scham und unter denen Achslen der Anfang der Krankheit ist, daß solche inficirte grosse Herzens Engigkeiten mit starker hitz klagen, welche, so annoch guter constitution die Beulen mit starken schmerzen heraus tretten, sich durch deren Hausmittelzeitigen und selbst öffnen, die werden conserviret, wo aber die Malignität gar zu starck, werden in unterschiedl. doch wenigen Tagen durch eberhauften tödlich zurfällen des Todes, die vergangene Nach seyend 11 gestorben an dieser Krankheit, und schätze deren gegen 20 annoch in Gefahr des Todes [...]." For a copy of this report from Stang to Pfefferkorn from 29 Sept. 1738 see OeStA KA HKR HR Akten 209, 1738-October-501, fol. 4rv.

29 Report of Pfefferkorn to Feldmarschallleutnant Marchese Guadagni, the head of the Landesdeputation in Esseg, from 30 Sept. 1738: OeStA KA HKR HR Akten 209, 1738-October-501, fol. 5r–6r.

30 Report of Pertsch to Feldmarschallleutnant Marchese Gudagni, from 30 September 1738: OeStA KA HKR HR Akten 209, 1738-October-501, fol. 7r–8v.

31 Note of the *Landesdeputation* in Esseg to the Aulic War Council from 2 Oct. 1738: OeStA KA HKR HR Akten 209, 1738-October-501, fol. 9r–13r.

32 Letter from Königsegg from the field camp in Zemun to the Aulic War Council from 4 Oct. 1738: OeStA KA FA AFA HR Akten 493, IX–XIII Türkenkrieg HKR, 1738-October-3.

33 For the field hospital 1738 see Kirchenberger, "Die älteste," 17–19, 36–39, 67–69, 75–77.

34 Kirchenberger, "Die älteste," 97–99.

35 For the plague in Transylvania see OeStA KA ZSt MilKom Sanitätshofkommission Bücher (1606–1752), fol. 35 and following folios.

36 Olaf Briese, *Angst in den Zeiten der Cholera: Über kulturelle Ursprünge des Bakteriums* (Berlin: Akademie Verlag 2003), 242–43.

37 In terms of the adjustable militarized cordons see the instruction of Charles VI to implement cordons on rivers from 23 May 1738: OeStA KA ZSt HKR HR Akten, No. 208, 1738-June-19, and the protocol of the Sanitary Court Commission OeStA KA ZSt MilKom Sanitätshofkommission Bücher (1606–1752) for the imperial resolution from 24 Mar. 1738 (article 810.) fol. 44 or concerning the alteration of the cordon because of the infection of a village near Szeged see the imperial resolution from 23 June 1738 (article 1) fol. 60–61.

38 "im Grünnischen Bataillon grassirenden epidemischen Kranckheit mit denen Beüllen." See Anton von Hammer, *Geschichte der Pest die von 1738 bis 1740 im Temeswarer Banate herrschte. Ein aus glaubwürdigen Quellen geschöpfter Beitrag zur Geschichte dieses Landes* (Temeswar: Verlag Joseph Beichel, 1839), 2–3; and for the letter of the commander of the fortress Scotti to the Aulic War Council from 19 Mar. 1738 see OeStA KA ZSt HKR HR Akten, No. 205, 1738-March-836.

39 Hammer, *Geschichte der Pest*, 3.

40 For §7 of the decree from 24 March 1738 see OeStA KA ZSt MilKom Sanitätshofkommission Bücher (1606–1752), fol. 43–4; Hammer, *Geschichte der Pest*, 4–5.

41 The journal of Tobias Dolfin and Marianus Caunes was sent from the fortress commander Scotti to the Aulic War Council on 19 Mar. 1738. See OeStA KA ZSt HKR HR Akten, No. 205, 1738-March-836.

42 "Beschreibung. Die auf Befehl eines löbl. Militar Ober Commando der auch Löbl. Grüenische Battaillon vor der Vorstadt über die Esplanade, die Quartier und Lazareth zugericht und abgetheilet worden, auch wie selbe durch die auß gestelte Militar und Bürger Posten verwacht werden." The plan was sent from Scotti to the Aulic War Council on March 19, 1738. See OeStA KA ZSt HKR HR Akten, No. 205, 1738-March-836.

43 Imperial decree from April 5, 1738. See OeStA KA ZSt MilKom Sanitätshofkommission Bücher (1606–1752), fol. 48.
44 Imperial decree from May 6, 1738. See OeStA KA ZSt MilKom Sanitätshofkommission Bücher (1606–1752), fol. 50.
45 Sabine Jesner, "The World of Work in the Habsburg Banat (1716–51/53): Early Concepts of State-Based Social and Healthcare Schemes for Imperial Staff and Relatives," *Austrian History Yearbook* 50 (2019): 58–77. In July 1738 was ordered to hinder contacts between the local population (the *Land-Volk*) and the army in the Banat. See OeStA KA ZSt MilKom Sanitätshofkommission Bücher (1606–1752), fol. 66.
46 Instruction of Charles VI from March 24, 1738. See Hammer, *Geschichte der Pest*, 36.
47 For the period of war, the protocol book of the Sanitary Court Commission offers hints for temporary quarantine stations (*Contumaz-hitten*) in Belgrade, Petrovaradin, Stari Slankamen, Pančevo, Szeged, Kecskemét, Somylo, Zilach, Szolnok, Cenad, Csongrád, Karlovac, Jagodina, Senta, Sânnicolau Mare, Ţaga, Baja, Vukovar, Dallia, and Zillaj. See OeStA KA ZSt MilKom Sanitätshofkommission Bücher (1606–1752).
48 For the mandatory quarantine see OeStA KA ZSt MilKom Sanitätshofkommission Bücher (1606–1752), fol. 58, 61–64, 113, and for the detailed printed instructions "Kontumaz- und Reinigungsordnung für die südlichen und östlichen Gebiete (Quarantine and Cleaning Order)" from 10 May 1738 see OeStA FHKA [Finanz- und Hofkammerarchiv] SUS [Sondersammung und Selekte] Patente 72.11.
49 For unauthorized travel see OeStA KA ZSt MilKom Sanitätshofkommission Bücher (1606–1752), fol. 122, 1738.
50 For the response of the Sanitary Court Commission from 6 Sept. 1738 see OeStA KA ZSt MilKom Sanitätshofkommission Bücher (1606–1752), fol. 171.
51 OeStA KA MilKom Sanitätshofkommission Akten (1738–1762), 19 October, 1738.
52 See for the instruction OeStA KA ZSt MilKom Sanitätshofkommission Akten (1738–1762), 1739-8; In April 1739, a separate directive was issued "Zu der Reinigung der inficirten oder suspecten Schiffen gehört," that elucidated the purification procedures of ships, including a prescription for the composition of the powder used in fumigation: OeStA KA ZSt HKR HR Akten, no. 211, 1739-April-50.
53 Königsegg from the field camp in Zemun on 4 Oct. 1738: "Alle üble Verhängnissen, wollen sich von Tag zu Tag über Eurer Kayl. May. Trouppen vermehren, es thun nit allein die ordinari Fieber und Durchbruch Krankheit von Tag zu Tag die Cavallerie und Infanerie an Officieres sowohl als Gemeine, und Bedienten sehr merklich zu Diensten vermindern, sondern muss ich leider anzeigen, das die Infection würklich bey einigen sowohl Infanterie als Cavallerie Regimentern eingedrungen habe, [...] Mann tut nun alles was die menschliche Vorsichtigkeit und Hülf dabey nur vermag, absonderlich ist mann sehr attent dergleichen Kranke als gleich zu separieren, bey der Artillerie habe sich auch schon etwas gezeiget, und die in größte Sorge das ob auch in die Provinz Häuser eingreisse, da wegen der sovill erkrankte und davon laufenden Leben, aus Noth mit Soldaten suppliert werden muss [...]." See OeStA KA FA AFA HR Akten 493, 1738, IX–XIII Türkenkrieg HKR, October, 1738-10-3.
54 Angeli, "Der Krieg," 430–31.
55 Erna Lesky, "Die österreichische Pestfront an der k.k. Militärgrenze," *Saeculum* 8 (1957): 82–106; Gunther E. Rothenberg, "The Austrian Sanitary Cordon and the Control of the Bubonic Plague: 1710–1871," *Journal of the History of Medicine and Allied Sciences* 28 (1973): 15–23; Ivana Horbec, Robert Skenderović, "The Quarantines of the Croatian and Slavonian Military Frontier and their Role in the 18th-Century Epidemic Control," in Černý and Horn, *Plague Between Prague & Vienna*, 190–230.

56   Concerning the Habsburg Military Border see among others Karl Kaser, *Freier Bauer und Soldat: Die Militarisierung der agrarischen Gesellschaft an der kroatisch-slawonischen Militärgrenze 1535–1881* (Vienna: Böhlau, 1997); Sabine Jesner, *Habsburgische Grenzraumpolitik in der Siebenbürgischen Militärgrenze (1760–1830): Verteidigungs- und Präventionsstrategien* (Ph.D. diss., University of Graz, 2013).

57   Sabine Jesner, "Habsburg Border Quarantines until 1837: An Epidemiological 'Iron Curtain'?" in *Medicalising Borders: Selection, Containment and Quarantine since 1800*, eds. Sevasti Trubeta, Christian Promitzer, and Paul Weindling (Manchester: Manchester University Press, 2020), 31–55.

58   For the censure concerning some delays during the erection of a military lazaretto (*Militar Lazareth*) in Timişoara see OeStA KA ZSt MilKom Sanitätshofkommission Bücher (1606–1752), fol. 99 and in terms of the plague column Horst Fassel, ed., *Die Dreifaltigkeits-oder Pestsäule in Temeswar. Stationen einer Wiederentdeckung* (Munich: Landsmannschaft der Banater Schwaben, 1996).

Further Reading

Černý, Karel and Sonia Hornová. *Plague between Prague & Vienna: Medicine and Infectious Diseases in Early Modern Central Europe*. Prague: Academia, 2018.

Hammer, Anton von. *Geschichte Der Pest die von 1738 bis 1740 im Temeswarer Banate herrschte. Ein aus glaubwürdigen Quellen geschöpfter Beitrag zur Geschichte dieses Landes*. Temeswar: Verlag Joseph Beichel, 1839.

Jesner, Sabine. "Habsburg Border Quarantines until 1837: An Epidemiological 'Iron Curtain'?" In *Medicalising Borders. Selection, Containment and Quarantine Since 1800*, edited by Sevasti Trubeta, Christian Promitzer and Paul Weindling, 31–55. Manchester: Manchester University Press, 2020.

Prinzing, Freidrich. *Epidemics Resulting from Wars*. Oxford: Clarendon Press, 1916.

Zimmermann, Wolfgang and Josef Wolf. *Die Türkenkriege des 18. Jahrhunderts. Wahrnehmen—Wissen—Erinnern*. Regensburg: Schnell & Steiner, 2017.

12 Shifting Power Relations along the Baltic

Poles, Lithuanians, and Russians in the Great Northern War

Mindaugas Šapoka

Poland-Lithuania and Russia: A Short History of a Long Relationship

In 1569, the Kingdom of Poland and the Grand Duchy of Lithuania signed the Union of Lublin, a constitutional settlement that established a single state, the multicultural and multireligious Commonwealth of Two Nations (also known as Poland-Lithuania, the Polish-Lithuanian Commonwealth or the Republic). From the end of the fifteenth century until the middle of the seventeenth century, Poland and Lithuania competed with the Grand Duchy of Moscow and its successor states (the Tsardom of Russia and the Russian Empire) for territory and power. Their relationship was marked by a long series of wars as their respective fortunes ebbed and flowed. In the middle of the sixteenth century along the Baltic, Poland and Lithuania won control of Livonia and much of Estonia. In 1610, a Polish-Lithuanian garrison entered Moscow following a period of crisis in Russia known as "The Time of Troubles." Success for Poland, though, was temporary. The conflict between the two states continued and culminated with the Russo-Polish War of 1654–67. Open hostilities officially ended in 1686 with the Treaty of Eternal Peace, an agreement that secured Russia's control of left-bank Ukraine. This treaty also marked a dramatic turn in relations between the two rivals. From an arch-enemy, Russia became an ally in the war against the Turks (1686–1700) and then in the Great Northern War (1700–21) against Sweden.

At the time of these new alliances Russia also began to interfere in the internal affairs of the Republic in an effort to influence domestic political processes in its favor. The growth of Russian influence in the Polish-Lithuanian Commonwealth was not a short-term phenomenon but an event of epoch-making significance in Eastern European history. At the end of the Great Northern War, Russia had become a major European power with its dominant army and navy, while Poland-Lithuania entered into a period of decline. Growing Russian influence in the Republic did not occur overnight, and, as Polish scholars have noted, it was not until the late 1720s that Russia had effectively established a protectorate over the Republic.[1] Thus the period of the Great Northern War is crucial to

DOI: 10.4324/9781003157700-16

understand how power relations between Poland-Lithuania and Russia shifted. How did the Polish nobility respond to these changes? This article examines the attitudes of these ruling elites towards Russia generally and Peter the Great in particular during the Great Northern War. It attempts to determine whether the Poles and Lithuanians trusted the Russians as their allies or whether they understood the danger of Russia's increasing influence within their country. These questions have received little attention. Andrzej Sowa has observed that the attitudes of the Polish-Lithuanian ruling elite were complicated, not least as the views of the same person were often inconsistent and contradictory. From the moment of the conclusion of the alliance with Russia against Sweden in 1704, the opinions of the elites oscillated between the conviction that it was necessary to maintain the alliance with Russia because of the benefits guaranteed under the treaty and distrust caused by the depredations of the Russian troops in the Republic.[2] It was the ministers of the Republic and their correspondence that primarily interested Sowa. In this chapter, I will examine the correspondence that Sowa did not use alongside official documents of Polish-Lithuanian institutions to determine if his conclusions can be applied not only to the government's ministers but also to the nobility more broadly. Additionally, to better understand the relationship between these two states, one also needs to examine the other side to determine what the Russian ruling elite thought of the Poles and Lithuanians.

Russian Policy Toward the Republic

The Polish king and the elector of Saxony, Augustus II, attacked Livonia, then ruled by Sweden, with his Saxon troops in February 1700. Augustus in this capacity was attacking not as a Polish king but as a Saxon duke, for even though he used the Grand Duchy of Lithuania as a base for his attack, Poland-Lithuania formally remained neutral in the Great Northern War. Thus, in the first years of the war the primary goal of Augustus II and Peter I, who had made an alliance in 1698 and reaffirmed it in 1701, was to drag the Republic into the war against Sweden. Few in the Republic, though, supported the fight against the Swedes. Many of the Poles would have preferred a war against Russia to re-conquer territories lost in the 1654–67 Polish-Russian conflict. Lithuanians, however, were traditionally inclined to have good relations with Russia as they would be the first to suffer if a war with their larger neighbor broke out. When the Great Northern War began in 1700, the terrible experiences of the 1654–67 war, when Russian troops occupied almost all of Lithuania, were still fresh in Lithuanian memory. It is a paradox, then, that Lithuania supported rapprochement with Russia for Russophobic reasons; they believed it was better to maintain friendly relations with Russia than to wage war against her. Regardless, the Lithuanian nobility viewed the conclusion of the treaties between Augustus and Peter with suspicion. In 1701, the local noble assembly (sejmik) of Brest declared that "the frequent correspondence of His Majesty and the Tsar of Moscow fills us with apprehension about our liberties." In the instruction for its envoys to the Sejm, the nobility asked the king "to establish confidence and friendliness with His Highness the Tsar of Moscow, thereby securing the Republic against future dangers which can arise from such a powerful neighboring sovereign."[3]

In the face of Saxon, Danish, and Russian attacks, Sweden and its army proved remarkably resilient. After landing a considerable body of his troops near Copenhagen, Charles XII of Sweden forced Denmark-Norway out of the war. He then defeated the Russians at Narva (November 1700) and the Saxon army at Riga in July 1701. Having occupied the vassal duchy of Courland, the main Swedish army stood at the borders of the Commonwealth whose internal affairs were turbulent. The Lithuanian nobility was embroiled in an internal conflict that essentially amounted to civil war. The powerful Sapieha family was challenged by a coalition of other magnate families. This alliance crushed the Sapiehas at the bloody battle of Olkieniki in 1700. After this defeat, the Sapiehas turned to the king and Polish magnates for assistance. Augustus was in a difficult position. He sought to pacify Lithuania and strengthen the defensive capabilities of the country as the Swedes were clearly ready to invade Lithuania at any time and indeed attacked in late 1701. To address this situation, Augustus II convened the Sejm of Poland-Lithuania.

When the Sejm considered Peter's offer of a military alliance against Sweden, the hostility of the Lithuanian warring factions manifested itself. The grand hetman of Lithuania Kazimierz Jan Sapieha opposed the Russian alliance and told the deputies that he regarded the Muscovites as the Republic's most dangerous enemy.[4] The Lithuanian Chancellor Karol Radziwiłł, however, was of a different opinion when he addressed Augustus:

> Your Majesty, turn the page and find the opposite, because it is the Swede who is ruining us with an intention to turn us into serfs, and yet we dispatch a diplomatic legation to him, and somebody speaks of him as a friend, not an enemy. I would not like to insult the mighty sovereign of Moscow, whom we truly need now, with the denial of an audience to his envoy. I would like us to have a diplomat permanently residing in Moscow to gain his [the tsar's] confidence.[5]

From the outset of the Great Northern War, Peter I's policy towards the Republic was determined by the fear of undermining the Eternal Peace of 1686. Despite the treaty, Peter and his court feared that Poland-Lithuania might resume the war for left-bank Ukraine. Having grown up in Moscow where memories of the Republic's former power were still potent, Peter developed a type of Polonophobia. In the initial period of his rule, he conducted his foreign policy as if there was no Republic at all.[6] The terrible start of the Great Northern War—Peter's defeat at Narva, the Saxon losses at Riga and Kliszów—obliged the tsar to change his attitude and seek rapprochement with the Republic, which formally remained neutral in the war, as his possible ally. In 1702, direct contacts were established by Peter's administration and the anti-Sapieha group which wanted the Republic to join the war against Sweden, not least because the rival Sapieha clan had asked for Swedish protection and thus legitimized the Swedish invasion of the Republic in 1701. The fear of damaging friendly relations with Russia combined with their concerns of the possible return of the Sapiehas left the Lithuanians with little room for manoeuvre. They sought the tsar's financial and military help.

There were still few supporters of the Russian alliance in Poland though. For this reason, Peter made three mutual assistance treaties with the anti-Sapieha group. Under the terms of these treaties, Lithuania, meaning Lithuania loyal to the anti-Sapieha faction, was to make war against Sweden even if Poland remained neutral. Formally, this was a breach of the Union of Lublin, which stipulated a common foreign policy for the kingdom and grand duchy. The Russians hoped that having dragged one part of the Republic into the war, the second part would soon follow fearful of breaking the union. The Russians, however, had to be very careful, as the new treaties with Lithuania promised Russian financial and military support to the anti-Sapieha faction. Additionally, any appearance of Russian troops on Lithuanian soil could cause Poland to accuse Russia of violating the Eternal Peace. Just as the separate treaties with Lithuania could accelerate Poland's entry to the war, the violation of the Eternal Peace could unite Poland and Sweden against Russia. Before the signing of the second treaty with the Lithuanians, the tsar voiced his concerns to the Lithuanian envoys:

> I have shown you much affection giving you money and auxiliary units, but you will follow the Swede and start a war against me.[7]

This, however, did not happen. The persistent efforts of the Swedish king Charles XII to remove Augustus from the Polish throne was the main reason why support for the Russian alliance grew in Poland in 1703–04. After all, Augustus was a legal king, and Swedish interference in the internal affairs of the Republic elicited significant discontent among the nobility. The Russians devised a cunning plan to make a treaty with the whole Republic against Sweden under conditions that would actually allow Russia not to honor its promises. In the fall of 1703, Johann Patkul, a former subject of the Swedish Crown now in the service of the Russians, drew up a falsified treaty between Russia and Saxony which was to be widely circulated in Poland and Lithuania. According to its conditions, Peter promised Augustus financial support only if the duke mortgaged a portion of his Saxon territory.[8] The design was to keep the real treaty secret and inflate the price of Russian support. The "falsified" treaty would delude the Poles and Lithuanians and reinforce the notion that

Russian assistance was not cheap. If the Republic wanted financial aid from the Russians, it should consider mortgaging a piece of its lands, just as Augustus had allegedly done.[9]

The Swedish invasion of the Republic highlighted the extent to which bitter internal rivalries within Poland and Lithuania had left them divided and incapable of mounting an effective resistance. Patkul was pleased to see the impotence of Polish-Lithuanian arms. "I would deceive myself if I believed that the Poles would ever enter into a reliable alliance with Your Tsarist Highness," he wrote. In his opinion, it was better to cooperate with Augustus in his capacity as Saxon elector and not to trust the Poles and Lithuanians. It was unwise to provide them with military and financial assistance "because now they are weak, and their idleness poses no danger to us; otherwise, they will become dangerous and firm enemies to be feared, just like the Swedes," Patkul continued.[10] Peter agreed that "it is impossible to agree with them [Poles and Lithuanians] on a good deal," but to increase the confidence of the Poles and Lithuanians he thought it was better "not to tell them that we regard the alliance with them as useless. We shall serve them the same sauce for their falsehood."[11] The treaty between Poland-Lithuania and Russia finally signed on 30 August 1704 at Narva fulfilled the Russian plan. It was a defensive and offensive military alliance valid until the end of the war. The tsar indeed pledged to pay the Republic an annual subsidy of 200,000 rubles, but the Republic had to muster an army of 48,000 men in return. The Russians were nearly certain the Republic would fail to conscript the required army thus freeing them of any obligation to pay subsidies.[12]

Russian Interference in Lithuania

In 1704–05, bitter disagreements arose between Michał Wiśniowiecki and Gregorz Ogiński, the hetmans of the Lithuanian army.[13] Consequently, Lithuanian troops were reduced to an auxiliary status with regard to the Russian army, which entered Lithuania in the spring of 1705. The conflict between the hetmans was significant as it allowed Peter to assume the role of mediator, not least as Augustus was absent having spent most of 1705 outside the Republic in Saxony. The hetmans used the tsar to harm each other. Initially, it was Ogiński who persuaded Peter that Wiśniowiecki was in secret correspondence with the Swedes. When Wiśniowiecki later met Peter, he persuaded the tsar that it was actually Ogiński who was negotiating peace with the Swedes and Sapiehas. The tsar's decision was to order Ogiński to obey the orders of Wiśniowiecki, who was then the grand hetman.[14]

The magnates benefited from cooperation with the tsar, either by enriching themselves or by strengthening their influence in the country, while the middling and petty nobility were of a different disposition. They viewed the Russian army in Lithuania with distrust if not open enmity. The Russians not only imposed an additional financial burden but were traditionally seen as eternal enemies. In February 1705, the English envoy of Queen Anne, Charles Whitworth, reported from Vilnius, the Lithuanian capital:

> He [Ogiński] therefore applauds himself for having introduced these aux-iliary forces hither [...] he made no mention at all of peace and probably has no inclinations that way, since his interests here, and the great figure he now makes, can only be maintained by the present troubles, or by the utter ruine of the potent house of Sapieha, whom it cannot be imagined the king of Sweden will ever so far abandon [...] they [Lithuanian nobil-ity] are highly discontented with the imperious carriage and exactions of their new guests, whom though they are bid to look upon as friends. They cannot so soon forget to have been their ancient and almost hereditary en-emies. This was the general language of the lesser nobility, which whom I had occasion to converse either here or on my road.[15]

Almost a year has passed, and in November 1705, Whitworth presented an en-tirely different assessment of the attitude of the great Lithuanian families to-wards the tsar:

> It is very probable that neither he [Ogiński], nor the prince [Wiśniowiecki] are too well inclined to the Czar's interests, for the great exactions of horses, carriages and provisions with other disor-ders, which are daily committed without any distinction of persons or estates, have entirely disgusted the chief families in this Duchy; nor do they dissemble their discontents in their private conversations, so that the complaisance they show in public certainly proceeds more from fear, then affection.[16]

Whitworth's observations proved accurate. Russian diplomats closely followed the sentiments of the leading magnates in the Republic although there is no documentary evidence to tell whether the Poles or Lithuanians knew they had been watched. The Russian plenipotentiary in the Republic Grigorii Dolgorukii was given the following instructions:

> We know that they [Lithuanians] can never fulfil their promises because they all lie. [...] Patkul wrote to the Great Sovereign [tsar] that there were some doubts in Lithuania about [the loyalty of] Wiśniowiecki and Ogiński; we think such assumptions are not valid, but if we find they are, we will treat them accordingly.[17]

There are very few sources about attitudes towards Russia among the non-noble population of the Republic. At the beginning of the war, the townsfolk of eastern

Lithuanian cities were more favorably inclined toward Moscow. A spy sent by Russian authorities with letters inviting burghers to move to Russia was welcomed in Połock and Vitebsk. The former declared their readiness to move in case of the danger of the Swedes, while the latter thanked the tsar for the care and his wish to "free them from the heavy yoke."[18] It is, however, impossible to determine whether these sentiments were genuine or carefully designed statements to placate the tsar and his ministers prompted by an underlying Russophobia. The situation with the nobility was somewhat different.

From the very moment the Russian army appeared in Lithuania, they complained about the arbitrary contributions raised by the Russians and by their conduct which reminded them more of conquerors than allies. They dispatched envoys from sejmiks to the Russian generals or the tsar with requests to reduce the size of contributions.[19] The Russians formally responded that they had come by invitation of the Republic, and thus their troops must be maintained by local resources.[20] Moreover, the tsar forbade the Lithuanian nobility to export their goods, mainly grain, via Riga, which was then controlled by Sweden. Russian troops were ordered to arrest anybody on their way to Riga with any goods whatsoever. The Lithuanians tried to persuade the tsar to rescind his order but to no avail.[21]

The success of Swedish troops against the Saxon, Russian, and Polish-Lithuanian armies forced Augustus to abdicate from the Polish throne in 1706 leaving the Russians in a challenging situation. Had the Swedish puppet king Stanisław Leszczyński, elected by a tiny minority of the Republic's nobility under the supervision of Swedish bayonets, been able to consolidate his power and win over those opposing Sweden to his camp, the Russians would have lost their only remaining ally in the war. Peter was fortunate that Charles XII's absolute disregard of Swedish opposition in the Republic precluded any chance of internal pacification. Nevertheless, the tsar was still compelled to make significant concessions to Poles and Lithuanians loyal to the Russian alliance while military discipline was increased in the Russian units. At the same time, Peter employed a series of tricks to temper Polish-Lithuanian opposition to the contributions. In 1707, Peter ordered his field marshal Boris Sheremetev to begin buying small quantities of provisions in many places simultaneously in order to give the impression that Russian troops actually did pay for their supplies:

> One thousand rubles are allocated to buy bread in two or three places. Firstly, when you receive this letter, send two or three persons immediately and proclaim that you have to provide me with as many provisions as soldiers under my command need [...] it is an order to buy in small quantities and continue to do so for a long time for the sake of glory. However, you must keep this order in secret.[22]

Such tactics were to demonstrate the "friendliness" of the Russian army and support their claims that they paid for their army's maintenance thus undercutting any complaint of the Poles and Lithuanians.

Double-Faced Russians

The crushing defeat of the Swedish army at Poltava in 1709 and the surrender of its remaining soldiers a few days later at Perevolochna allowed Augustus to return to the Polish throne. The Republic's nobility confirmed him as legal king the following year. At the same time, the destruction of the Swedish army also allowed the Russians to return to the Republic. Just as Charles XII proclaimed he was defending the Republic's constitution from a king who had broken the laws by starting a war nobody had wanted, Peter now declared that by defeating the Swedes he had defended the Republic's constitution from Swedish aggression and the illegal election of Stanisław Leszczyński. The main aim of this rhetoric was to use the Republic as a base for taking the offensive against Swedish Livonia and Pomerania.[23]

After Poltava, the Russians governed Lithuania with a firm hand. Former Swedish supporters were in a complicated situation because, in order to avoid arrest, they had to regain the confidence of the tsar and seek his pardon. When the Russian general Samuel von Rentsel arrived in Nyasvizh the center of the estates owned by the Lithuanian Chancellor Karol Stanisław Radziwiłł who in 1707 defected to Stanisław Leszczyński's camp, he asked the servant where his master was. The servant answered that Radziwiłł was in Biała, one of his estates. The Russian retorted:

> Not true, you are a liar! Your master is with Stanisław and the Swedes. If you do not provide my unit with provisions immediately, I will order them to take them from you, and you will have to transport it 100 miles. I have an order to punish you with the collection of provisions as an enemy servant.[24]

Radziwiłł asked Ludwik Pociej, the newly appointed Lithuanian grand hetman, to help him reconcile with the tsar. Pociej spoke about Radziwiłł with Peter and later in his letter to Radziwiłł he quoted the tsar's words, "Write in my name to the duke chancellor that he will receive an amnesty, but he will need to furnish new uniforms to my Preobrazhensky regiment."[25] In 1713, Radziwiłł wrote a letter to Peter claiming that he had already proved his loyalty and asked for the tsar's favor.[26]

Courtesy was expressed in direct correspondence with the tsar, but the private correspondence between the Poles and Lithuanians was full of complaints about the treacherous conduct of the Russians. In the next years of the war, the Russian army marched through the Republic without permission from the Republic's authorities, and collected arbitrary contributions, while Russian merchants did not pay customs duties on the Republic's border. The grand marshal of Lithuania Marcjan Wołłowicz wrote:

> Field marshal Sheremetev [of the Russian troops] arrived in Lithuania demanding colossal contributions [...] The leaders of the districts visited him and presented the deplorable situation of the districts where many people had died because of famine and plague. Asking him to reduce the numbers of households [on which the contributions were imposed], he replied, "I do not care about it. You must give me as many provisions as soldiers under my command need."[27]

Requests for compensation for lands devastated by the Russians did not bring any tangible results. When a member of the Sapieha family, the most prominent supporters of Leszczyński now granted amnesty by Peter, asked the tsar to consider reimbursement, Peter delivered an indifferent answer, "We were at war. I have lost many of my soldiers and burned my own country; it was impossible to do it any other way."[28] The Danish envoy at the Russian court commented from the sidelines concerning the issue of Russian contributions:

> The Poles [...] are more opposed to the Russians than to the Swedes. Though the Swedes had come as enemies and raised immense contributions, they allowed the Poles to control their lands peacefully. Meanwhile, the Russians had come as friends but also demanded colossal contributions, but having received them, they continue to rob, burn, and occupy their lands and estates. Even during Advent when Poles do not eat, the Russians slaughter animals only to sell their skins and feed the meat to dogs; the Poles also complain about their king who allows the tsar to govern in Poland as he pleases.[29]

It would be a mistake to claim that the Poles and Lithuanians did not understand that they were being exploited and cheated by the Russians. In the summer of 1710, when the fortress of Riga, then controlled by the Swedes and promised to the Republic by the treaty of Narva of 1704 but besieged by the Russians in late 1709, was about to fall, the Russian Chancellor Golovkin sent letters to Polish and Lithuanian hetmans demanding that they send infantry to help the Russian army.

> Because His Tsarist Highness is besieging the fortress not for himself but according to the alliance agreement and knowing Your Lordship's zeal for the Fatherland and our common interests, he has ordered me to inform you, Honorable Sir, and ask you to call up the regular forces of the Crown and infantry to the aforementioned fortress as soon as possible so that it can be captured more easily. The infantry is very much needed for the attack and to beat off any relief force were it to appear.[30]

The Lithuanian grand hetman Pociej could not hide his surprise:

> Chancellor Golovkin of His Tsarist Highness wrote me demanding in the name of his sovereign that I come to Riga with my army. He informed me that the tsar was besieging Riga for us, but this letter is dated late, and I am surprised that he did not use an express post for such a delicate and urgent matter [...] I replied to him that in this matter they should have used a faster post, and at the moment it was better if I watched His Tsarist Highness's flank. This letter, however, may be useful as evidence that Riga was taken for us.[31]

A week after Riga fell to Russian troops, Golovkin sent new letters to the hetmans informing them that the city had surrendered, and thus Peter was no longer in need of reinforcements at Riga.[32] Further events showed that Pociej was right. When the grand marshal of Lithuania Marcjan Wołłowicz was dispatched as the Republic's extraordinary envoy to the tsar's court in late 1710, one of the most critical points of his instruction was to demand the return of Riga to the Republic. The Russians, however, refused to hand over control of the city, for they claimed that the Republic's forces were too weak to defend it in case of a Swedish attack. The Russians, in fact, raised an additional complaint:

> His Tsarist Highness had spoken about the preparation of an army of five thousand in such way that it could be sent as reinforcements for the army of His Tsarist Highness to help him capture the fortresses of Livonia. This request was made in the letters to His Royal Highness and the hetmans, but the army was never sent.[33]

Disagreements with the Russians continued during the Russo-Turkish War of 1711. Here the principal aim of Peter was to drag the Republic into the war against Turkey, but Polish-Lithuanian unwillingness to help the Russians proved to be a tough nut to crack. The Poles and Lithuanians had had enough of the tsar's empty promises: the subsidies promised at Narva in 1704 had not been paid; Livonia and the right-bank Ukraine, which the Russians had captured from the rebellious Cossacks in 1703, had not been returned. Those in leadership feared that the presence of Polish and Lithuanian troops in the Russian camp could provoke the Ottomans to declare war on the Republic.[34] Thus the Lithuanian grand hetman Pociej did not lift a finger to help the Russians:

> Moscovite gentlemen, according to their habit, do not keep their word. I have not received a pfennig from the promised sum. [...] I am marching towards the border; if they do not send me money there, my troops will not be able to continue the march.[35]

After Russia concluded the Pruth River treaty with the Turks in 1711, Peter met the Polish-Lithuanian ministers in Zhovka. He communicated the conditions of the treaty and the news that the Swedish king was about to return to Sweden via the Republic with a Turkish escort. Pociej was unhappy to hear that the tsar refused to order Field Marshal Sheremetev, the commander of the Russian troops in the Republic, to unite his troops with the Polish-Lithuanian armies. When asked about the payment of subsidies, the tsar reproached the Poles and Lithuanians, "Why haven't you been with me?" prompting Pociej to conclude, "The burden of the threat of Swedish attack has been hung on our necks."[36] In 1712, when the war moved to Swedish Pomerania, the Russians asked the Poles to furnish the Russian troops in Pomerania with provisions. The Polish Chancellor Jan Szembek refused, but Duke Menshikov, the tsar's favorite, offered the Poles the return of the Elbląg fortress, which was in Polish territory but had been captured by the Russians from the Swedes in 1710, a prize they had dangled before the Poles on a number of occasions. These Russian maneuverings, though, were proving counterproductive. The Poles were on the verge of losing their temper. Stanisław Denhoff, the Lithuanian field hetman commented that "the pressing demand by Duke Menshikov to the chancellor affirms and demonstrates that nation's usual manner of dealing with us."[37] The Russians, however, were quick to notice the annoyance of their allies. One commander of the Russian army in the Republic suggested to his government:

> Because of the current affairs with the Turks, it would be better for us to buy the provisions and incline the Poles to our side [...] if we make peace with the Turks we will take back twice as much from the Poles.[38]

Polish-Lithuanian and Russian interests intersected not only in Livonia but also in the Duchy of Courland, which was a vassal duchy of the Republic located between Livonia and Lithuania on the Baltic Sea. In June 1710, a marriage between Anna, the daughter of Peter's half-brother, and Frederick William, duke of Courland, was finalized. Frederick accepted the marriage contract which was dictated by Peter as he planned to become the Russian governor of Livonia, now conquered from Sweden by Peter's troops.[39] These plans, however, did not materialize. In January 1711, a few months after his marriage, the duke died on his way from St. Petersburg to Courland. The reins of government passed to Anna, who was to rule Courland for almost twenty years. This, in reality, turned the duchy, billeted with Russian troops, into a Russian protectorate even though formal supremacy of the Republic remained. Polish-Lithuanian ministers viewed the expanding Russian bridgehead in Courland with apprehension. They had no illusions about the future of Courland. The Polish treasurer called it a catastrophe:

> I am afraid of the marriage of the prince of Courland which had to take place in St. Petersburg, for the tsar can bargain with him on the sovereignty of Courland [...] thus we need to be very cautious so that we do not give in or pay for anything and that we come out of this catastrophe skilfully.[40]

The Lithuanian Chancellor agreed that the marriage was tantamount to the annexation of Courland by Russia:

> I am sending you the news from Riga [...] the duke of Courland has completed the marriage contract with the tsar's family [...] and so it seems we have lost Courland.[41]

The Lithuanian grand hetman was afraid that the Curonians would not be able to serve in the Lithuanian army—a substantial portion of its officers in the western-style infantry and dragoon units had come from Courland in the past—while the actual presence of Russian troops in Courland posed a direct military threat to the Republic itself:

> Duke Menshikov has marched to Prussia through Courland imposing hefty contributions there. We ought to help Courland not only because it belongs to us but also from Christian compassion. I am sure Your Ducal Highness, according to the duties of your office, will not fail to work for the benefit of that province at His Majesty's court. If we, God forbid, lose it, this will create a thousand dangers for us; not only we will lose all our army officers, but we will also receive a whip to be cracked on us.[42]

Peter the Defender of Noble Liberties

However, there was a much more severe problem in Russo-Polish-Lithuanian relations than the Russian contributions or the Russian refusal to surrender Livonia. That problem was the Polish king Augustus who had returned to the throne in October 1709 but who was feared by the Polish-Lithuanian nobility, and not without reason, to be plotting against the laws and constitution of the Republic, the preservation of which was a guiding principle of the nobility. Significantly, Stanisław Denhoff, one of the most loyal supporters of Augustus and the Russian alliance, secretly offered Peter to guarantee the Republic's constitution before the return of Augustus.[43] The offer was a personal initiative of Denhoff, and there is no evidence that it had any immediate consequences. The mistrust of Augustus, however, was great in the following years, and the nobility looked to the tsar as a guarantor that the king would not violate the noble freedoms. Peter did not forget Denhoff's offer and was only too willing to offer his services to restrain the Polish king.

In 1713 Augustus took advantage of the Turkish military threat and redeployed his Saxon troops in Poland and Lithuania. He hoped to use the presence of his troops to force a measure that would secure the succession of the Polish throne for his son, reorganize legislative and executive procedures, and secure a series of fiscal reforms.[44] Because Augustus refused to convene the Sejm to discuss the reform openly and arbitrarily imposed contributions for the maintenance of the Saxon troops—a breach of a fundamental right of the nobility as only the Sejm could agree on any extraordinary taxation—the Poles and Lithuanians formed a noble confederation to fight the Saxon troops and resist the coup they thought the king was planning.[45] When it became clear that the tsar would not support either side, the confederation asked Peter to mediate in the peace talks, as the nobility needed assurances that Augustus would withdraw the Saxon army. They believed that such a guarantee could only be given by the Russian tsar. Negotiations between the king's court and the confederation lasted six months and illustrated how great the distance between the two sides were. Both parties agreed to sign a peace agreement only when another corps of the Russian army had entered the Republic. By asking the tsar to mediate the conflict with the Polish court, the nobility actually demonstrated that they trusted Peter more than their own king. The ink was barely dry on the agreement, however, when the nobility and king's representatives began drawing up plans to remove the mediators with flintlock muskets. One participant in these negotiations observed:

> The representatives of the confederation asked the help of the king's plenipotentiaries to invoke the clemency of His Majesty so that their wishes would be accepted and that His Majesty would help to evacuate the Muscovite [troops] in an efficacious way, and that the Saxons would observe the armistice. Again and again, measures of how to get rid of the Muscovites were weighed. It was argued that the best cure is unity, and when that was restored, the king, the primate, and the Republic would send legations to His Grace the Tsar. [...] And later there will be other ways which must be kept silent now.[46]

The final crisis during the Great Northern War occurred in 1719 when Augustus as the elector of Saxony signed the Treaty of Vienna which established an anti-Russian alliance of Great Britain, Hannover, and the Holy Roman Empire. These powers viewed the growing might of Russia with unease, and it was thought that the Republic would also soon join the anti-Russian alliance. Sweden as well had to join the coalition in one form or another, for the ultimate goal of this design was to re-conquer the eastern Baltic provinces from the Russians and return them to Sweden. The Republic was to be awarded Kyiv and Smolensk, lost to Russia in 1654–67.[47] Though some in Poland may have been tempted by this plan, many harbored doubts. In the eyes of the nobility, Augustus during his twenty-two years as king had endeavoured to weaken their status. His aspirations to strengthen royal power alongside his adventurous foreign policy in the Great Northern War, the results of which were catastrophic for the Republic's economy, elicited the mistrust of the nobility. Though after 1710 no fighting from the Great Northern War took place on the Republic's territory, all was not well. In 1715 the king's controversial policies prompted a civil war between the

Saxon army and the nobility. Now four years later, the king's actions threatened the Republic with a new war, this time against Russia. There was a great fear of fighting Russia among the nobility, and for itself, Russia sought to prevent the Republic from joining the Vienna Treaty at all costs. Noble supporters of the Russian alliance from the difficult days of 1707–08 were ready to offer the tsar their services once more. They disrupted the Sejm and did not allow the Republic to join the anti-Russian alliance.

In retrospect, this moment may have been one of the best opportunities for the nobility to have challenged emerging Russian supremacy over the Republic. This chance, however, was missed. It is important to realize, though, that the nobles did not take advantage of this situation because they were subservient to Peter. On the contrary, they thought they were patriots of their homeland. By resisting plans to confront Peter directly, they thought they were protecting the Republic from a greater evil, a war against Russia, which in their estimation would have inevitably resulted in the destruction of the country.

Did They Trust the Russians and Peter?

So in the end, did the Poles and Lithuanians trust the Russians and Peter? The short answer is yes and no. The Russians, however, did not trust the Poles and Lithuanians. Arch enemies could hardly become reliable allies in just a single generation. On the other hand, distrust of allies is a standard feature of politics at all times. Despite all the disagreements and mistrust, however, the Republic's nobility was beginning to see Russia and Peter as a friendly force. Both sides regarded each other as essential allies in the war against Sweden. If the Republic was useful for the Russians as a base to supply and maintain its army, the Poles and Lithuanians needed Russia to defend the Republic's constitution: first, from the Swedes and their lackey Leszczyński and then from Augustus and his dream of royal absolutism. The only occasion when the nobility showed absolute confidence in Peter was with his mediation of the 1715–17 crisis triggered by Augustus's policies and the Saxon army. Yet this short-term alliance only emerged as the nobility alone was unable to compel the king to uphold the law. The Polish-Lithuanian nobility turned to the Russians as a more powerful elder brother who could defend the younger sibling. When they needed the tsar to fight an internal opponent or to counter the king's attempts to strengthen royal powers, they would ask Peter's help without hesitation. They would also seek the tsar's favor at every opportunity. However, when the Poles and Lithuanians were asked to pay for Russian services, for instance, to maintain Peter's army, the nobility began to protest just as a younger brother may resent a more dominant sibling.

Hindsight makes it easy to spot the danger of growing Russian interference in the Republic's internal affairs. At the time, however, the vast majority of the nobility saw cooperation with Russia as beneficial for the Republic. They were fighting the king's attempts to strengthen and centralize royal power. Hindsight makes it easy to portray such actions of the Poles and Lithuanians as simple folly. Still, one must take into account that their priority was the preservation of the Republic's constitution based upon the Roman republican ideal of citizenship. If the tsar offered his assistance, he could be trusted, and such help was appreciated. It is more than a bit ironic, however, that the most absolutist of rulers, Peter, was seen as the best protector against the absolutism of the Saxon king.

One of the great examples of such a mentality is the story of Michał Puzyna. A middling noble from central Lithuania, he held the honorary post of court ensign. He was a member of the anti-Sapieha coalition and a loyal supporter of the Russian alliance throughout the war. In 1712, the Sejm appointed him envoy to Peter's court. Puzyna's mission was critical as he was instructed to demand the immediate return of Livonia promised to the Republic by the Narva treaty and the withdrawal of Russian troops from the Republic. However, instead of negotiating these matters seriously, he harried the Russians instead with his request to be awarded a portrait of Peter for "his faithful services to His Tsarist Majesty."[48]

The story of Michał Puzyna is indicative of wider changes transforming the political landscape of the region. For generations, Polish and Lithuanian nobility had used the term "Muscovite" as a type of sneer. It was a way to indicate their superiority to the Russians, an unwavering belief that the political system and culture of the Republic were in every way superior to that of their eastern neighbor. An important shift, however, was now occurring. In 1716, the tsar informed Polish-Lithuanian ministers to no longer use the term Muscovy when referring to Russia. From now on, it must be called Russia and its inhabitants the Russian nation.[49] Significantly, several years later after the Nystad peace (1721) which ended the Great Northern War, a similar demand was submitted to the Swedish State Council, which accepted the Russian request with no objection.[50] The efforts of the tsar and his entourage to discontinue the use of this older demeaning term reflected a changing balance of power in Eastern Europe. The time had at long last come to respect a new European power, Russia.

Notes

1   Józef A. Gierowski, *Traktat przyjaźni z Francją w 1714 r* (Warsaw: PWN, 1965); idem, "Europa wobec unii Polsko-Saskiej," in *Na szlakach Rzeczypospolitej w nowożytnej Europie*, ed. Andrzej Link-Lenczowski (Cracow: Księgarnia Akademicka, 2008), 287–300. Jacek Staszewski pushed back the boundaries of the Republic's international decline from the mid-1720s, as suggested by Gierowski, to the 1740s, Jacek Staszewski, "Stosunki polsko-rosyjskie w pierwszej połowie XVIII w. na tle międzynarodowego położenia Rzeczypospolitej," in *Przemiany w Polsce, Rosji, na Ukrainie, Białorusi i Litwie (druga połowa XVII—pierwsza XVIII w.)*, ed. Juliusz Bardach (Wrocław and Warsaw: Zakład Narodowy im. Ossolińskich, 1991), 9–21.

2 Andrzej Sowa, *Świat ministrów Augusta II* (Cracow: Biblioteka Jagiellońska, 1995), 113–27.
3 The instruction of the sejmik of Brest, 18 April 1701, *Akty izdavaemye Vilenskoiu arkheograficheskoiu kommissieiu* 4 (Vil'na, 1870), 309.
4 P. Smolarek, ed., *Diariusz Sejmu Walnego Warszawskiego 1701–1702* (Warsaw: Państwowe wydawnictwo naukowe, 1962), 341.
5 Ibid., 345.
6 Andrzej Kamiński, "Zagadka rosyjskiej bezczynności w trakcie bezkrólewia po śmierci Sobieskiego," *Sobótka* 15 (1982): 391–93.
7 Account on the negotiations with the tsar, July 1703, Lietuvos mokslų akademijos Vrublevskių bibliotekos Rankračių skyrius (LMAVB RS) f. 17 b. 177, 372.
8 Treaty with the Polish king Augustus, 27 November 1703, *Pis'ma i bumagi imperatora Petra Velikogo* (*PiB*) 2, (St. Petersburg, 1889), 288–91.
9 Patkul to Peter, 23 October 1703, Ibid., 676.
10 Patkul to Peter, 29 September 1709, Ibid., 679–80.
11 Golovin to Patkul, 1 November 1703, Ibid., 687.
12 Vladimir Korolyuk, "Vstuplenie Rechi Pospolitoi v Severnuiu voinu," *Uchenye zapiski instituta slavianovedeniia* 10 (1954): 283–87; Treaty with Poland, 30 August 1704, *PiB* 3, St. Petersburg, 1893, 129–35.
13 J. Bartoszewic, ed., *Pamiętniki Krzysztofa Zawiszy, wojewody mińskiego (1666–1721): wydane z oryginalnego rękopismu i opatrzone przypiskami* (Warsaw, 1862), 236; Whitworth to Harley, 3 June 1705, *Sbornik imperatorsko russkogo istoricheskogo obshchestva* 39 (St. Petersburg, 1884), 97.
14 Whitworth to Harley, 7 November 1705, Ibid., 182–84.
15 Whitworth to Harley, 30 January 1705 (Old Style?), Ibid., 20–21.
16 Whitworth to Harley, 7 November 1705, Ibid., 183–84.
17 Golovin to Dolgoruki, 8 February 1704, *PiB* 3, 548.
18 Gintautas Sliesoriunas, "Wywiad moskieski o wydarzeniach w Wielkim Księstwie Litewskim na przełomie wieku XVII i XVIII," in *Między Zachodem a Wschodem. Etniczne, kulturowe i religijne pogranicza Rzeczypospolitej w XVI–XVIII wieku*, eds. K. Mikulski, A. Zielińska-Nowicka (Toruń: Wydawn. Adam Marszałek, 2006), 115.
19 One of the many examples: Decision of the sejmik of Brest, 3 January 1706, *Akty izdavaemye Vilenskoiu arkheograficheskoiu kommissieiu* 4, 347.
20 Peter to Augustus, 5 May 1704, *PiB* 3, 51–52; Peter to Augustus, 24 January 1705, Ibid., 222–25.
21 Peter to Menshykov, 8 April 1705, Ibid., 302–03; Peter to Augustus, July 1705, Ibid., 377.
22 Peter to Sheremetev, 13 March 1707, *Sbornik imperatorsko russkogo istoricheskogo obshchestva* 25 (St. Petersburg, 1878), 20.
23 Already in 1703 in his manifesto to the Republic the tsar declared that the Republic must be sure of "our true intention to diligently preserve [the Republic's] freedoms in an irreplaceable state," 20 April 1703, *PiB* 3, 147.
24 Tauryłowicz to Radziwiłł, 27 September 1709, Archiwum Główne Akt Dawnych (AGAD), Archiwum Radziwiłłów (AR) 16163.
25 Pociej to Radziwiłł, 12 September 1709, Ibid. 11913.
26 Radziwiłł to Peter, 3 December 1713, Rossiiskii gosudarstvennyi arkhiv drevnikh aktov, f. 79 op. 1 1713 No. 13, 1–2.
27 Wołłowicz to Sieniawski, 1711, Biblioteka im. Ks. Czartoryskich w Krakowie (BCzar) 5988 No 48437.
28 Antoni Sapieha's letter, 30 July 1716, LMAVB RS f. 139 b. 4064/1.
29 Iu. Shcherbachev, ed., *Zapiski Iusta Iulia datskogo poslannika pri Piotre Velikom (1709–1711 gg.)* (Moscow, 1900), 380.

30 Golovkin to Sieniawski, 8 July 1710, BCzar 5819 No. 12464.
31 Pociej to Radziwiłł, 11 August 1710, AGAD AR V 11913.
32 Golovkin to Sieniawski, 21 July 1710, BCzar 5819 No 12468.
33 Reponse by the tsar's ministers, 13 March 1711, AGAD, Archiwum Koronne Warszawskie 54/3, 6.
34 Józef Feldman, *Polska a sprawa wschodnia 1709–1714* (Cracow: Polska Akademja Umiejętności, 1926), 64–66.
35 Pociej to Szembek, 3 July 1711, BCzar 500 No. 46.
36 Pociej to Sieniawski, 26 August 1711, BCzar 5916 No 30033.
37 Denhoff to Sieniawski, 30 June 1712, BCzar 5791 No. 7703.
38 Dolgorukii to Golovkin, 4 May 1712, *PiB* 12, part 1 (Moscow: Nauka, 1975), 388.
39 Bogusław Dybaś, "Inflanty a polsko-litewska Rzeczpospolita po pokoju oliwskim (1660)," in *Między Zachodem a Wschodem. Studia z dziejów Rzeczypospolitej w epoce nowożytnej*, ed., J. Staszeski, K. Mikulski, J. Dumanowski (Toruń: Wydawn. Adam Marszałek, 2002), 120–21.
40 Przebendowski to Sieniawski, 11 August 1710, *Listy Jana Jerzego Przebendowskiego podskarbiego wielkiego koronnego do Adama Mikołaja Sieniawskiego wojewody bełskiego i hetmana wielkiego koronnego z lat 1704–1725*, ed. A. Perłakowski (Cracow: Księgarnia Akademicka, 2007), 62.
41 Radziwiłł to his wife, 17 July 1710, AGAD AR IV, 19–236
42 Pociej to Radziwiłł, 18 April 1712, AGAD AR V, 11913.
43 Józef Andrzej Gierowski, *W cieniu Ligi Północnej* (Wrocław: Zakład Narodowy im. Ossolińskich, 1971), 88.
44 Jacek Staszewski, "Pomysły reformatorskie czasów Augusta II," *Kwartalnik Historyczny* 82 (1975): 750–61; Józef A. Gierowski, *W cieniu*, 106–10, 135–38
45 Mindaugas Šapoka, *Warfare, Loyalty and Rebellion: The Grand Duchy of Lithuania and the Great Northern War, 1709–17* (London; New York: Routledge, 2018), 134–87.
46 Diary of the Negotiations, 13 November, 1716, LMAVB RS f. 9 b. 3116, 544–45.
47 Lucjan Lewitter, "Poland, Russia, and the Treaty of Vienna of 5th January 1719," *Historical Journal* 13 (1970): 3–23.
48 Puzyna to Peter and Golovkin, 6 October 12, *Rossiiskii gosudarstvennyi arkhiv drevnikh aktov*, f. 79 op. 1 1712, No. 34, 106 7.
49 Dunin to Szembek, 4 April 1716, BCzar 471 No. 67.
50 B'ern Asker, "Natsiia, kotoroi nado pokazyvat' zuby," *Poltava. Sud'by plennykh i vzaimodeistvie kul'tur*, ed. T. Toshtendal'-Salycheva, L. Iunson (Moscow: Rossiiskii gosudarstvennyi gumanitarnyi universitet, 2009), 372.

Further Reading

Gierowski, Józef Andrzej. *W cieniu Ligi Północnej*. Wrocław: Zakład Narodowy im. Ossolińskich, 1971.
Korolyuk, Vladimir. "Vstuplenie Rechi Pospolitoi v Severnuiu voinu." *Uchenye zapiski instituta slavianovedeniia* 10 (1954): 239–347.
Lewitter, Lucjan. "Poland, Russia, and the Treaty of Vienna of 5 January 1719." *Historical Journal* 13 (1970): 3–30.
Šapoka, Mindaugas. *Warfare, Loyalty and Rebellion: The Grand Duchy of Lithuania and the Great Northern War, 1709–17*. London; New York: Routledge, 2018.
Sowa, Andrzej. *Świat ministrów Augusta II*. Cracow: Biblioteka Jagiellońska, 1995.

13 Silent Victims

The Hidden Costs of War in Brandenburg, 1648–1700[1]

Mary Lindemann

More than forty years ago, the historical demographer, Myron Gutmann, published a study of rural life in the Basse-Meuse (today eastern Belgium), a region "almost always involved in the thick of early modern war." In discussing the European conflicts of the 1670s, 1680s, and early 1690s, he stressed the consequences of troop movements on agriculture, concluding that even temporary breakdowns "accumulated into long-term agricultural stagnation." Gutmann's analysis also described an extended post-war period of inaction and desolation punctuated by what he termed "aborted recoveries." Recently, Emmanuel Kreike has advanced a much broader idea of *environcide* wars that "[consist] of intentionally or unintentionally damaging, destroying, or rendering inaccessible environmental infrastructure through violence."[2] Like the Basse-Meuse, Brandenburg suffered a series of wars in the mid to the late seventeenth century with obvious results in the destruction of sacred and secular buildings, the misery of people harmed or displaced, the crushing financial exactions, the brutal robberies, and the vandalism that stripped whole villages bare. Yet the hidden costs—the damage to the land that fighting, troop movements, and prolonged neglect caused—proved just as determinant in shaping a recovery that profoundly altered social, economic, communal, and even political life. Simply put, the Brandenburg that emerged by the early eighteenth century differed substantially from the Brandenburg of the early seventeenth century. That transformation, however, can neither be attributed solely nor even principally to the activities of a dirigist state, nor subsumed within a story of the rise of Brandenburg-Prussia or the creation of absolutism. Rather, we need to look more deeply at the transformations in the land that restructured post-1648 Brandenburg.

Much historical scholarship on late seventeenth-century Brandenburg, however, has focused on other subjects, two in particular. The first involves the long-standing debate over the character of a free versus an unfree peasantry (*Leibeigenschaft, Erbuntertänigkeit,* or *Schollenbindung*). The historiography of the last several decades has done much to explode the myth of "the second serfdom" and to emphasize the existence of variations and mixed forms of economic relationships shaped by chronologies and locations.[3] The second is a still

DOI: 10.4324/9781003157700-17

older interpretation that, however, continues to demonstrate considerable stay-ing power. In this narrative, generations of historians placed Friedrich Wilhelm ("The Great Elector") as the first in a "legendary series of four hyperactive rul-ers," who lifted Brandenburg out of the wartime wreck and initiated the "rise of Prussia."[4] These analyses have largely turned on top-down political initiatives as forms of state-building (not state-*formation*) and administrative centraliza-tion, often gathered under the now shop-worn idea of absolutism.[5] A study that takes the land or the landscape *(Landschaft)* seriously moves the discussion in a decidedly different direction. *Landschaft* can mean many things: scenery, land-scape, countryside. However, these definitions capture only part of the greater analytical significance embedded in the richer idea of a social or a vernacular landscape that is "above all a space shared by a group of people." Accordingly, "landscape is not a natural feature of the environment," but rather a *"synthetic space*, a man-made system of spaces superimposed on the face of the land, func-tioning and evolving not according to natural laws but to serve a community."[6] The actions of human beings, including war, repeatedly remolded *Landschaf-ten* yet never fully obliterated older ones. In her study "The Reformation of the Landscape," Alexandra Walsham points out that a variety of scholars have "compared [landscape] with a parchment and palimpsest, a porous surface upon which each generation inscribes its own values and preoccupations without ever being able to erase entirely those of the preceding one."[7] The interpretation this article advances rests on an understanding of Brandenburg as an equally malle-able landscape.

In older historiography, the Thirty Years War has held a special place in the catalog of wartime horrors as an "all-consuming fury." Indeed, the Thirty Years War became a metonym for war.[8] More recent historical investigations have con-siderably modified and refined this interpretation. The results have done much to alter the older picture of a conflagration that ravaged all the German lands equally by stressing the differential and sporadic impact of the war. Nonetheless, older views of destructive fury and widespread desolation possess a good deal of validity for many places in Brandenburg. An early nineteenth-century histo-rian noted that the Thirty Years War was "nowhere more destructive than in the Mark Brandenburg." For example, the location of the small town of Rathenow "could not have been more unfortunate," lying as it did athwart the Havel Pass. It thus provided "a convenient passageway for enemy and friendly troops [alike] as they marched back and forth," and was repeatedly besieged and occupied by Danish, Swedish, Saxon, and Imperial troops. The resulting destruction in the town was frightful, but few historians have explored the effects on the land it-self.[9] Indeed, for the environmental history of the early modern world, the Thirty Years War remains a "gap [and] perhaps the most important one (and hence also an area of greatest promise)."[10]

Moreover, often little separated times of war from non-war. Peace remained a relative term and an elusive goal. After 1648, provisions for the withdrawal and demobilization of troops were not finalized until the Nuremberg Execution Congress (*Exekutionstag*) of 1649–50.[11] Even thereafter, peace proved transitory and ephemeral. Several conflicts between 1648 and 1721 raged in and around Brandenburg, involving the territory or parts of it either directly or indirectly. Historians have come to understand this series of conflicts cumulatively as the Northern Wars among the several powers—Sweden, Poland-Lithuania, Denmark, Russia, and Brandenburg (Prussia)—that vied for the control of territory along the Baltic.[12] One might reasonably see these repeated struggles as something like a "second hundred years war." Like the conflicts of the thirteenth and fourteenth centuries, this newer version was equally episodic but hardly less destructive. Likewise similar was the massive disruption wrought by troop movements, especially through marches, even when the theater of war lay elsewhere.

After the Peace of Westphalia, a brief hiatus of relative calm reigned until 1655, when the outbreak of the Second Northern War (1655–60) involved Brandenburg and struck areas that had only begun to recover from the Thirty Years War. Of course, here, too, war affected some areas more dramatically and longer than others. Then, on Christmas Day, 1674, the Swedes attacked the Uckermark and Havelland precipitating the Swedish-Brandenburg War (1674–79). The consequences of these two wars "obliterated all progress at rebuilding made during the interwar period."[13] The more westerly province of Ruppin suffered as well and reports tallied burned-out buildings and the seizure by Swedish troops of all food, domestic animals, and stores of grain. In short, "the most gruesome horrors of the Thirty Years War" returned.[14] For these wars as well, archives document the less well-known, but quite extensive, damages to the land and landscape: crops untimely reaped, sown fields trampled, gardens stripped of their ripe and unripe fruits, bridges, dams, and causeways demolished. Typically, little differed between the destruction caused by enemy or friendly troops.

Despite the wealth of existing evidence to draw on, historians have tended to ignore these voiceless victims of war and even the most comprehensive works on the territory do not directly address the land and landscape, natural or built.[15] Furthermore, for the early modern period, the scholarly literature has mostly ignored the extended chronology of war damage.[16] Historians of the modern world, however, have produced numerous treatments of the deeper effects of wars on the environment and the landscape, particularly for World War I and the American Civil War. Several works have analyzed the global effects of the American Civil War that created an "empire of cotton" and described how World War I shifted the global flows of material and foodstuffs, thus permanently altering pre-war patterns of production and consumption.[17]

Early modernists, however, have been less active in exploring similar topics, perhaps because of the faulty perception that more recent conflicts, especially those categorized as "total wars," proved far more destructive with longer-lasting effects. The ravaged landscapes of trench warfare, the jagged ruins produced by sustained shelling, and the aerial bombardment of cities seemingly wreaked exponentially greater devastation. Nonetheless, one can point to examples of almost unfathomable physical ruin in the early modern world as well: the virtual leveling of Magdeburg in 1631 and the massacre of its inhabitants, or the double destructions of Heidelberg in 1688 and then again in 1693 during Louis XIV's campaigns in the Palatinate.[18]

Seventeenth-century documents, however, say a great deal about these lesser-known and silent victims of war. The inhabitants of Guben (a small Lusatian town bifurcated by the Neisse River) suffered from "innumerable *excursions*" by Swedish soldiers who repeatedly "infested the entire province." Besides confiscating farm animals, the marauders prevented "anyone from working in the fields [and] we could bring nothing under the earth."[19] In 1639, Jobst von Dalchaw petitioned the electoral government for permission to fell wood from his estate (*Lehngut*). Dalchaw told a familiar tale, portraying himself, not unbelievably, as "a nobleman impoverished ... by the persistent warfare in this place." His estate had remained "desolate" for years and its restoration would require "great expenditures." The only resource left to him was his woodlands. The sale of wood to Hamburg lumber merchants would allow him to rebuild and reclaim fields long left "unprepared and empty (*verwuestet*) ... unplowed and choked with weeds."[20] Dalchaw was not exceptional in seeking to tap the few resources remaining to him. Villages and especially the smaller towns (*Ackerbürgerstädte*) that retained a pronounced rural character, possessed similar resources and likewise sought to convert them into cash. War raised prices for lumber, encouraging agents from Hamburg, Lüneburg, Lübeck, and sometimes even farther afield, as from Amsterdam, to swarm into Brandenburg looking to make deals. Lumber was also needed locally for the reconstruction of secular and sacred buildings, including the profitable wind and water mills that proved prime targets of destructive actions in war. Throughout the prolonged period of depression and rebuilding that lasted at least until the end of the seventeenth century, the commercialization of forests, if by no means new, became a more essential part of a cash-generating economy throughout Brandenburg. Deforestation could be one result; monoculture of more useful trees, such as oak or fir, was another. The rebuilding of demolished mills, the erection of new ones, the replacement of water mills with more expensive and complicated windmills, the amalgamation of several mills or different types of mills into a mill complex, and the construction of larger mills with more axles, as well as profitable commercial enterprises such as sawmills and oil-pressing mills characterized the economic initiatives of many estate holders (*Gutsherrn*), but also of many towns, villages, and individual entrepreneurs that sought to expand their income and cover their debts.[21]

These thousands of specific examples taken together create the basis for a longitudinal analysis of how wars and the process of overcoming war-damages physically and topographically shaped Brandenburg after 1648. War-caused damage and recovery were processes, not events. As these petitions, surveys, and visitation reports abundantly document, the immediate destruction formed a middle point of the story, not its beginning. Transformations of the land were never merely the result of war, of course. Since prehistoric times, humans and animals have acted in deliberate and unintended ways to alter their living spaces. The concept of a global "great acceleration" has come to dominate much historical discussion of a changing environment.[22] The interpretation presented here, however, does not address global change or, for that matter, even large regional changes, nor does it consider the long-term shifts that environmental historians have studied extensively: warm and cool cycles, deforestation, desertification, irrigation-caused salinization, species extinctions or near-extinctions, or soil erosion. Rather, I argue that the wars of the mid to late seventeenth century generated and accelerated major historical transformations in particular if by no means unique ways, not only in the landscape, but in how inhabitants of this landscape came to understand their place in the world, contoured or re-contoured their identities, and reshaped their social and political interactions. Perhaps the trickiest analytical difficulty here lies in the *post hoc, ergo propter hoc* fallacy; coincidences are not causes. Moreover, pinpointing turning points or locating origins are fraught historical exercises. In evaluating the effects of "changes in the land," such interpretive hazards loom especially large.[23]

Nonetheless, it is possible to explicate how wars and their aftermath affected communities by homing in on quotidian affairs and by focusing on the mostly small-scale initiatives that individuals and groups undertook to recover or meliorate their surroundings. A multiplicity of efforts "from below" proved more successful than initiatives guided by the visible hand of the electoral government.[24] None of these actors—estate owners, district officers (*Amtmänner*), village communes, individual entrepreneurs, single peasants, or groups of them—enjoyed unrestricted freedom of action, yet they were not totally constrained by their legal and social situations. The sum of their efforts substantially re-shaped a disrupted agricultural world and redesigned social and communal traditions linked to the land. Historians have not often looked at this level of activity that was, admittedly, mostly scattered or sparse, frequently inchoate, and sometimes virtually imperceptible. Small, even insignificant, though these changes may appear individually, in aggregate they proved important and deserve to be studied alongside the better-known and better-documented initiatives that have most often caught the historical eye, such as the initiatives launched by the Great Elector Friedrich Wilhelm that included attempts to improve river navigation, immediately following the Thirty Years War and, in the eighteenth century, those of his grandson, Friedrich Wilhelm I, and his great-grandson, Friedrich II.[25] Even in strictly administrative and political terms, "changes in the longer-term structures of the Mark Brandenburg [were] certainly not accomplished by government-initiated measures."[26]

Population loss provides a rather obvious point of departure. For all the areas touched by the war that ended in 1648, global estimates range between an older value of forty percent and more recent estimates of between fifteen percent and twenty percent. Brandenburg provinces as a whole, however, experienced a striking demographic decline and large parts of Brandenburg remained underpopulated for the rest of the century. The urban population declined from about 113,500 to 34,000 and the rural population from 300,000 to 75,000, leaving about half the villages thinly populated at best. If scholars continue to argue about the degree of loss, and the relationship between loss and mortality, the fact itself appears indisputable.[27] Later wars only exacerbated the problem.

Within Brandenburg substantial regional and chronological variations existed; some places were considered totally obliterated and continued to be designated *wüst* (abandoned, deserted, destroyed, fallow, unusable, unused, or uninhabitable) for decades thereafter; other places remained untouched by war for years only to be mercilessly ravaged later in the conflict.[28] Resettlement proceeded slowly; the number of peasant holdings that were occupied (*besetzt*) in Ruppin, for example, only exceeded its 1624 total (354) in 1717 (452).[29] Yet caution is required here. Many places calendared as *wüst* in the late seventeenth century resulted not only, or perhaps even principally, from the havoc the Thirty Years War caused; rather, these frequently dated from the economic and demographic decline of the later Middle Ages.[30] Moreover, *wüst* did not invariably mean "abandoned" or "uncultivated." Contemporary usage defined a *wüst* farm as one on which no peasant resided; the land itself did not necessarily remain barren.[31] Despite the prevalence of centuries-old *Wüstungen*, the wars of the mid to late seventeenth century created a new group beginning in the 1620s whose character differed in that, for example, they had been inhabited in the not-so-distant past.

In 1652, the electoral government ordered a visitation throughout Brandenburg in order to survey damages and assess what had been accomplished in rebuilding since the end of the war. These reports convey a vivid picture of dire circumstances in many places.[32] No area in Brandenburg had suffered more in these years than the Uckermark. The visitors there described rural areas that were "completely plundered and their inhabitants driven out." In the absence of seed and labor, the fields remained untended and went wild. Forests encroached on, and sometimes engulfed, inhabited and cultivated areas. For example, the fields of the village of Günterberg had disappeared; once fertile lands "had become [newly] forested and were no longer recognizable." Not until the end of the century were the tenants able to clear and cultivate these fields.[33] Few towns in the Uckermark were even moderately large, but the Thirty Years War emptied them of people; some never recovered. Even somewhat bigger towns, such as Prenzlau and Angermünde, only slowly revived. "In surveying this city," the 1652 visitation of Angermünde listed 114 houses abandoned; "the war, as also the plague ... had caused this [disaster]." Another 102 dwellings in Angermünde languished in various states of disrepair; seventy-five had to be pulled down and only sixty-six were occupied.[34] Even areas spared the worst excesses regained their pre-war size only gradually. More than four years after the Peace of Westphalia, in the town of Müllrose on the Schlaube River (part of the Oder River system), the *Pfarr-Acker* ("the Parson's Piece") was still "overgrown with scrub and scraggly trees."[35]

Population loss thus possesses a strong, if often unrecognized and under-valued, spatiality and spatial significance. In the wake of the Thirty Years War, and for decades thereafter, many rural habitations were simply empty of people and domestic animals. In the early modern world, physical safety and a sense of well-being depended on the presence of others, not only family and kin but also neighbors and the beasts of the field and farmyard. People lived in villages, rarely in isolation. Villages were, of course, not happy lost worlds of close and warm communal relations; such communities often seethed with sharp rivalries and deep hatreds. However, communities ran on human relations that laws and traditions sustained and when significant numbers of inhabitants, or people central to the social, economic, and cultural webs of life, were removed or moved, the fabric of fundamental human and social relationships frayed. Physical markers of possession and boundaries, such as stones, fences, or slashed trees were often gone as well. Thus, much had to be renegotiated.

Wars also distorted the social organization of agriculture. Villagers tilled fields in common and village councils determined when to plow, plant, tend, and harvest. In Brandenburg, the three-field system dominated although with many variations and one can rarely document the existence of a "rigid and intensive form" in practice.[36] The model theoretically involved a precise rotation: planting one field in the fall with hardy winter wheat or rye that would be harvested in late summer, another planted in spring with summer grain (oats or barley) and usually harvested in July and a third left fallow. Fallow, however, did not necessarily mean unused; villagers frequently planted fallow fields with leguminous crops or turned them into temporary pastures for animals that ate weeds and manured the soil. Differing micro-climates, soil compositions, and water levels determined the alteration of crops and affected the timing of sowing and harvesting. Such agricultural systems demanded regular attention and involved rotation of plowing and harrowing to loosen and aerate the soil, thus allowing plants to root properly.[37] The specifics of what needed to be done, and when, varied from region to region; agricultural traditions, human memory, and orally transmitted knowledge determined routines. None of this, however, mandated a calcified system. Yet whatever the local or even micro-local differences, war inevitably disrupted agricultural cycles; the lack of seed reduced yields, the absence of horses frustrated plowing, and sheer fear drove peasants from their fields.

The resultant neglect not only upended agricultural routines; it also upset all the normal structures of communal life that were embedded in agriculture. Seasons and weather largely directed agricultural cycles and demanded moments of greater and lesser activity. The early authors of agronomic texts and husbandry manuals emphasized the need for diligence and regularity and fashioned model calendars listing the tasks to be accomplished month-by-month or even week-by-week. Although cheaply printed almanacs that specified times for sowing and reaping, for instance, became commoner in the eighteenth century, they already existed in the sixteenth and seventeenth centuries.[38] Far more agricultural wisdom, however, was enshrined in the memory of the oldest villagers who were called upon to say what constituted an (usually undesirable) innovation or not.[39] Armed conflicts disrupted all these routines. Diligence perforce gave way to neglect and memory thinned then sifted away entirely as its human repositories perished or migrated. The results were dramatic; wartime upheavals prevented labor in the fields and reduced harvests to levels where not even seed sufficient for the next year's planting was recovered. Faced with starvation, whole villages took to the roads, seeking shelter in towns or crossing the borders to where they hoped to find safety and perhaps a new home.

Demographic loss, whether from mortality or migration, gouged deep rifts in communal organizations and traditions that formed the basis not only for agriculture but also for village politics and socio-cultural life. The cultivation and management of arable fields (*Äcker*) in a landscape that typically lacked clearly demarcated roads or paths proved possible only if all participants adhered to the good discipline that constituted the basis of an always contentious "ordering of village lands" (*Ordnung der Flur*) and that included communal control over planted fields, hay-meadows, pasturage, common land, and forest rights.[40] Thus, as historians of the agricultural world have noted, all those with an "interest in the village" (*Interessengemeinschaft*) had to accept "neighborly law" (*Nachbarrecht; Flurzwang*). The village council determined virtually everything about cultivation and punished those who failed to obey. Additionally, the council determined boundaries and access rights. The inner working of these councils is, however, a tough nut to crack; their proceedings left few written records.[41] Often only when conflict spilled out of the village and when internecine quarrels led to disputes that drew external agents into the fray does one glimpse what was at stake and who the stakeholders really were. The demographic collapse caused by war ripped gaping holes in communal life and upset its internal regulatory mechanisms. Yet, while war foreclosed opportunities for many, it could create more favorable situations for others, including women, younger sons, and newcomers.[42] Changes in the land could produce both a sense of despair and glimmers of hope; they also intensified tensions between proponents of restoration or innovation.

The after-effects of cumulative damage to the natural and built environments persisted long after the Thirty Years War had stuttered to a close. Troops continued to destroy fields and interrupt the normal flow of the agricultural year.[43] Crafting new agricultural regimens took time and rarely proceeded harmoniously. The impact of the damage in the long term, however, became apparent early. By 1632, the agricultural town of Wolmirstedt had already suffered severely and the rural areas surrounding the city perhaps even more: "for several years our fields have lain fallow and ponds once rich in fish have dried out." All the horses had either been stolen, died from disease, or killed. Three cows, seven chickens, and a few pigs—"two of which had been eaten by wolves"—completed the animal inventory. The only grain harvested in three consecutive years (1631–33) was the meager amount that had self-germinated.[44]

Seventeen years after the Peace of Westphalia, the war's pernicious effects were still all too obvious in Zossen, a Lusatian town lying about twenty miles south of Berlin. When the district clerk (*Amtsschrieber*), Christoph Vogelsmark, arrived there in 1661, the outlying fields (*Vorwercke*) of the nearby village of Werben were still overgrown. In the two-and-a-half years since then, they had been partly cleared of brush and weeds. Prolonged flooding, however, had "soured" the soil of the *Vorwerck* Cammerforst to the extent that it could not support even a crop of the relatively non-demanding barley. Moreover, ash trees, birches, large pines, and tree shoots covered most arable land. Near the village of Lüdersforst, the district possessed a large hay-meadow but in the decades "since the great devastation" of the Thirty Years War, it had lain unused because straggly trees and bushes had taken over. The process of reclaiming fields was complicated and Vogelsmark's description vividly portrayed the interrelated problems that frustrated attempts to return land to productivity. Two villages, Zossen and Trabnin, once shared the hay crop grown on their abutting meadow but now lacked the labor necessary to clear it. The debris that clogged the drainage and boundary ditches (*Gräben*) alongside the meadow needed to be cleared out and then the ditches had to be retrenched in order to provide proper drainage and irrigation. To do so would require employing a specialized engineer (*Teichgräber*) for weeks or even months; Vogelsmark doubted that the still-suffering villages could afford such "great expense."[45]

The electoral government, of course, had a vested interest in seeing peasants resettled and farming resumed. The taxation system rested on payments in *naturalia* or cash; thus untilled lands were unprofitable in more than one sense. Estate owners depended on such income to cover their own expenses. In the wake of each war, the electoral government issued new ordinances or revived older ones, instructing local officials and those holding vassal lands (*Lehngüter*) to clear fields and resettle peasants on *wüst* lands. The electoral government also introduced a range of other measures to repair dams, roads, and causeways damaged by the war, and to remove brush and other debris that impeded river traffic, hindered log driving or rafting, and interfered with fishing.[46] Of course, a dearth of labor and a lack of money frustrated these initiatives, but conflicting ideas about usage and the expectations brought to the table by the various interested parties (*Interessanten*) proved equally obstructive. Resistance, however, cannot simply be attributed to a pig-headed reluctance to change on the part of peasants or the self-interested greed of grasping landlords. Problems that had developed during and because of war persisted not only due to the legacy of long-term neglect but also because ideas diverged about what should be done, by whom, and when. Admittedly, difficulties about the allotment of responsibilities for the upkeep of infrastructures and how to pay for them were hardly new in the later seventeenth century; still, the dimensions of the problems burgeoned in the wake of war.

The recovery of land "gone wild" and fields "grown over" demanded investments of time and labor that impoverished and depopulated villages could rarely muster. In many places, fields remained uncultivated until late in the seventeenth century. In 1672, a report from the small town of Bötzow bemoaned its continued poverty despite the considerable money the elector's wife and Bötzow's proprietress, Louise Henriette, had poured into it.[47] Likewise, in Ruppin, many fields still remained uncultivated in 1680. On the Kahlenheüdsche and Franckendorffische fields, for example, nothing grew except thickets of puny firs and stunted oaks.[48]

In 1696, the electoral government ordered a survey of three uncultivated plots in the district of Amt Zehdenick in anticipation of transferring these to Groß Woltersdorf where fertile land was scarce and where "each peasant farms no more than a single holding (*Hufe*). The cottars possess just a few acres each and the whole village lacks meadowland."[49] In reviewing the history of the village, the report dated "the great devastation" from 1637 when soldiers had torched the buildings and driven the people out. Some returned after 1648, only to succumb to the plague. Those who survived soon fled again and often never returned. Even when the village finally regained some of its lost population, the land remained overgrown, crop yields low, barns leaky or destroyed, and houses ramshackle. As the village repopulated, problems persisted and indeed a very welcome influx of newcomers and returnees exacerbated them as supplies of food and fodder limped behind.[50] If weed-choked fields permitted only minimal cultivation, and if labor shortages short-circuited the work of reclaiming them, these still were not the most intransigent obstacles; the damage ran deeper and affected the local ecosystem. As trees, wild grasses, and weeds, as well as swarming pests—mammalian, avian, and insectoid—invaded overgrown fields, they attracted more fearsome predators; wolves reappeared and lurked around houses and barns.

To comprehend the fullness of the damage agriculture suffered in these decades, however, requires an understanding of how other environmental factors, in particular errant waters, affected the land and the *Landschaft*. Brandenburg was—and remains—one of the water-richest areas in Germany. Productive farming required control of water; too much, too little, and at the wrong times diminished yields. Taming the waters required the erection, maintenance, and coordination of a system of structures, mechanisms, and procedures that harnessed water to useful purposes and prevented it from running wild. The story of the waters of Brandenburg is a hugely important and complex topic that cannot be fully discussed and analyzed here.[51] Suffice it to say that the many large and small hydraulic measures developed over centuries to control water proved particularly susceptible to war-damage. The earthen dams that predominated in Brandenburg required continuous attention and timely repairs, all of which depended on gangs of locals working with picks and shovels. Neglected upkeep caused frequent dam-breaks, but armies, too, destroyed dams and dikes. In 1636, for example, when the city of Cüstrin denied passage to the Swedish field marshal Johan Banér, in retribution he dispatched 200 Finnish soldiers to cut through the dike along the Oder River. The subsequent flooding allowed shallow-draft boats to sail across the low, swampy plain (*Oderbruch*) while also isolating Cüstrin from Berlin.[52]

Cüstrin's experience was only one of many that resulted from the flooding of fields caused by breached dams or clogged channels, rivers, and drainage ditches. A short foray into the discipline of soil science highlights their prolonged adverse effects on agriculture. Two branches of the subject bear special relevance here: pedology, involving the chemistry and morphology of soil, and edaphology, the study of soil-dependent uses and, in particular, soil's interactions with plants. Most people are familiar with some positive effects of flooding, especially the regular inundations that deposit silt (alluvium) and increase fertility. But, besides the immediate harm done to people and structures, as well as run-off erosion, flood waters profoundly damaged crops and soil. Floods washed away seeds and caused the roots of germinated plants to tear loose, rot, or mold. Water that stood on fields for weeks (or even months) reduced the amount of oxygen in the soil and altered it in ways that prevented soil aggregation that allowed for the retention of the gases and water critical to plant health and crop yield. The wetter the soil, the less air it retains. Any imbalance modifies soil chemistry and affects the soil's ability to nourish plants.

The process of compaction has an equally injurious effect on plant life. Today, most compaction results from carelessly driving heavy machinery over saturated ground. Horses and cattle, however, also damage the soil in ways that reduce its crop-bearing potential. Army horses pastured in fields could quite effectively compress the soil.[53] Wars that robbed communities of their plow animals and decimated the laboring population vastly reduced the possibility of rectifying compaction by plowing. Opportunistically rooting trees and shrubs prevented deep-plowing and bent moldboards and plowshares. In many places, and often for years after 1648, farmers who lacked draft animals had to replace the sturdier plows (*Beetpflug* or *Kehrpflug)* that turned over soil and facilitated re-aeration and desirable particle aggregation with an older, lighter type of plow (*Hakenpflug*) that only scratched the surface. In some places, planting by digging sticks (*Grabstock*) proved the only feasible solution, yet it further diminished productivity.[54]

Of course, neither peasants nor early agricultural writers knew anything about pedology or edaphology but they understood how flooding reduced soil productivity. Early experts on agronomy and husbandry, as well as villagers who petitioned for relief from taxes or resolution of conflicts, spoke quite knowledgeably about immediate issues—the inability to sow flooded fields and the corrosive and erosive effects of rapidly advancing and receding waters, for example— but also about prolonged damage to the soil that they described as a "sanding" or "souring" (acidification). Both diminished fertility. Floods also swept away boundary and field markers and triggered long, bitter, and even violent disputes within and between villages. The demands for the reallotment of fields threatened to disturb older socio-political structures within villages because one's status derived directly from the amount of land farmed and, often, its proximate location.

Not only actual warfare produced flooding, of course. The neglect of old, often ancient, hydraulic systems, and the inadequate or nonexistent maintenance of parahydraulic structures, such as dams and drainage ditches, also profoundly disrupted agriculture and formed one of the most prevalent and insidious costs of war. Moreover, damages were always synergistic. Besides the need to redraw obliterated boundaries, these problems included the rebuilding of transportation links, especially local ones, like the simple stone causeways laid across low-lying swampy areas and the dredging and clearing of rivers, fish ponds, and mill-streams. The loss or deliberate destruction of documents that specified rights, duties, and privileges, as well as those that delineated the specifics of property holdings, threw up additional stumbling blocks that thwarted recovery.[55] But if many inhabitants attempted to return to a well- or poorly-remembered past, others embraced change; it would be folly to believe that only the visible hand of the state was active in reconstruction. Grandiose projects often never got off the drawing board, although not all were failures; one can point to the building and improvement of a canal system in Brandenburg that all Europe came to admire.[56] But if we look only at that large-scale, seemingly planned, level of activity, we miss the efforts of thousands of others who transformed the land and the landscape and with it the communal and individual lives of the territory's inhabitants. War, despite all its horrible consequences, opened up room for the many changes that made eighteenth-century Brandenburg a different place than it was a hundred years before. It remains to be seen to what extent Brandenburg's experiences during and after the wars of the mid to late seventeenth century resembled those in other parts of the German territories. Numerous factors— topographical, political, economic, social, and cultural—could certainly have wrought very different results. Nonetheless, a consideration of the impact of war and its aftermath on land and landscape cannot be separated from the more general history of those fraught decades and the generations that followed.

Notes

1 A part of this article was presented as a keynote address at the *Frühneuzeitinterd-isziplinär* conference on "Rethinking Europe: War and Peace in the Early Modern German Lands," March 8–10, 2018, Washington University, St. Louis as "Picking Up the Pieces: Rebuilding Brandenburg After the Thirty Years War."

2 Myron Gutmann, *War and Rural Life in the Early Modern Low Countries* (Princeton, NJ: Princeton University Press, 1980), 7, 76, 106; Emmanuel Kreike, *Scorched Earth: Environmental Warfare as a Crime against Humanity and Nature* (Princeton, NJ: Princeton University Press, 2021), 3.

3 Sean A. Eddie, *Freedom's Price: Serfdom, Subjection, and Reform in Prussia, 1648–1848* (New York: Oxford University Press, 2013); William W. Hagen, *Ordinary Prussians: Brandenburg Junkers and Villagers, 1500–1840* (Cambridge: Cambridge University Press, 2007); Jan Peters, *Märkische Lebenswelten: Gesellschaftsgeschichte der Herrschaft Plattenburg-Wilsnack, Prignitz 1550–1800* (Berlin: Berliner Wissenschaftsverlag, 2007), 599–634.

4 The tendency to cast Friedrich Wilhelm in a heroic role began in his lifetime, continued through the nineteenth and twentieth centuries, and seems to be alive and well today. Johannes Burkhardt coined the phrase "four hyperactive rulers" in *Vollendung und Neuorientierung des frühmodernen Reiches 1648–1763* (Stuttgart: Klett-Cotta, 2006), 175. This positive view is enshrined in older works such as Ferdinand Schvelli, *The Great Elector* (Chicago: University of Chicago Press, 1947) and Derek McKay, *The Great Elector: Frederick William of Brandenburg-Prussia* (London: Longman, 2001). Christopher Clark in his revisionist study of Prussia also gives Friedrich Wilhelm principal credit for rebuilding in *Iron Kingdom: The Rise and Downfall of Prussia, 1600–1947* (Cambridge, MA: Harvard University Press, 2006), 38.

5 Absolutism is a fraught term, of course. See Peter H. Wilson, *Absolutism in Central Europe* (London and New York: Routledge, 2000), 1–9. According to Wilson, state-building "emphasises the conscious and deliberate actions behind political change" (21). A significant analytical difference separates state-*building*, understood as a process initiated from above, from state-*formation*, a term that recognizes and allows for the play of multiple factors, multiple groups, and multiple forces and acknowledges that, even if plans existed, results often deviated significantly from them. It is now often argued that the evolution of the state was a far more contingent process than the term state-*building* suggests.

6 John Brinckerhoff Jackson, *Discovering the Vernacular Landscape* (New Haven and London: Yale University Press, 1984), 5, 7.

7 Alexandra Walsham, *The Reformation of the Landscape: Religion, Identity, and Memory in Early Modern Britain and Ireland* (New York: Oxford, 2011), 6.

8 Anuschka Tischer, "Kriegstyp 'Dreißigjähriger Krieg' und seine unterschiedlichen Typologisierungen von 1618 bis zur Gegenwart," in *Diplomatie, Medien, Rezeption: Aus der editorischen Arbeit an den ACTA PACIS WESTPHALICAE*, eds. Maria-Elisabeth Brunert and Maximilian Lanzinner (Münster: Aschendorff, 2010), 2.

9 Samuel Christoph Wagener, *Denkwürdigkeiten der churmärkischen Rathenow: Nicht bloß für Rathenower, sondern für Geschichts-und Vaterlands Freunden überhaupt* (Berlin: Matzdoff, 1803), 219, 233, 252–53.

10 Shepard Krech III, J. R. McNeill, and Carolyn Merchant, eds., "War," in *Encyclopedia of World Environmental History*, 3 vols. (Great Barrington, MA: Berkshire Publishing Group, 2004), 3: 1286; Verena Winiwarter, "Land Use and Agrarian Knowledge as Topics of Early-Modern Environmental History," in *An Environmental History of the Early Modern Period: Experiments and Perspectives*, eds. Martin Knoll and Reinhold Reith (Vienna and Berlin: LIT, 2014), 57. But, see Kreike, *Scorched Earth*.

11 Bernhard R. Kroener, "Der 'Zweiunddreißigjährige Krieg'—Kriegsende 1650: Oder wie lange dauerte der Dreißigjährige Krieg?" in *Wie Krieg Enden: Wege zum Frieden von der Antike bis zur Gegenwart*, ed. Bernd Wegner (Paderborn: Ferdinand Schöningh, 2002), 67–93; Peter Wilson, *Europe's Tragedy: A History of the Thirty Years War* (London: Allen Lane, 2009), 762–64, 769–71; Antje Oschmann, *Der Nürnberger Exekutionstag 1649–1650: Das Ende des Dreißigjährigen Krieges in Deutschland* (Münster: Aschendorff, 1991).

12   Robert I. Frost, *The Northern Wars: War, State and Society in Northeastern Europe, 1558–1721* (Abingdon: Routledge, 2000).

13   Lieselott Enders, "Wiederaufbau nach dem Großen Krieg: Ein Ansiedlungsvertrag von 1655 in der Uckermark," *Mitteilungen des Uckermärkischen Geschichtsvereins zu Prenzlau* 11 (2003): 36.

14   Johannes Schulze, *Die Herrschaft Ruppin und ihre Bevölkerung nach dem 30j. Kriege* (Neuruppin: n.p., 1925), 31.

15   As Tait Keller observed in "Destruction of the Ecosystem," *International Encyclopedia of the First World War*, an online publication at http://www.1914-1918-online.net/. Several authors have contributed detailed studies of regions of Brandenburg that, while they do not directly engage issues of the *Landschaft*, provide extensive background and material for creating such a history. The three massive volumes by Lieselott Enders: *Die Prignitz: Geschichte einer kurmärkischen Landschaft vom 12. bis zum 18. Jahrhundert* (Potsdam: Verlag für Berlin-Brandenburg, 2000); *Die Uckermark: Geschichte einer kurmärkischen Landschaft vom 12. bis zum 18. Jahrhundert* (Berlin: Berliner Wissenschafts-Verlag, 2008); and *Die Altmark: Geschichte einer kurmärkischen Landschaft in der Frühneuzeit (Ende des 15. bis Anfang des 19. Jahrhunderts)* (Berlin: Berliner Wissenschafts-Verlag, 2008) are essential works; her use of *Landschaft* means territory. Equally valuable are several studies by Jan Peters, especially his *Märkische Lebenswelten.*

16   Gutmann, *War and Rural Life*, is one exception; another is Maren Lorenz's study of military violence in Swedish occupied areas after 1648, *Rad der Gewalt: Militär- und Zivilbevölkerung im Norddeutschland nach dem Dreißigjährigen Krieg (1650–1700)* (Cologne: Böhlau, 2007); and, recently, Kreike, *Scorched Earth.*

17   Keller, "Destruction"; Selena Daly, Martina Salvante, and Vanda Wilcox, eds., *Landscapes of the First World War* (New York: Palgrave Macmillan, 2018); Lisa M. Brady, *War Upon the Land: Military Strategy and the Transformation of Southern Landscapes during the American Civil War* (Athens: University of Georgia Press, 2012); Brian Allen Drake, ed., *The Blue, the Gray, and the Green: Toward an Environmental History of the Civil War* (Athens: University of Georgia Press, 2015); Sven Beckert, *Empire of Cotton: A Global History* (New York: Knopf, 2014).

18   Emmanuel Kreike emphasizes that early modern wars also caused "the indiscriminate and simultaneous destruction of society *and* the environment." *Scorched Earth*, 1.

19   From 11 May 1642, Geheimes Staatsarchiv, Preußischer Kulturbesitz [hereafter: GStAPK], I. HA Rep. 24c Nr. 23 fasc. 12, fol. 65.

20   GStAPK, I. HA Rep. 9 (AV) S3b, from [9?] July 1639.

21   Werner Peschke, *Das Mühlenwesen der Mark Brandenburg: Von den Anfängen der Mark bis um 1600* (Berlin: VDI, 1937), 16–18. Besides mills, the years after the Thirty Years War saw the establishment, or re-founding, of a number of other enterprises that had their own peculiar, albeit often dramatic, effects on the *Landschaft*: brick works, tar and pitch refineries, glassworks, and limeworks.

22   Scholars disagree as to when to date the advent of the anthropocene. Some argue for a "great acceleration" in the mid-twentieth century. Others take a longer view, arguing that the anthropocene began 200, 500, or even 2000 years ago. Erle C. Ellis, *The Anthropocene: A Very Short Introduction* (Oxford: Oxford University Press, 2018), 53, 72–74, 101; John R. McNeill, *Something New Under the Sun: An Environmental History of the Twentieth-Century World* (New York: W. W. Norton, 2000).

23   William Cronon discussed this problem in his influential work on *Changes in the Land: Indians, Colonists, and the Ecology of New England* (New York: Hill & Wang, 1983), 9.

24 In his four-volume work on Prussian *Landwirtschaft,* August Meitzen observed that agricultural improvements in this period "were undoubtedly the work of private [groups and individuals]." *Der Boden und die landwirtschaftlichen Verhältnisse des Preussischen Staates nach dem Gebietsumfange vor 1866* (Berlin: Wieghardt & Hempel, 1868–71), 1: 140. In describing the devastation in Pomerania, Herbert Langer noted that one could not speak of a state-driven repopulation policy. "Rather, it was mostly the inhabitants themselves who initiated and carried out [programs of] reclamation and repopulation." "Pommern nach dem Dreißigjährigen Krieg," in *Rechtsprechung zur Bewältigung von Kriegsfolgen: Festgabe zum 85. Geburtstag von Herbert Langer,* ed. Nils Jörn (Hamburg: Dr. Kovac, 2012), 49; Wolfgang Neugebauer, *Zentralprovinz im Absolutismus: Brandenburg im 17. und 18. Jahrhundert* (Berlin: Arno Spitz, 2001), 68–71.

25 Ernst Friedlander, "Beiträge zur Geschichte der Landesaufnahme in Brandenburg-Preußen unter dem Großen Kurfürsten und Friedrich III./I.," *Hohenzollern-Jahrbuch* 4 (1900): 336–59; Rudolph Stadelmann, *Friedrich Wilhelm I. in seiner Thätigkeit für die Landescultur Preussens* (Stuttgart: Hirzel,1878); Christa Kouschil, *Landesbau in der Neumark unter Friedrich II* (Buskow bei Neuruppin: edition bodoni, 2012).

26 Neugebauer, *Zentralprovinz,* 72.

27 Wilson, *Europe's Tragedy,* 786–89. See also John Theibault, "The Demography of the Thirty Years War Re-Visited: Günther Franz and his Critics," *German History* 15 (1997): 1–21.

28 Several areas in Brandenburg, such as the Altmark, Uckermark, Ruppin, and Prignitz, were especially hard hit; Neugebauer, *Zentralprovinz,* 62–65. See also Enders: *Prignitz*; *Altmark*; and *Uckermark.*

29 "Tabelle, Des Königl. Ambts Ruppin, worinnen verzeichnet, wie viel Höfe in diesen Ambtsdörffern vor ao. 1624 gewesen, wie viele anietzo würcklich besetzet seyn,. . ." appendix to the printed order from "Königliche verordnete Curmärckische Ambts-Cammer," from 30 June 1717 in Brandenburgisches Landeshauptarchiv [hereafter: BLHA], Rep. 7 (Alt) Ruppin 494.

30 Günter Mangelsdorf, *Die Ortswüstungen des Havellandes: Ein Beitrag zur historisch-archäologischen Wüstungenskunde der Mark Brandenburg* (Berlin: Walter de Gruyter, 1994). In the Herrschaft Boitzenburg (Uckermark), for example, the fifteenth century was the major period of *Wüstungen,* not the mid to late seventeenth century, Wilhelm Abel, *Die Wüstungen des ausgehenden Mittelalters,* 3rd ed. (Stuttgart: Gustav Fischer, 1976).

31 Meta Murjahn, "Die gutsherrlich-bäuerlichen Verhältnisse des 17. Jahrhunderts im nördlichen Domanialgebiet des Landes Stargard (in den Ämter Stargard, Broda, Nemeow, Wanska)," *Jahrbuch des Vereins für Mecklenburgische Geschichte* 98 (1934): 9n14.

32 Not all areas in Brandenburg were visited (the Neumark, for example, was excluded) and many of the reports have been lost. Some still can be found in BLHA, especially in Rep. 7 (Landesherrlichen Ämter) and Rep. 37 ([Adlige] Herrschafts-Guts-und Familienarchive). Older authors often used them, see J. M. de la Pierre, *Ausführliche Geschichte der Uckermark* (Prenzlau: Carl Vincent, 1847), 215–16. Johannes Schulze published a complete one for Ruppin: *Die Herrschaft Ruppin und ihre Bevölkerung nach dem 30jähr. Kriege* (Neu-Ruppin: Historisches Verein der Grafschaft Ruppin, 1925).

33 Ernst Fidicin, *Die Territorien der Mark Brandenburg,* vol. 4: *Der Kreis Prenzlau, Der Kreis Templin, Der Kreis Angermünde* (Berlin: J. Gutentag, 1864), xi.

34 GStAPK, I. HA. Rep. 21 Nr. 2 Paket 1, from 15 May 1652.

35 GStAPk, I. HA. Rep. 63 Nr. 630; quote from Hermann Trebbin, *Müllrose: Aus den Schicksalen und Kämpfen einer märkischen Landstadt* (Frankfurt a.d. Oder and Berlin: Trowitzsch & Sohn, 1934), 42.

36 Anneliese Krenzlin, *Dorf, Feld und Wirtschaft im Gebiet der großen Täler und Platten östlich der Elbe: Eine siedlungsgeographische Untersuchung* (Remagen: Amt für Landeskunde, 1952), 51.

37 Rolf Kießling, Frank Konersmann, and Werner Trossbach, *Grundzüge der Agrargeschichte*, vol. 1: *Vom Spätmittelalter bis zum Dreissigjährigen Krieg* (Göttingen: Böhlau, 2016), 55–62; Rainer Beck, *Ebersberg, oder das Ende der Wildnis: Eine Landschaftsgeschichte* (Munich: Beck, 2003), 26–30; idem, *Unterfinning: Ländliche Welt vor Anbruch der Moderne* (Munich: Beck, 2004), 120–38.

38 Mix von York-Gothart, ed., *Almanach- und Taschenbuchkultur des 18. und 19. Jahrhunderts* (Wiesbaden: Harrassowitz, 1996).

39 For example, Abraham von Thumbshirn, *Oeconomia Oder Nothwendiger Unterricht und Anleitung wie eine gantze Hauß Haltung am nützlichsten und besten (so ferne GOTTes Seegen und Gedeyen darbey) kan angestellet werden* (Frankfurt a.M. and Leipzig: Hennige Grossens Seel. Buchladen, 1677[1616]) in Gertrud Schröder-Lembke, ed., *Martin Grosser, Anleitung zu der Landwirtschaft, Abraham Thumbshirn, Oeconomia: Zwei frühe detusche Landwirtschaftsschriften* (Stuttgart: Gustav Fischer, 1965), 83.

40 Kießling, et. al., *Grundzüge*, 1: 77–80; Beck, *Unterfinning*, 82–95; idem, *Ebersberg*, 25–26.

41 Fritz Schröer, *Das Havelland im Dreissigjährigen Krieg: Ein Beitrag zur Geschichte der Mark Brandenburg* (Cologne: Böhlau, 1966); Thomas Rudert and Hartmut Zückert, eds., *Gemeindeleben: Dörfer und kleine Städte im östlichen Deutschland (16–18. Jahrhundert)* (Cologne: Böhlau, 2001).

42 Krenzlin, *Dorf, Feld und Wirtschaft*, 51–53. In his study of *Rural Society and the Search for Order in Early Modern Germany* (Cambridge: Cambridge University Press, 1989), Thomas Robisheaux points out that even in times of crisis certain "groups carved out places of power. . . [including] widows, even on occasion youths, craftsmen, and the village poor." (11). In her book, *Coping with Life during the Thirty Years War (1618–1648)* (Leiden: Brill, 2021), Sigrun Haude discusses how people could exploit opportunities even in the worst situations of the war.

43 Thumbshirn, *Oeconomia*, for instance, offered a month-by-month description of what a household had to do.

44 *1000 Jahre Warmerstidi—Wolmirstedt, 1009–1009* (Wolmirstedt: Köthen, 2009), 97.

45 Report from 13 May 1663 in BLHA, Rep. 2 D21539, fols. 1–3.

46 Christian Otto Mylius, *Corpus Constitutionum Marchicarum Oder Königl. Preußische in der Chur-und Marck Brandenburg, auch incorporirten Landen publicirte und ergangene Ordnungen, Edicta, Mandata, Rescripta etc.* (Berlin and Halle: Waisenhaus, 1737–1755), Th. 4, 1, 2 (forest, wood); 4, 2,4 (dikes, dams,drainage ditches); 5, 2, 1 (trade, transportation), and 5, 3, 2 (clearing fields, surveying); Ernst Opgenoorth, *Friedrich Wilhelm: Der Große Kurfürst von Brandenburg;* vol. 1: *1620–1660* (Göttingen: Musterschmidt, 1971), 173–80, 255; vol. 2: *1660–1688* (Göttingen: Musterschmidt, 1971), 50–51, 301–303, 347–49.

47 GStAPk, I. HA Rep. 36, Nr. 3038; Toni Saring, *Luise Henriette von Oranien: Die Gemahlin des Großen Kurfürsten* (Göttingen: Deuerlichsche Verlagsbuchhandlung, 1941); Ulrike Hammer, *Kurfürstin Luise Henriette: Eine Oranierin als Mittlerin zwischen den Niederlanden und Brandenburg-Preußen* (Münster: Waxmann, 2001).

48 January or February 1680, GStAPK, I. HA. Rep. 55 Ruppin Nr. 5 fasc. 2, fol. 18.

49 *Hufe* was an imprecise measure. A *Hufe* was the amount of land that a family could farm and from which it could subsist. In Brandenburg, 1 *Hufe* usually equaled a little more than 17 hectares. Helmut Kahnt and Bernd Knorr, *Alte Maße, Münzen und Gewichte: Ein Lexikon* (Mannheim: Bibliographisches Institut, 1987), 126.

50 GStAPk, I. HA. Rep, 21 Nr. 13b1 fasc. 4, fol. 51 (1672).

51 The most influential work on the subject to date is David Blackbourn, *The Conquest of Nature: Water, Landscape, and the Making of Modern Germany* (New York: W. W. Norton, 2007).

52 Johann Carl Brandt, *Aus der Geschichte der Stadt Lebus* (Görlitz: Max Kretschmer, 1926), 17.

53 Donald P. Franzmeier, William W. McFee, and Helmut Kohnke, *Soil Science Simplified*, 5th. ed. (Long Grove, Illinois: Waveland Press, 2016), 1–3, 26–29, 145–152; Graham Stirling, Helen Hayden, Tony Patterson, and Marcelle Stirling, *Soil Health, Soil Biology, Soilborne Diseases and Sustainable Agriculture: A Guide* (Clayton South [Australia]: Csiro Publishing, 2016), 20–22, 55, 90. William Cronon noted the impact of English horses confined on relatively small pastures in colonial New England: "their weight had the effect of compacting soil particles so as to harden the soil and reduce the amount of oxygen it contained." *Changes in the Land*, 146–47, 151.

54 Kießling, et. al., *Grundzüge*, 1: 62–67; Michael Koch, *Traditonelles Arbeiten mit Pferden* (4th. ed.; Stuttgart: Eugen Ulmer, 2012), 39–48.

55 In the unquiet times before 1618, villagers in Forst-Pförten, fearing what war would mean for important documents, petitioned their *Herrschaft* to make and keep copies of the many papers, ordinances, and *Statuta*, to be preserved in view of "the often occurring plundering" that threatened to wipe out the rules of their lives. Dated 14 July 1614, BLHA, Rep. 37 Forst-Pförten.

56 Werner Michalsky, *Zur Geschichte des Lebuser Landes: Der Friedrich-Wilhelm-Kanal* (Seelow: Rat des Kreises Seelow, 1984). Less successful were early attempts to stimulate agriculture and manufacturing by importing migrants, for improving tolls on waterways, and attempts to facilitate international trade. Opgenoorth, *Friedrich Wilhelm*, 1: 173–77, 180, 2: 52–53, 348.

Further Reading

Gutmann, Myron. P. *War and Rural Life in the Early Modern Low Countries*. Princeton, NJ: Princeton University Press, 1980.

Hagen, William W. *Ordinary Prussians: Brandenburg Junkers and Villagers, 1500–1840*. Cambridge: Cambridge University Press, 2002.

Kreike, Emmanuel. *Scorched Earth: Environmental Warfare as a Crime against Humanity and Nature*. Princeton, NJ: Princeton University Press, 2021.

Peters, Jan. *Märkische Lebenswelten: Gesellschaftsgeschichte der Herrschaft Plattenburg-Wilsnack, Prignitz 1500–1800*. Berlin: Berliner Wissenschafts-Verlag, 2007.

Scott, James C. *Seeing Like a State: How Certain Schemes to Improve the Human Condition Have Failed*. New Haven: Yale University Press, 1998.

Index